IMF AND WORLD BANK SPONSORED STRUCTURAL ADJUSTMENT PROGRAMS IN AFRICA

IMF and World Bank Sponsored Structural Adjustment Programs in Africa

Ghana's experience, 1983-1999

Edited by
KWADWO KONADU-AGYEMANG

Routledge
Taylor & Francis Group

LONDON AND NEW YORK

First published 2001 by Ashgate Publishing

Reissued 2018 by Routledge
2 Park Square, Milton Park, Abingdon, Oxon OX14 4RN
605 Third Avenue, New York, NY 10017

Routledge is an imprint of the Taylor & Francis Group, an informa business

Publisher's Note
The publisher has gone to great lengths to ensure the quality of this reprint but points out that some imperfections in the original copies may be apparent.

Disclaimer
The publisher has made every effort to trace copyright holders and welcomes correspondence from those they have been unable to contact.

A Library of Congress record exists under LC control number: 2001088797

ISBN 13: 978-1-138-63429-9 (hbk)
ISBN 13: 978-0-415-79283-7 (pbk)
ISBN 13: 978-1-315-21041-4 (ebk)

Table of Contents

Contributors

Joe Amoako-Tuffour is an Associate Professor of Economics at St. Francis Xavier University, Antigonish, Nova Scotia, Canada. He holds a Ph.D. degree in Economics from the University of Alberta, and has interests in financial economics, public finance and applied econometrics. Dr. Amoako-Tuffour has published in the American Journal of Agricultural Economics, Review of Economics and Statistics, and the Journal of African Economies. Since 1995, he has been a regular visitor to the Center for Policy Analysis (Ghana) as a Research Fellow.

Nicholas Amponsah received his Ph.D. in Political Science from Claremont Graduate University, California, in 1999. Amponsah was a Visiting Assistant Professor of Political Science as well as a Research Fellow at the Center for International Security and Strategic Studies, Mississippi State University at Mississippi in 1999/2000. He is currently a lecturer in the Political Science Department at the University of Ghana, Legon. His research interests include political and economic liberalization reforms in Africa and the Third World.

Siaw Akwawua is currently an Assistant Professor of Geography at the University of Saskatchewan, from where he received his Ph.D. degree. Prior to his current appointment, he taught at the University of Regina. Dr. Akwawua's main research focus is in human/economic geography, and is particularly interested in population, migration, spatial interaction modeling, transportation, quantitative methods, spatial analysis and culture.

Kojo Appiah-Kubi is an economist and a research fellow at the Institute of Statistical, Social and Economic Research (ISSER) of the University of Ghana. He attended Johannis Gutenberg Universität in Mainz in Germany, where he earned his Ph.D. degree. His area of research includes public finance, fiscal policies, taxation and public institutions.

Charles Anyinam is a geographer. He received his master's degree from Carleton University, Ottawa, and his doctorate degree from Queen's University, Kingston, Ontario, Canada. Dr. Anyinam has published extensively in various geographical, health and environmental journals. His expertise is in spatial dimensions of development, human-environment relations, health and health care, solid waste management, as well as geographic information systems (GIS).

Robert Armstrong works as a consultant in the Operation Evaluation Department at the World Bank in Washington, D.C. He was a Lead Economist and Economic Adviser for the Africa Region. He first visited Ghana in March 1957 and has since visited the country many times. He holds degrees from Yale, Johns Hopkins (School of Advanced International Studies) and Northwestern in history, international studies and economics, respectively. He has taught at Northwestern, Williams College and the American University in Cairo, and worked for the Ford Foundation. He is the author of Ghana Country Assistance Review. A Study in Development Effectiveness published (World Bank, 1996) a summary of which is presented in chapter 18 of this book.

Samuel Aryeetey-Attoh is Professor and Chair of Geography and Planning at The University of Toledo. He received his Ph.D. from Boston University and his M.A. from Carleton University, Ottawa. He also earned a B.A. degree with Honors from the University of Ghana, Legon. Dr. Aryeetey Attoh's research and teaching interests are in Urban and Regional Planning, Urban Geography, Housing and Community Development, and the Geography of Africa. He has several publications and reviews in journals such as Urban Geography, Local Environment, Tidjschrift voor Economische en Sociale Geografie, Professional Geographer, Journal of Cultural Geography, Regional Science Review, Applied Geography Proceedings, Cities and Villages, Conference of Latin Americanist Geographers Yearbook, and Modeling and Simulation Proceedings.

Eric Asa is a Ph.D. candidate in Mining Engineering, Department of Civil and Environmental Engineering, University of Alberta. He holds Masters degrees in Mining Engineering, and Mineral Economics from the Colorado School of Mines. Prior to pursing graduate studies at Colorado, he worked working for Ghana's Minerals Commission as a Monitoring and Evaluation Officer. He has also worked for Agri-Petco/Primary Fuels (GH) Limited, Ashanti Goldfields Corporation, and BHP World Minerals. His research interests include artificial intelligence algorithms applicable to mining engineering problems, computerized open pit and underground mine design and optimization, mine management in the information age, strategic management, economic development and mineral economics.

Kwame Boafo-Arthur holds a Ph.D. degree in Political Science, and is a senior lecturer in the Political Science Department at the University of Ghana, Legon. Dr. Boafo-Arthur was a Fulbright Senior African Research Scholar at the

James S. Coleman African Studies Center, UCLA in 1998. He has written extensively on Ghana's political economy, international economic relations, African development, and foreign policy.

Francois K. Doamekpor received his Ph.D. from the University of Pittsburgh. He is an associate professor in the Department of Public Administration and Urban Studies, University of Akron, Ohio, U.S.A. He teaches Public Budgeting, Statistics, Public Sector Economics, Quantitative Techniques for Decision Making and, Project Design and Management. His current research interests focus on the development of performance measures in public sector organizations, the changing factors affecting public revenue and expenditure forecasting and management, and comparative public management.

Noble T. Donkor is a Ph.D. Candidate (Wildlife Ecology and Management) at the University of Alberta, Edmonton, Alberta, Canada. He holds a MSc. Degree in Wildlife Ecology and Management from University of Guelph, Guelph, Ontario, Canada. His research interests include environmental impact assessment, wildlife productivity and Natural Resource Management.

Peter Fuseini Haruna is an Assistant Professor of Public Administration at Texas A and M International University, Laredo, Texas. Dr. Haruna graduated from the University of Akron with a Ph.D. degree and his research interests include governance structures and processes, as well as administrative leadership.

Kwadwo Konadu-Agyemang is an Assistant Professor in the Department of Geography and Planning at The University of Akron, Akron Ohio. He earned his Ph.D. degree from Monash University, Melbourne, Australia in 1991. His research interests include housing and urban development in Africa, spatial organization in developing countries, and the role of NGOs and international organizations in development. Dr. Konadu-Agyemang has published scholarly papers in *The Professional Geographer, Urban Studies, The Canadian Geographer* and *Ghana Studies*. He is the author of *The political economy of housing and urban development in Africa: Ghana's experience from colonial times to 1998* (Praeger, 2000).

Joseph R. Oppong is an Associate Professor of Geography at the University of North Texas. He has a Ph.D. and M.A. from the University of Alberta in Edmonton, AB, Canada and a B.A. (Hons) degree from the University of Ghana, Legon. His major research interest focuses on Medical Geography and includes the geography of disease and the geography of health services. Dr. Oppong is currently studying HIV-AIDS, Buruli Ulcer and Tuberculosis in Africa, and HIV-AIDS and Cancer in Texas.

Kwaku Osei-Akom earned his undergraduate degree in History, Economics and Education from the University of Cape Coast in 1975, and his MA degree from SUNY(Albany) in 1987. He earned his Ph.D. degree in African Studies from Howard University in 1999. His Ph.D. dissertation, which involved extensive fieldwork among cocoa farmers in the Tano District in the Brong Ahafo Region, was on the impact of Structural Adjustment Programs on cocoa production in Ghana. Dr. Osei-Akom is currently studying the impact of the 1992 and 1996 Multi-Party Elections on Ghana's Economic Recovery Program. His research interest includes the impact of economic restructuring, economic history of Africa, and problems of African economic development.

Kwamina Panford is a Professor of African Studies at Northeastern University, Boston. He earned his Ph.D. degree from Cornell University, specializing in International development with emphasis on the roles of international organizations including the International Labor Organization, the Bretton Woods Institutions and the World Trade Organization. His on-going research is on Globalization and Labor Laws, policies and practices in Africa.

Andy C.Y. Kwawukume holds a BSc (Political Science) and M.Phil. (Administration and Organization) degrees from the University of Bergen, Norway. He was working as a researcher at Chr. Michelsen Institute in Bergen, Norway, when he wrote his contribution to this book. His specialty is microcredit organizations, participation and rural development. He currently lives and works in London, England.

Baffour Kwaku Takyi is currently an Assistant professor of Sociology at the University of Akron, Ohio. He holds a Ph.D. and Masters degree in Sociology from the State University of New York at Albany (SUNY). Prior to his current position, he worked as a policy research analysts for the New York State Department of Social Services, and later Health between 1987 and 1997. His research interests are primarily in the sociology of African societies, family dynamics, reproductive related behavior, and health and illness behavior in Ghana.

Ian E.A. Yeboah is an Associate Professor of Geography and Black World Studies at Miami University, Oxford, Ohio. He received his Ph.D. from the University of Calgary, in Canada, in 1994. His research and publications focus on the economic status of urban residents and spatial structure of urban places in Sub-Saharan Africa. He has extensive research experience on Ghana.

Preface and Acknowledgments

Since the early 1980s, the World Bank and the International Monetary Fund (IMF) have been actively involved in the management of the economies of many developing countries, including more than forty in Africa. Almost invariably, their prescription for economic development has been in the form of Structural Adjustment Programs. This book seeks to appraise the impact of the World Bank and IMF programs in Africa, focusing on Ghana's 16-year experience. Although the original intention was to use case studies from several African countries undergoing economic restructuring under the auspices of the Bretton Woods Institutions, no researchers with expertise on those countries responded to our call for papers. This notwithstanding, we believe that many of the countries undergoing Structural Adjustment Programs in Africa and elsewhere can identify with most of the issues pertaining to Ghana's adjustment programs that we have presented in this volume. Moreover, since the World Bank and IMF have often touted Ghana as the best example of successful economic restructuring in Africa, we believe that the lessons derived from the experience of this West African nation could, with appropriate caveats, be used to guide development efforts elsewhere.

The book is the result of the untiring efforts of many individuals whose names cannot be mentioned here. However, I am extremely grateful to Linda Bussey of the Geography and Planning Department at the University of Akron. Linda applied all her skills and learned all the new tricks available to make the camera-ready version of the book possible. I am also grateful to the my former department chair, Dr. Chuck Monroe, and the current one, Dr. Rob Kent, for allowing Linda to spend many hours on this project. Linda's student assistants, Justin and Dawn, also deserve to be thanked for preparing the index and some of the charts and tables.

I also wish to express my appreciation to Dr. Allen Noble for his comments on the draft proposal, and some of the chapters; to Wendy Risner for assisting me with the editing; my graduate assistants Louis Machado and Onai Chitiyo for their contribution in the form of maps and charts. I am also indebted to the contributors without whose effort 'this huge elephant' could not have been killed, let alone dragged home.

I also wish to acknowledge the Publications Department of World Bank for permitting me to use a summary of Armstrong (1996) presented in Chapter 18.

Finally, I must applaud the efficiency and cooperation of the editors and production staff at Ashgate Publishing without which this book project would not have become a reality.

The contributions of other people to the production of this book notwithstanding, none of them is to blame for any fault, error or unfairness of opinions and argument. The editor and the contributors take responsibility for all these defects.

Kwadwo Konadu-Agyemang

1 An Overview of Structural Adjustment Programs in Africa

KWADWO KONADU-AGYEMANG

1:1 Introduction

The 1970s and early 1980s saw the economies of almost all African countries teetering towards complete collapse, due to several factors that included economic mismanagement, political instability, and unequal exchange. In order to redress the situation and improve their living conditions, most countries asked for help from the World Bank/IMF to restructure their economies, and almost invariably, the prescription was in the form of Structural Adjustment Programs (SAPs). What do we know about the impact of these adjustment programs in Africa? Have the Structural Adjustment Programs been totally beneficial to Africa as is often claimed by the IMF and the World Bank?

Structural Adjustment is the process whereby economic policies and relevant institutions are reformed with a view to enhancing economic growth, improving resource allocation, increasing economic efficiency and increasing the economy's resilience to changes in its domestic or global market. This process is normally supported by policies designed to achieve sustainable deficit reduction and to reduce the rate of price inflation (Gilpin, 1994:547). Designed by the World Bank and IMF, Structural Adjustment Programs are imposed on debtor countries as a pre-condition for debt relief, acquiring new loans, as well as attracting foreign investment. Structural Adjustment Programs often consist of two parts: shorter-term stabilization and longer-term adjustments. While stabilization measures are primarily designed to reduce short term imbalances between demand and supply, which are normally manifest in balance of payment and budget deficits, structural adjustments seek to address a wider range of obstacles to growth, many of them limiting the ability of the economy to increase supply (Duncan and Howell, 1992:5). But in view of the overlapping

1

nature of stabilization and structural adjustments, SAP is used as an umbrella term to cover all of them.

Structural Adjustment Programs were first introduced in Turkey in 1980 and have since become the cornerstone of World Bank and IMF development policies in the so-called Third World countries. By 1992, 64 countries had been 'adjusted' through 187 separate lending operations totaling $28.5 billion (Jayarajah and Ghani, 1996). Countries ranging from Argentina, Bangladesh, Bolivia, Brazil, Ghana, Nigeria, Peru, Zambia to Zimbabwe have come under Structural Adjustment Programs in one form or the other during the last 18 years. During the last 3 years, the IMF has also prescribed SAPs as the solution to the economic malaise afflicting South Korea, Thailand, Indonesia along with other Asian economies and Russia. As of June 30, 1998, 70 countries had various Enhanced Structural Adjustment Facilities (ESAF) arrangements with the IMF in place. These arrangements involved a colossal sum of $43,502.50 million, which these countries would eventually owe the IMF.

From the perspectives of the Bretton Woods Institutions, and other supporters of Structural Adjustment Programs, the problems of underdevelopment and inefficiencies in the developing countries are primarily generated and sustained by internal factors. These factors such as state interference in the workings of the price mechanism, over-bloated public service, exchange control, corruption, state ownership of manufacturing enterprises and investment in social services only serve to clog the wheels of efficiency (World Bank, 1994; Green, 1987). According to neo-classical economic theory that provides the theoretical and philosophical foundation for Structural Adjustment Programs, unbridled market forces represent the only tested and proven path for the elimination of bottlenecks, and any attempt to manage the economy through state interference only results in unwarranted distortions in the economy, and breeds inefficiency. Therefore, economic efficiency and rational management driven by market forces, they argue, should constitute the sole scale upon which all national considerations of production, administration and culture are measured (Ould-Mey, 1996:14).

The Bretton Woods agencies claim that the implementation of Structural Adjustment Programs almost invariably leads to poverty reduction and bridges the gaps between the rich and the poor, and between the rural and urban areas, and that countries that adjust tend to be better off than the non-adjusting ones (World Bank, 1994). Three main arguments are often presented in support of this claim.

- Failure to adjust will, ultimately, impose huge costs on the poor, with unsustainable budget and trade deficits leading to hyperinflation, currency instability, and economic collapse (Watkins, 1995).
- Structural Adjustment Programs where properly implemented have not only created the conditions for growth, but for growth which is pro-poor. The key contention is that state intervention in the rural sector, where a vast majority of the poor live, has lowered prices and reduced market opportunities, and thereby depressed household income. Rural market de-regularization, however, raises prices and creates rural employment. In the urban sector, import liberalization makes local industries more competitive, by allowing them to take advantage of imported technology. Liberalization, together with labor market deregulation then, it is argued, will lead to thecreation of more jobs (Watkins, 1995:78; World Bank, 1991).
- Structural Adjustment Programs incorporate 'social conditionality' and provisions that aim at protecting welfare service delivery in areas of concern to the poor (World Bank, 1993; Watkins, 1995).

Indeed, the World Bank claims that its central goal is poverty reduction as is evidenced in a speech delivered to the board of governors in 1988 by its then president, Barber Conable. The address focused on

> the central goal of the bank: the reduction of poverty. Poverty on today's scale prevents a billion people from having even minimally acceptable standards of living. To allow every fifth human being on our planet to suffer such existence is a moral outrage. It is more: it is bad economics, a terrible waste of precious development resources. Poverty destroys lives, human dignity, and economic potential (quoted in Nafziger, 1990:99).

Barber Conable's declaration was reiterated in a recent World Bank Report that asserted that its fundamental objective is to help client countries to reduce poverty and improve living standards through policies that promote broad-based, labor-demanding growth and increasing the productivity and economic opportunities for the poor (World Bank, 1999).

The implementation of Structural Adjustment Programs, especially in Africa, has generated much research and controversy. Not surprisingly, most of the researches that are either sponsored by the World Bank/IMF, or produced by their own staff give high accolades to the programs. These researchers, almost invariably, take the view that there are no viable alternatives to Structural Adjustment Programs, and therefore developing countries are better off with Structural Adjustment Programs than they will be without them. Furthermore, they point to growth in gross domestic product, increase in exports, drops in inflation level, and other macro level changes as evidence of development under Structural Adjustment Programs (World Bank, 1994; Husain, 1994 a/b; Hadjimichael *et al.*, 1992; Kapul *et al.*, 1991; Zulu and Nsouli, 1985). On the other hand, most of the independent views from academia, charitable NGOs and UN institutions such as UNESCO tend to question the benefits of Structural Adjustment Programs. Concerns about impacts of adjustments on the more vulnerable elements in society, social welfare, spatial inequalities, the environment, labor, gender, and on farmers rhyme through the literature (e.g. Cornia *et al.*, 1987; 1988; Weissman, 1990; Gladwin, 1991; Development Gap, 1993; Loewenson, 1993; Mackenzie, 1993; Anyinam, 1994; Watkins, 1995; Stewart, 1995; Brydon and Legge, 1996; Panford, 1994; Konadu-Agyemang, 1998).

1:1.2 Structural Adjustment Programs in Ghana: Background

Ghana launched its far reaching Structural Adjustment or Economic Reform Program (ERP) in 1983, and has since experienced a dramatic socio economic transformation that includes a GDP growth of 4-6 percent over most of the past 16 years, increase in export revenue and the revamping of industrial capacity. As a result of these achievements, the World Bank/IMF and the Western countries have hailed Ghana as the most successful case of Structural Adjustment Programs in Africa. But is this really the case?

1:1.2.1 Ghana's Slide from Glory to Dust

The country of Ghana came into being when the British granted political independence to their then colony, the Gold Coast. The British had ruled or controlled the country in one form or another since 1844, and had through colonial plunder turned their so called model colony into a "periphery" country, unscrupulously exploited by the "core" country (Howard, 1978). The result of the exploitative activities of the colonial government itself and the European firms (cocoa buying, banking, shipping and mining especially) was that when the British finally pulled out of Ghana in 1957, they left behind a country, which bore all the most important features of underdevelopment. In particular, it suffered from an acute form of structural dislocation. Production and consumption were not integrated within the country, but through external trade. Thus the colony exchanged commodities such as cocoa and gold, which it produced but did not consume, for mainly manufactured goods that it did consume, but in many cases could not produce itself (Kay, 1972:xv).

All the exploitation and symptoms of underdevelopment notwithstanding, the British probably left Ghana in a much better shape than any other country in Sub-Saharan Africa. Ghana's primary material export-based economy had been well established. At independence in March 1957, Ghana was the world's leading producer and exporter of cocoa, exported 10 percent of the world's gold and had foreign reserves of £200 million (equivalence of 3 years imports). Furthermore, it had one of the best infrastructure systems in Africa, the most educated, skilled and experienced workforce in Sub-Saharan Africa, and well-formulated development plans (Huq, 1989; Werlin, 1994; Roe and Schneider, 1992). By all criteria, Ghana was probably much better endowed than most third world countries, and certainly had better opportunities to develop than any other country in Sub-Saharan Africa. Naturally, Ghana's per capita income of £50 in 1957 put it at par with South Korea (Werlin, 1994), and was comfortably placed among the club of middle income countries. In 1960, Ghana's per capita national income of £70 was still higher than that of Nigeria (£29), India (£25) and Egypt (£56), and the annual growth rate was around 6 percent (Huq, 1989). As Apter (1972) aptly put it,

> Richer than most (African) countries, carefully groomed for independence, with trained cadres exceeding those of far larger countries, without racial minority problems, having inherited a good and expanding educational system, Ghana is regarded as having the resources, manpower and moral and spiritual qualities to set the pace and tone of political development in all Africa (p. 337).

The bright and promising star of Africa, however, did not maintain its luster for long. By the middle of the 1960s, Ghana was teetering towards bankruptcy. Growth rate of GDP had fallen to 0.4 percent; foreign reserves had dried up, and the nation was in serious debt (US $1 billion); real value of the minimum wage had dropped by 45 percent; public sector earnings and industrial earning had fallen by 20 percent and 25 percent, respectively (Rimmer, 1992; Loxley, 1991; Huq, 1989). Furthermore, the producer price paid to cocoa farmers had been decreased by 66 percent (Fitch and Oppenheimer, 1966), and the lower half of income earners shared one-third of national income as against one-half in 1956.

Ghana's economic woes continued through the rest of 1960s into the 1970s and early 1980s, orchestrated by political instability and corruption. Between

1970 and 1983, import volumes declined by over 33 percent, real export earnings fell by 52 percent while domestic savings and investments dropped from 12 percent of GDP to almost zero, and an unprecedented number of Ghanaians - artisans, teachers, medical professionals, as well as the unskilled - left the country. By the early 1980s inflation rate was in excess of 100 percent, the per capita GDP had fallen from its 1960 level of US \$1,009 to \$739, and the nation was going through one of the worst droughts and famines in its history. As if these troubles were not enough, the Nigerian government suddenly repatriated well over 1 million Ghanaians whose arrival worsened the already chaotic socio-economic environment (Huq, 1989; Chazan, 1991; Rimmer, 1992; Werlin, 1994).

Researchers have suggested a number of factors behind Ghana's fall from a promising middle-income economy to a country teetering towards bankruptcy. These include economic mismanagement-over-regulation, inability to control inflation, productivity disincentives, an over-bloated and mismanaged public sector, over-subsidized social services (education and health in particular), overvalued currency; political instability, corruption and inept leadership, as well as acts of God, unfavorable terms of trade, and clientelism (Huq, 1989; Chazan, 1991; Rothchild, 1991; Rimmer, 1992; Werlin, 1994; Brydon and Legge, 1996). It was this bleak and dismal economic situation that led to the "invitation" of the twin Bretton Woods "doctors", the IMF and World Bank, by the Ghanaian government in 1983 to provide solutions.

1:1.2.2 The IMF Diagnosis and Treatment

The IMF and the World Bank came on the scene February 1983, and predictably, their prescription was *Rx Structural Adjustment*. Indeed the IMF/World Bank's solution to economic maladies in Third World countries since the late 1970s has always, without exception, been in the form of Structural Adjustments, a new euphemism for the modernization and growth at all cost thinking that had been "abandoned" in the 1970s (see Chapter 2, this volume). IMF/World Bank inspired structural adjustment programs often consist of a package of actions that include currency devaluation, reducing inflation, downsizing the public service, drastic cut back on government expenditure on education, health and welfare; financial reforms, privatization of public enterprises, export promotion, and other policies geared to enhancing economic growth.

According to the IMF and other proponents of Structural Adjustment Programs, the economic problems of developing countries have nothing to do with exogenous factors as is often asserted by neo-Marxist theorists (see Amin, 1974; Frank, 1975), but are rooted in endogenous factors that serve as obstacles to development, and therefore need to be removed to pave the way for economic growth. These so called obstacles to development such as unwarranted state interference in the workings of the price mechanism, over-bloated public service, exchange control, state ownership of manufacturing enterprises and investment in social welfare only serve to clog the wheels of efficiency (World Bank, 1981). African countries in particular have been singled out as carrying an unnecessary and excess baggage of price control, overvalued currencies, political instability, inward-looking trade policy and heavy government spending. These factors, it has been argued, constitute a drain on the efficiency of the market, and have caused Africa's dismal economic performance, *vis a vis* Asian and Latin American countries, in the past three decades (Roe and Schneider, 1992; Werlin, 1994; Sahn, 1994; World Bank, 1994). The only path to the elimination of economic bottlenecks, and hence fostering of development, is to allow the invisible hands of market forces to determine economic activities unfettered (Roe and Schneider, 1992; Ould-Mey, 1996).

In April 1983, Ghana's economic recovery program designed by the IMF was put into action. The implementation of Structural Adjustment Programs in Ghana over the past 16 years has consisted of compressing government expenditure through massive cuts in social services and retrenching public sector workers; adjusting the exchange rate through discrete devaluation of Ghana's currency, the *cedi*; abolishing domestic price controls, mobilization of government revenue through broadening of the tax base, and the strengthening of tax administration. Other strategies include the privatization of state-owned enterprises, promoting the efficient allocation of resources for growth, rehabilitation of economic infrastructure, increased reliance on market-based monetary policy; encouraging private sector development. There has also been a massive export drive that encourages the production and export of both traditional exports (cocoa, mineral resources, timber) and non-traditional exports–crafts, foodstuff, vegetables and the like (Anyinam, 1994; World Bank, 1995; ISSER, 1997; Ghana Government, 1997;1998). Ghana's economic reform programs have been undertaken with World Bank and other multilateral loans that total more than $6 billion (Armstrong, 1996; IMF, 1999a/b).

1:1.3 Objectives and Structure of This Book

What have been the full implications of these programs for socio-economic development in Ghana (and for Africa)? What are the opportunities? What are the constraints? What lessons can be learnt? The principal objective of this book is to assess the ramifications of these programs since their inception in 1983. The chapters seek to critically evaluate the impact of Structural Adjustment Programs. The book brings together geographers, planners, political scientists, economists, rural development specialists, bankers, public administrators and other development experts to assess the impacts of Structural Adjustment Programs from a multi-disciplinary perspective. While the IMF and the World Bank evaluate the success of their programs by comparing the economy under Structural Adjustment Programs to others which have not been under Structural Adjustment Programs (Loxley, 1986), the chapters in this volume will primarily focus on the "before and after the fact". This study is intended to throw some light on the nature of Structural Adjustment Programs and how the programs have impacted on Ghana.

The book consists of 20 chapters and is divided into 8 parts. Following the introductory chapter, Kwadwo Konadu-Agyemang and Baffour Takyi discuss the political economy of development and underdevelopment in Ghana. They revisit the age-old nagging questions about what constitutes development and what the aims and objectives should be. They then evaluate Ghana's Structural Adjustment Programs in the light of what true development should aim at. They conclude that despite the fact that the programs have resulted in impressive GDP growth, better economic management and engendered international confidence in Ghana's economy, among other things, they have also created large scale unemployment, and perpetuated poverty and inequalities. Consequently, they question the assertion by the sponsors of the program that Ghana has experienced massive development during the last 16 years.

One of the most significant developments over the adjustment period has been Ghana's ability to borrow capital from international sources. Indeed, between 1983 and 1995, Ghana borrowed to the tune of $2 billion from the World Bank alone, and Ghana's external debt now amounts to more than $6 billion. In chapter 3, Joe Amoako-Tuffour examines the impact of such borrowing. He points out that Ghanaian public finance history has been characterized by two major developments: (1) an increase in the size of the government sector, and (2) an uninterrupted string of budget deficits that have led to an increase in the ratio of public debt to total national output from under 10 percent in the first half of the 1960s to almost 100 percent in the mid 1990s. In this chapter he analyzes thirty-eight years of government budget deficits and public debt from 1960-1998, with particular focus on the last two decades.

In Chapter 4, Kojo Appiah-Kubi discusses fiscal management, one policy measure that has been at the core of Ghana's financial and structural adjustment. The chapter evaluates the success of adjustment in Ghana with particular reference to fiscal policies: tax and expenditure reforms and their relationship to other macroeconomic indicators such as savings and investment. It also attempts to assess Ghana's past fiscal policy in the context of the efficacy of mobilization of resources for development.

In Chapter 5, Peter Haruna examines changes in public administration since 1983. The chapter describes the changing role of public administration and of the state in Ghana under current reforms, economic recovery, and structural adjustment program. The central thesis is that the role of public administration and of the state has transitioned from their developmental focus during the independence period to a managerial role under the IMF/World Bank sponsored reform, economic recovery, and structural adjustment.

The cocoa industry is unrivalled in its contribution to Ghana's political economy. How has the goose that has been laying the golden eggs for Ghana since the 1890s fared under the Structural Adjustment Programs? In chapter 6, Kwaku Osei-Akom provides an insight into changes in cocoa production under Structural Adjustment Programs with a case study from one of the most productive cocoa growing areas in the country.

In Chapter 7, Eric Asa discusses the interconnection between economic restructuring and the mining industry. Indeed, the success of Structural Adjustment Programs can be directly related to the success of the mineral industry, especially gold mining ventures. Under Structural Adjustment Programs, the Government of Ghana established new institutional and legislative frameworks resulting in the promulgation of a plethora of laws guiding the industry, and the issuing of unprecedented number of mineral exploratory and mining licenses to various national and international companies. Asa argues that Ghana might have unknowingly fallen prey to the resource curse paradigm. The benefits from the present strategies may be short-lived and may not address the long-term needs of the country.

In Chapter 8 Charles Anyinam discusses the controversial issues of development versus the environment. Can we have real development while safeguarding the environment? This chapter contributes to the discussion of the environmental impacts of Structural Adjustment Programs by examining mining development in Ghana in the last 16 years and the role that the current upsurge in minerals production has been playing in the deterioration of the country's ecosystems. It is argued that ample evidence shows that recent mineral developments have intensified the rate of pollution, land degradation, and environmental destruction of mining areas in Ghana.

Noble Donkor, in Chapter 9 continues with the environmental theme examined by Anyinam by discussing the impact of Structural Adjustment Programs on forests in Ghana. The massive export drive engendered by the adjustment programs has led to an unprecedented harvesting of timber, fuel wood and other forest produce consequently diminishing Ghana's forest resources. This chapter provides a preliminary insight into how Structural Adjustment Programs might have stimulated or discouraged sustainable management, deforestation and degradation of forests in Ghana.

Kwamina Panford, in Chapter 10, deals with labor and employment matters and focuses on two basic issues emanating from Ghana's structural Adjustment Program in the last ten years: how organized labor is meeting the dilemmas of a shrinking formal sector as a result of privatization and trade liberalization and other components of Structural Adjustment Programs, and how the SAP in Ghana is adversely impacting human resource development and utilization? Emphasis is on the status of women.

In Chapter 11, Boafo Arthur tackles the issues of adjustment, democratization and the politics of continuity. Is it possible to successfully pursue democratization and structural adjustment programs at the same time? Given the "mechanisms" adopted by governments with the support of the World Bank and the IMF to impose adjustment policies on many developing countries prior to political liberalization, and given the underpinnings of democracy, some argue that the two processes are antithetical. Boafo-Arthur challenges the seemingly embedded determinism of the incompatibility thesis and argues it is possible for adjustment and democratization to be pursued *pari passu*.

In Chapters 12 and 13, respectively, Andy Kwawukume and Siaw Akwawua discuss rural development issues. Kwawukume argues that, while theoretically, Structural Adjustment Programs aim at shifting the focus of development from the urban to the rural areas where more than 60 percent of the population live, in practice the urban areas still continue to receive a lion's share of investment capital and services. As a result, the rural areas suffer from an acute form of underdevelopment, and poverty among rural people is still a major problem, and seems to be getting worse. One approach being used by the government to address the urban-rural dichotomy is to improve access of rural people to credit facilities. Kwawukume examines the impact of the Financial Sector Adjustment Program (FINSAP) under the Rural Finance Project (RFP), with special focus on the Rural Banks of Ghana, and developments in rural credit intermediation.

Siaw Akwawua continues the discourse and discusses the strategies used by rural households to cope with the impacts of economic restructuring. He argues that while rural-urban and international migration have been taking place in Ghana since the 1950s, the volume of migration increased to unprecedented after the introduction of SAP in the country in the early 1980s. He sees the unprecedented rural-urban and international migration among rural people, and the associated remittances targeted to rural areas as strategies adopted by rural people to deal with the devastating impact of SAP. This chapter describes a case study based on a 1999 household survey data collected in a rural settlement in Southeast Ghana. Akwawua concludes that improvements in rural household living standards are attributable more to household survival strategies than to any direct beneficial effects of Structural Adjustment Programs.

Among the litany of ills that characterize African cities are housing shortages, deteriorating and inadequate infrastructure, ineffective and chaotic public transit systems, as well as poor land use planning. Chapters 14 and 15 deal with the implications of restructuring for urban planning and management. Ian Yeboah discusses the relationship between globalization, Structural Adjustment Programs as a development alternative in Sub-Saharan Africa (SSA), and emerging urban form. He argues that increasingly Structural Adjustment Programs in SSA are seen as an extension of globalization forces (albeit, peripheral), and that relationships exist between Structural Adjustment Programs and urban places, especially urban expansion and spatial form. Based on mapping the expansion of Accra and relevant literature, this chapter focuses on the emerging spatial form of Accra under Structural Adjustment Programs. It also examines planning and management implications of Accra's expansion for the new millennium.

Samuel Aryeetey-Attoh continues the planning/urban management theme and management issues. He argues that while at the international scale, aid to Sub-Saharan Africa has taken several forms; the IMF and the World Bank's Structural Adjustment Programs have conditioned technical assistance since the 1980s. The Structural Adjustment Programs were initiated to adjust malfunctioning economies and promote greater economic efficiency and economic growth in order to make Sub-Saharan economies more competitive in today's global economy. Within this context the chapter examines the impacts of Structural Adjustment Programs on planning and urban management implementation in Ghanaian cities with emphasis on Accra. It focuses specifically on the effects of currency devaluation, cuts in subsidies, and cost recovery programs on the delivery of urban services, the segmentation of housing markets, and the overall management of urban areas.

In Chapter 16, Joseph Oppong examines health issues. The organization, management and delivery of health care services in almost all African countries follow a common pattern. While a few private and parochial health care facilities exist, the government, through its ministry of health pays for and operates most hospital facilities, often at no cost to the users. Consequently, government health policy changes tend to have wider ramifications for the health sector. As part of the structural adjustment programs, a radical restructuring of health delivery has taken place. The changes include the introduction of "pay as you use". How has the health care sector suffered and/or benefited from the programs? How have they affected the quality of care and access to services, especially for the poor and rural population?

Under Structural Adjustment Programs, the private sector is expected to be the main propulsive power that drives development. Indeed, most analysis on Ghana's limited adjustment reforms since 1983 admit that one of the reasons for the lack of significant success is the unfavorable response of private economic actors in Ghana, especially Ghanaian private entrepreneurs. Nevertheless, no in-depth studies on why domestic entrepreneurs have reacted coldly to the reforms have been done. Is it because Ghanaian, or African private entrepreneurs in general, simply lack the entrepreneurial acumen that made private business elsewhere in East Asia served as the engine of growth? Using empirical field survey data on 448 Ghanaian Private Entrepreneurs, Amponsah argues in Chapter 17 that the reasons why private entrepreneurs in Ghana have failed to respond favorably to reforms is the poor institutional environment.

Chapter 18 provides insights into a World Bank study that evaluates the sustainability of Ghana's economic restructuring. Armstrong raises the very important, yet often overlooked question: Is growth sustainable under the present economic regime? He points out that Ghana's economic progress, that has become possible with more than $2 billion in World Bank loans, will not be sustainable unless Ghana speeds up the implementation of a large unfinished agenda of policy reform. He argues that the strategy has not fostered the response needed from the private sector, nor has it raised agricultural productivity as expected. He concludes that the bank needs to frame its assistance strategy to help Ghana to achieve participatory development, and self-reliant private sector led growth.

Reform programs in several countries have reduced the dominant role once played by central governments. Through these programs, the extent of government economic activity has been drastically reduced, including the dominant role once played in certain key sectors of the economy. In Chapter 19, Francois Doamekpor proposes an alternative model that could be used to ascertain the effectiveness of economic restructuring programs not only in Africa, but also in other developing countries. This chapter is devoted to a review of some of the methods for measuring government conduct of work. It is written to provide a rationale and broad overview of some of the measures used to gauge performance. It is not intended to prescribe country-specific techniques or approaches for assessing organizational performance. As a result, no country data were used. This was done on purpose to facilitate a general but an in-depth presentation of an integrative model that incorporates various performance measures and indicators.[1]

Chapter 20 represents a synthesis of the key issues addressed in the chapters. In this chapter, Kwadwo Konadu-Agyemang highlights the costs and benefits of the economic restructuring programs, and suggests lessons that policy makers at the World Bank and in adjusting countries may have to learn to make the structural adjustment programs more sustainable.

Note

1   Performance measures and indicators are sometimes used interchangeably, although their technical definitions are different. Measures define program output and desired outcomes or results. They relate directly to desired outcomes or results. Indicators, on the other hand, are surrogates for expected performance results and are used when measures are difficult to determine or define.

References

Amin, Samir (1974). *Accumulation on a World Scale: A Critique of the Theory of Underdevelopment*. New York: Monthly Review Press.

Anyinam, Charles (1994). "Spatial Implications of Structural Adjustment Programs in Ghana". *TESG* 85 (5): 446-450.

Apter, David E (1972). *Ghana in Transition*. Princeton, N.J: Princeton University Press.

Brydon, Lynne and Karen Legge (1996). *Adjusting Society: The World Bank, IMF and Ghana*. London: Tauris Academic Studies.

Chazan, Naomi (1991). "The political transformation of Ghana under the PNDC" in *Ghana: The Political Economy of Recovery*. ed. Donald Rothchild. Boulder, CO: Lynne Reinner Publishers.

Cornia, C.A., R. Jolly and F. Stewart (1987). *Adjustment with a human face* Vol. I. Oxford: Clarendon.

Cornia, C.A., R. Jolly and F. Stewart (1988). *Adjustment with a Human Face* Vol. II. Oxford: Clarendon.

Development GAP (1993). *The other side of the story: The real impact of World Bank and IMF Structural Adjustment Programs*. Washington DC: Development GAP.

Duncan, Alex and John Howell (eds.) (1991). *Structural Adjustment and the African Farmer*. London: Zed for ODI.

Fitch, Bob and Oppenheimer, Mary (1966). *Ghana: End of an Illusion*. New York: Monthly Review Press.

Frank, Andre Gunder (1975). *On Capitalist Underdevelopment*. Bombay: Oxford University Press.

Ghana Government (1997). *Budgetary Statement*. Accra: Government Printer.

Ghana Government (1998). *Budgetary Statement*. Accra: Government Printer.

Gilpin, Raymond (1994). "Pill or Poison?". *West Africa*. 28th March - 3 April.

Gladwin, C.H. (1991). "Introduction" in *Structural Adjustment and African Women Farmers*. C.H. Gladwin (ed.) Gainsville, Fl : University of Florida Press.

Green, Reginald H (1987). *Stabilization and Adjustment Policies and Programs. Country Study I: Ghana.* Annankatu, Finland: WIDER.

Hadjimichael, M.T., T. Rumbaugh and E. Verreydt (1992). *The Gambia: Economic Adjustment in a small open economy.* Washington, DC: IMF.

Howard, Rhoda (1978). *Colonialism and Underdevelopment in Ghana.* London: Croom Helm.

Huq, Mohammed M (1989). *The Economy of Ghana: The First 25 Years.* London: Macmillan.

Husain, I (1994a). *Macroeconomics of adjustments in Sub-Saharan African Countries: Results, Lessons.* Washington, DC: World Bank.

Husain, I (1994b). *Why do some countries adjust more successfully than others: Lessons from seven African countries.* Washington, DC: World Bank.

IMF (1999a). *Ghana – Enhanced Structural Adjustment Facility: Economic and Financial Framework Paper,* 1998-2000. Washington DC: IMF.

IMF (1999b). *Ghana.* Selected Issues. Staff Country Report No. 99/3. Washington: IMF.

Institute of Statistical, Social and Economic Research (ISSER) (1997). *The State of Ghana's Economy in 1996.* Legon, ISSER.

Jayarajah, C. and Ghani, E (1996). *Trade Reform, Efficiency, and Growth.* Washington, D. C: World Bank.

Kapul, I., M.T. Hadjimichael, P. Hilbers, J. Schiff, P. Szymczak (1991). *Ghana: Adjustment and Growth,* 1983-91. Washington, DC: IMF.

Kay, Godfrey (1972). *The Political Economy of Colonialism in Ghana: A Collection of Documents and Statistics.* Cambridge: Cambridge University Press.

Konadu-Agyemang, Kwadwo (1998). "IMF Sponsored Structural Adjustment Programs and the Perpetuating of Poverty in Africa: Ghana's Experience Revisited". *Scandinavian Journal of Development Alternatives,* Vol. 17: 3-4 127-144.

Loewenson, Rene (1993). "Structural Adjustment and Health Policy in Africa". *International Journal of Health Services,* Vol. 23 (4) 717-730.

Loxley, J. (1986). *Debt and Disorder: External Financing and Development.* Westview Press, Boulder Colorado and North South Institute.

Loxley, J. (1991). *Ghana: The Long Road to Recovery 1983-1990.* Ottawa: North South Institute.

Mackenzie, Fiona (1993). "Exploring the Connections: Structural Adjustment, Gender and the Environment". *Geoforum,* Vol. 24 (1) pp. 71-87.

Nafziger, E. Wayne (1990). *The Economics of Developing Countries.* Englewoods, N.J: Prentice Hall.

Ould-Mey, Mohammed (1996). *Global Restructuring and Peripheral States: The Carrot and the Stick in Mauritania.* Lanham, MD: Rowman and Littlefield.

Panford, Kwamina (1994). "Structural Adjustment, the State and Workers in Ghana". *Africa Development,* XIX (2) 71-95.

Rimmer, Douglas (1992). *Staying Poor: Ghana's Political Economy 1950-1990.* Oxford: Pergamon Press.

Roe, Alan and Hartmut Schneider (1992). *Adjustment and Equity in Ghana.* Paris: OECD.

Rothchild, Donald (1991). *Ghana: The Political Economy of Recovery.* Boulder, CO: Lynne Rienner Publishers.

Sahn, David E. (1994). *Adjusting to Policy Failure in Africa.* Ithaca: Cornell University Press.

Stewart, Frances (1995). *Adjustment and Poverty.* London: Routledge.

Watkins, Kevin (1995). *The Oxfam Poverty Report.* Oxford: Oxfam.

Weissman, S.R (1990). "Structural Adjustments in Africa: Insights from the Experience of Ghana and Senegal". *World Development,* Vol. 18 (12), 1621-1634.

Werlin, Herbert H (1994). "Ghana and South Korea: Explaining Development Disparities". *Journal of African and Asian Studies*, 3-4: 205-225.

World Bank (1981). *Accelerated Development in Africa: An Agenda for Action*. Washington DC: World Bank.

World Bank (1991). *Poverty Handbook*. Washington DC: World Bank.

World Bank (1993). *Ghana: 2000 and Beyond*. Washington DC: World Bank.

World Bank (1994). *Adjustment in Africa: Reforms, Results and the Road Ahead*. New York: Oxford University Press.

World Bank (1994). *Adjustment in Africa: Reforms, Results, and the Road Ahead*. New York: Oxford University Press for the World Bank.

World Bank (1995). *Country Briefs*. Washington DC: World Bank.

World Bank (1999). World Development Report. New York: Oxford University Press.

Zulu, J.B. and M. Nsouli (1985). *Adjustment Programs in Africa: The Recent Experience*. Washington, DC: IMF.

2 Structural Adjustment Programs and the Political Economy of Development and Underdevelopment in Ghana

KWADWO KONADU-AGYEMANG AND BAFFOUR KWAKU TAKYI

The questions to ask about a country's development are therefore: What has been happening to poverty? What has been happening to unemployment? What has been happening to inequality? If all three of these have declined from higher levels, then beyond doubt this has been a period of development for the country concerned. If one or two of these central problems have been growing worse, especially if all three have, it would be strange to call the result 'development', even if per capita income doubled (Dudley Seers).

2:1 Introduction

The decades since political independence have not been kind to many sub-Saharan Africans. Indeed, African countries emerged on the world scene following centuries of colonial rule brimming with hope and optimism for a new era of structural and social changes, otherwise called development. Well into the twentieth century, however, these high hopes of the original nationalist leaders such as Kwame Nkrumah of Ghana and Sekou Toure of Guinea and the masses have faded as these societies are yet to see the fruits of the development they were hoping to get during the liberation struggle. What happened? Why has development by-passed the majority of Africans as we begin the third millennium? These are among the issues we address in this chapter.

17

2:2 The Issue of Development

Social scientists have long debated the question of development, a concept widely understood to mean a transition or transformation from one state of growth to another. What constitutes development, and what is the ultimate goal of the development process? The quest for development has been at the root of the socio-economic and political initiatives of most nations over the years. Yet still, what constitutes development and what its ultimate goal should be has never been fully answered. While most post-World War II developmental experts equated development with modernization or westernization, the problems associated with such interpretation have been the subject of a considerable number of debates (Goulet, 1971; Seers, 1969; 1972; 1977; Toye, 1987; Hettne, 1990).

From the viewpoint of economists, development is synonymous with growth. Economic growth, according to economic thinking entails a sustainable rise in the standard of living of the people, which is often measured by such indicators as increases in per capita income (real income), the gross national product (GNP) and the gross domestic product (GDP). A related theme associated with development is that of changing economic structures as epitomized by increasing urbanization and industrialization.

While economic interpretation of development has guided most of the programs in post-independent Africa, and those of other developing countries, starting in the mid 1970s, some social scientist started to challenge this conceptualization of development. Rather than view development primarily in terms of increases in aggregate measures such as GNP, the sociological perspective goes beyond economic growth. Acknowledging that increases in output could in turn help improve the situation of the masses, the sociological approach to development looks at how goods and services are distributed among different groups and sub-populations, the alleviation of poverty, and an overall improvement in the life chances of the majority of the people. A more recent variant of this approach can be seen in the UNDP's Human Development Index that quantify progress differently from those relied on by traditional economists.

Our conceptualization of development is in line with the recent approaches. Borrowing from the works of Todaro (1996:16) and consistent with recent trends and theory on socioeconomic development, we argue that development entails a sustained elevation of an entire society and social system towards a better and humane life. In other words, development to us should include the benefits that are associated with "development". This would include an increased availability of basic life sustaining goods such as food, shelter, health and protection; raising the levels of living, including higher incomes, the provision of more jobs and better education, and expanding the range of economic choices available to individuals and nations by freeing them from servitude and dependence not only in relation to other people and nation states, but also the forces of ignorance and human misery (Todaro, 1996:18). We submit that if a program of action does not achieve these objectives, then its characterization as "development" will, in our opinion, be dubious.

2:3 From Modernization to Structural Adjustment Programs

The question of development and which approaches should be used in the African context has been a subject of debate over the years. Starting in the late 1950s, and intensifying soon after political independence in the1960s, modernization theory gained prominence, as a theoretical paradigm in explaining the state of underdevelopment in Africa, pointing to what Africans should do in order to develop. Indeed, social science thinking during the early period of independence argued that the only path to development that would move Africans from their supposed underdeveloped state, alleviate poverty and improve the quality of life was the one successfully taken by the Western European countries and the US. The "modernization" paradigm saw development as a linear path through which societies must travess in order to develop. This implied that African societies would follow the same patterns that had been pursued in earlier decades by North America and Western Europe, and had resulted in significant changes in these societies (Slater, 1986). The process of development was thus portrayed as unilinear, requiring the then developing countries to jettison certain behavioral, social and cultural traits that were considered to be antidevelopment, in favor of European traits. This would enable them, the argument went at that time, to achieve the characteristics of modernity. As Toye (1987) expresses it,

> Modern society supposedly had typical social patterns of demography, urbanization and literacy; typical economic patterns of production and consumption, investment, trade, government finance; and typical psychological attributes of rationality, ascriptive identity and achievement motivation. The process of development consisted, on this theory, of moving from traditional society, which was taken as the polar opposite of the modern type through a series of stages of development (p. 11).

This unilinear approach, while making some allowances for individual countries and regions, describes contemporary world history as the progression of each country from underdeveloped or traditional to developed and modern and postulates a series of two or more stages through which all countries are sooner or later alleged to pass. Thus the more "traditional" of the dual segments is seen as historically more archaic or less advanced, and it can see the image of its own future in the less traditional of the two segments. Nowhere is this approach to development as well exemplified as in the work of Walter Rostow. In his stages of growth model of development, he argued that all societies could be classified as lying within one level of five stages of economic development: the traditional society, the preconditions for take-off into self-sustaining growth, the take-off, the drive to maturity and the age of high consumption (Rostow, 1962). Most of the developing economies, including those in Africa, according to Rostow's model, were still in the traditional or pre-take off stages. Since the modernization paradigm was the dominant approach espoused by the development planners at the time, it became the basis of development.

The basic approach was to seek economic growth at all cost by concentrating on increasing the gross domestic product (GDP) and the Gross National Product (GNP) with the hope that the benefits thereby accrued would "trickle-down" to the whole society at large in the form of jobs and other economic opportunities, or create the necessary conditions for the wider distribution of the economic and social benefits of growth. Poverty reductions, income redistribution, access to services and improved quality of life were thus seen as byproducts of the development process (Todaro, 1996:14).

By the late 1960s, it was apparent that the so called development decade during which the newly independent countries were to be transformed into modern industrial societies had been a dismal failure, and neither the modernization experience nor the long awaited "trickle-down" benefits had materialized. When it became clear that the "Promised Land" was no where in sight, and that many of the developing economies that sought to follow the linear model were not progressing, the theory and practice of development based on this model came under serious criticisms (see for example, Wallerstein, 1974; Toye, 1987; Todaro, 1996; Bradshaw and Wallace, 1996).

Following the abysmal growth experienced by several Third World countries during the 1970s and 1980s, attempts at explaining what had happened during the early post-independence era in Africa emerged. This new thinking suggested that in large part, the problems that these LDCs had found themselves in were primarily their own making. Rather than see the exploitative relationship between the core industrialized nations and the peripheral raw material producing nations as a major cause of the "underdevelopment" in Africa, the new orthodoxy was aimed at "blaming the victims" for their own calamities. In this case, references were made to mismanagement, over-centralization of government as well as governmental intrusion into the operations of the market economy (Toye, 1987; Hettne, 1990). These arguments were given more credence by the political and civil instability that has been associated with post-independent Africa.

What was the next option available to cash-strapped African countries such as Ghana? Many LDCs felt they had been given the wrong prescription for their problems and called for a new international economic order (NIEO). The response of the World Bank and other institutions engaged in international development was in the form of a new philosophy of development centered on basic needs. The basis of this new approach was that the direct provision of goods and services could help alleviate poverty at a much faster rate than policies that seek to accelerate economic growth by pushing up the GDP at all cost. The key arenas tackled under this approach included primary education, shelter, rural development, nutrition and health (Thirlwall, 1995). The essence of basic needs strategies was not, however, to completely replace the growth at all cost paradigm, but to "supplement it by commitment of funds to the poorer sections of the community who were becoming worse-off despite a wider process of aggregate national growth" (Higgott, 1984:60). This new approach to development that was actively supported by the World Bank, the major industrialized nations and aid granting agencies, however, was short-lived and emphasis shifted back to policies aimed at achieving economic growth measured in terms of GNP and GDP, and resultant "trickle-down" benefits. The goddess of growth at all cost was thus resurrected, returned to its pedestal (Higgot, 1984) and re-christened "Structural Adjustment Programs" (SAPs).

Structural Adjustment Programs is the umbrella term used to describe a set of comprehensive economic policies aimed at altering the nature of a country's economy and the role of its government. Designed by the World Bank and the IMF, Structural Adjustment Programs are literally shoved into the throats of countries that are in dire need of international financing (The Development GAP, 1995). The programs have two major components, namely

= Stabilization of prices through balanced budgets, and
= Market liberalization/deregulation plus public sector reform as the environment for free-markets (Rojas, 1997; Gilpin 1994:547; Duncan and Howell, 1991:5).

Structural Adjustment Programs involve large-scale reorganization of the economic system from state control to private sector management and control. Closed markets are opened up and the forces of demand and supply allowed to not only determine the prices of goods and services, but also access to them. Thus it involves extensive divestiture of government owned businesses, retrenchment of workers employed by the state and quasi-state institutions, cut back on government expenditure on social services, and massive currency devaluation (Rojas, 1997). Since the early 1980s, almost all developing countries, particularly those in Latin America and Africa have been undergoing Structural Adjustment Programs. Although the monetarist principles involved in Structural Adjustment Programs have been applied by the World Bank and the International Monetary Fund on developing countries since the late 1950s, it was not until the late 1970s that the Bretton Woods Institutions adopted them as their principal policies toward economic development in the so called Third World (Rojas, 1997). Following the Mexican debt crisis of 1982, Structural Adjustment Programs became practically synonymous with lending from international financial institutions like the World Bank and the IMF, and virtually any country that wants low-interest loans or debt rescheduling must implement a Structural Adjustment Program (Jubilee 2000, 1998).

Given that most of the countries requesting financial aid from international banks are often in desperate situations, the terms offered to them by the sponsors border around what Adedeji (1995) has described as "swallow-or-starve". These conditions almost invariably include trade liberalization, privatization of state enterprises, deregulation, credit reduction, wage suppression, currency devaluation and massive cuts in government expenditure on health, education and other social services. It also includes shifting agricultural and industrial production from staple foods and basic goods for domestic use to exportable commodities. In effect, like modernization, the development approach taken by Structural Adjustment Programs seek to get rid of, or minimize the impact of, the so called internal constraints or bottlenecks that are perceived as being the major factors responsible for the state of underdevelopment, and to openly embrace capitalism as the ideal path to development.

2:4 Structural Adjustment Programs in Ghana: A Path to Development?

Following several years of near economic collapse dating from the mid-1960s, but especially during the late 1970s, Ghana launched an economic recovery program (ERP) in 1983. Developed in close collaboration with the World Bank and the IMF, this program was aimed at reversing the economic decline which was evidenced by deteriorating real export earnings, very low domestic savings and investments, inflation rate in excess of 100 percent, lax financial management, and extensive government involvement in the economy (IMF, 1999a; Dordunoo and Nyanteng, 1997). Although some scholars have argued that the programs carried on between 1983 and 1986 (i.e the stabilization period) should not be considered as part of the Structural Adjustment Programs that actually took off in 1986 (see Adepoju, 1993), technically speaking they represent the first phase of the programs. Therefore for the purpose of this discussion, the ERP and the "real adjustments" are lumped together as Structural Adjustment Programs.

Ghana's economic restructuring programs have been implemented in six phases:

Phase 1: 1983-1986 Period of Economic Stabilization

The principal objectives of phase one were to compress government expenditure from the 1982 level of 10.2 percent to 8.6 percent of GDP, adjust the exchange rate through discrete devaluation of the cedi, abolish domestic price controls, mobilize government revenue through broadening of the tax base and strengthening of tax administration (Dordunoo and Nyanteng, 1997; World Bank, 1995a).

Phase 2: 1987-1989

Phase 2 also, among other things, aimed at initiating structural changes to address the deep-rooted causes of imbalances in the economy and rebuilding the productive base; a comprehensive liberalization of the exchange rate and trade system; privatizing state enterprises; reforming and reducing the size of the public service; reforming and improving the tax system and its administration; institutional and financial reforms to strengthen the domestic banking system; and the rehabilitation of economic infrastructure, key export industries and the public sector (Anyinam, 1994; Dordunoo and Nyanteng, 1997).

Phase 3: 1990-1993

The third phase which started in 1990, aimed to build upon the two previous phases by implementing structural and institutional reform. Decisive steps were taken towards a market economy by completing exchange rate reforms; increasing reliance on market-based instrument of monetary policy; concentrating on, civil service reform, public expenditure and poverty alleviation programs.

Furthermore, tax policy changes and cocoa sector policy reform were undertaken with the aim of achieving an average annual real GDP growth of at least 5 percent; reducing the average rate of inflation from 37 percent in 1990 to 5 percent in 1993; and generating an overall balance of payments surplus of at least $90 million a year (World Bank, 1995b; Anyinam, 1994:451).

Phase 4: 1994-1996

In Phase 4 that began in 1994, the primary objectives of government programs, among other things, were to reduce inflation, support a realistic exchange rate, complete the privatization of State enterprises, and promote the efficient allocation of resources for growth (IMF, 1998; Armstrong, 1996).

Phase 5: 1995-1997

The program for 1995-97 sought to accelerate real GDP growth to 5.5 percent by the end of the period, and to 6 percent by the end of the decade, while reducing inflation to 20 percent and the external current account deficit to 2.7 percent of GDP by 1997. It also aimed at improving the environment for private savings and investment by pursuing sound financial policies consistent with reducing the inflation rate and maintaining a stable exchange rate. The structural policies were directed at providing a supporting environment to private sector initiative and foreign investment through curtailing public sector involvement in the economy (IMF, 1999a).

Phase 6: 1998-2000

Under the 1998-2000 Program, the government's overall objective was to secure a stable macroeconomic environment that supports economic growth led by the private sector, thereby creating jobs, increasing incomes, and reducing poverty. The key macroeconomic targets of this period were (1) to achieve annual real GDP growth of 5.6, or 2.5 percent on a per capita basis; (2) to reduce annual inflation from 20.8 percent in 1997 to 11 percent by the end of 1998, and further halve it to 5.5 percent by the end of 1999; and, (3) to contain the current account deficit at 7.3 percent of GDP while maintaining gross official reserves at 2.7 months of imports (IMF, 1999a).

2:4.1 Accomplishments of Economic Restructuring Programs in Ghana

By all accounts, a dramatic transformation has indeed taken place in Ghana over the past 16 years. Ghana's dire economic situation in the pre-1983 era has changed for the better, judging by standard economic indicators. Among other things, the implementation of Structural Adjustment Programs has resulted in a consistent GDP growth of 4-6 percent, a tremendous increase from an annual average of 1.5 percent in 1970-83. Even with an average population growth of 2.6 percent over the past 16 years, Ghana has been experiencing real per capita growth rate of more than 2 percent per year. The fact that the rate of per capita GDP growth has remained continuously positive since 1984 (with the exception of 1990) is also impressive. In addition to the impressive growth in GDP, real incomes per head have also grown by an average of 2 percent. The Structural Adjustment Programs era has also seen major improvement in infrastructure services, and a shift towards developing non-traditional export goods like arts and crafts and horticultural produce. Furthermore, industrial capacity expanded from about 25 percent of installed capacity before 1984 to 35-46 percent in the 1990s, while export goods production more than tripled between 1986 and 1998. More significant, annual inflation also dropped from 123 percent p.a. in 1984 to 32 percent in 1991, 34 percent in 1994; and 29 percent in 1997 and 10 percent in 1999 (IMF, 1999b; ISSER, 1999; GSS, 1998; Armstrong, 1996; Anyinam, 1994; Sowah, 1993; Boachie-Danquah, 1992; Kapul *et al.*, 1991; Jonah, 1989). The programs have also boosted international confidence in Ghana's economy and therefore helped to attract foreign capital especially in the areas of telecommunication, banking, mining and infrastructure.

Considering all these apparently impressive improvements, the World Bank and IMF, and indeed the Western industrialized nations, have often touted Ghana's case as the most successful economic restructuring program in Africa, a model for others to emulate. But impressive as these achievements are, can we say with any certainty that development has taken place? Should the growth in GNP and GDP, drop in inflation, attracting foreign investment, and increasing industrial productivity be seen as conclusively engendering development?

2:4.2 The Cost of Structural Adjustment Programs: A Bitter Pill

Ghana's development programs in the past 16 years have involved much borrowing to the extent that the country is listed among the 41 highly indebted poor countries (HIPC) in the world. Its total stock of external debt amounted to about US$7,510 billion in 1999, 75.6 percent of which are multilateral debts. In addition, Ghana's total debt has been growing at an average of 7 percent each

Table 2:1 Inflation Table, 1970-1999

Consumer Price Index and Rate of Inflation, 1970-1999 (1977=100)			
		Inflation	
Year	Combined	Food	Non-Food
1970	3.85	2.39	1.94
1971	8.76	6.20	5.32
1972	10.07	16.06	10.11
1973	17.68	18.87	10.82
1974	18.13	20.11	23.96
1975	29.82	33.48	26.25
1976	56.08	44.22	34.40
1977	116.45	128.83	40.65
1978	73.30	59.40	81.10
1979	54.24	61.67	67.92
1980	50.09	52.31	59.52
1981	116.53	111.16	62.96
1982	22.30	35.67	41.28
1983	123.00	144.98	76.85
1984	40.16	11.07	78.63
1985	10.37	-11.16	29.27
1986	24.57	20.27	25.33
1987	39.82	38.48	37.56
1988	31.36	34.11	30.25
1989	25.23	25.09	24.65
1990	37.26	40.14	37.37
1991	18.07	8.97	25.21
1992	10.02	10.37	9.51
1993	24.96	24.96	26.50
1994	24.87	25.88	23.21
1995	59.47	62.21	56.40
1996	46.56	35.77	53.21
1997	29.00	20.30	31.10
1998	15.20	21.00	13.20
1999	15.70	7.71	16.20

Source: *Ghana Statistical Service; 1989, 1999 CEPA, 1999*

year since 1987 and now amounts to over $350 per person (Department of International Development, 1999). As a result of extensive borrowing under the restructuring programs, debt service now absorbs a huge chunk of Ghana's export revenues. Debt service ratio increased from only 3.7percent of export goods in 1977 to 45 percent in 1987. Although the debt-service ratio dropped to 32.6 percent in 1997 and again to 28 percent in 1998, the debt commitment is far too high and has greatly eroded the ability of the state to cater for educational, health and other needs. As seen in Table 2:2, Ghana's total debts/GDP ratio also doubled from 33.4 in 1977 to 64.6 in 1987 and climbed further to 93.5 percent in 1998 (World Bank, 1999; CEPA, 1999; Bank of Ghana, 1997, 1998).

The IMF alone receives close to one-third of Ghana's expenditure on external debts, and the Fund together with the World Bank and other multilateral organizations based in the G-7 nations receive more than 70 percent. The rest is owed to these powerful nations who also dominate the IMF and the World Bank. The dominance of both Bretton Woods Institutions, the architects of the adjustment programs, by the G-7 nations gives them a chance to promote open and free markets, which is profitable for the developed countries, but not for the fragile economies of many developing countries. The powerful countries also benefit in a more direct way from the loans given by the World Bank to the developing countries. In 1992, for instance, more than half of the World Bank's loans were used by the recipient countries to purchase foreign goods and services. About seventy percent of these moneys went to the rich donor nations (Hittle, 1992, p.13). Looking at the evidence, one cannot help but to agree to with Hittle's assertion that

> The World Bank is a body whose loans allow a country to go into debt in order to purchase developed country goods or developed country advice. The Bank allows donors to create the appearance of providing "billions of dollars" to economically stricken areas, when in fact, they are simply providing the billions to their own contractors to do work in less developed regions. The real transfer is, in effect, between the public and the private sectors of the developed countries (Hittle 1992, p.14).

Added to the huge foreign debt is the increasing central government domestic debt that rose from 2.8 percent of GDP in 1990 to 15.6 percent in 1997. Domestic debt servicing now accounts for 20 percent of government expenditure. The huge external indebtedness and the very high percentage of export earnings apportioned for debt servicing have had grave ramifications for adjusting countries like Ghana who have lost their ability to use their earnings to service their own socio-economic programs and to make long term investments. Ghana now spends about four times more on debt servicing than on health care (Unicef/Oxfam, 1999).

Another salient feature of Structural Adjustment Programs that has had severe consequences for the economy is the massive devaluation of the Ghanaian currency, the *cedi*. Due to Structural Adjustment Program's mandated currency devaluation, the value of Ghana's currency *vis a vis* the US dollar, has been reduced by over 120,000 percent in the past 16 years (i.e. from 2.75 cedis =US$1 in 1983 to 3400 cedis =US $1 in 1999) as shown in Table 2:3. The government first adjusted the exchange rate in April 1983 as a first step in the adjustment process, and the value has since been falling. Within the first three years alone, the currency was devalued by 3,172 percent (i.e. from 2.75 to 90 cedis =$1). While the principal purpose of currency devaluation is to stimulate exports by making export products cheaper, this strategy has not worked well for Ghana since the country's primary produce have to compete with those of other developing countries producing similar commodities, and using the same devaluation strategy. Indeed, the IMF stimulates all the poor countries to simultaneously lower prices and to produce more for a market, which is already saturated. As a result, the country faces the prospect of having to exploit and export more resources in order to reach or exceed previous earning levels. Moreover, devaluation leads to higher import prices, which is often harmful for developing countries as imports often make up a large share of the economy. Ghana's dependence on imports for essential supplies such as fuel, medicines and machinery has made Structural Adjustment Programs unconditional devaluation a serious burden. The effects of higher import prices has been to increase the cost of living through its effects on the prices of all items that are either imported or is manufactured or serviced with imported components and fuels. In this case virtually everything bought or sold in Ghana is affected.

2:4.3 Unemployment and Underemployment

The Structural Adjustment Programs period has also seen considerable cut backs in both government and private sector employment. A key conditionality of Structural Adjustment Programs requires the governments of adjusting countries to cut back on public sector employment and thereby reduce government expenditure and improve public sector performance. Between 1983 and 1992, 200,000 public sector employees were retrenched. The Cocoa Marketing Board (CMB) alone retrenched 40,000 workers in 1985, and an additional 12,000 were

Table 2:2

Ghana's Debt									
	1977	1987	1992	1994	1995	1996	1997	1998	1999
External Debt	1,067	3,287	3969	5,022	5,074	4,347	6,345	6,900	7,510
External debt/GDP (%)	33.4	64.6	60.7	97.1	82.1	92.2	92.0	93.5	93.1
Total debt service/exports	3.7	45.8	25.1	27.1	35.8	35.7	32.6	28.0	24.5
Total debt service	3.7	45.8	25.1	27.1	35.8	35.7	32.6	28.0	24.5
Included IMF (US$ M)	n/a	38	279	376	566	473	536	552	572
Excluding IMF (U S$ M)	n/a	n/a	193	282	447	337	363	390	n/a
IMF Share debt service (%)	n/a	31	25	21	29	32	27	22	13.7
Multilateral (incl. IMF)	n/a	n/a	n/a	n/a	n/a	n/a	74.1	75.0	75.6
Bilateral	n/a	n/a	n/a	n/a	n/a	n/a	25.3	24.4	24.4

Sources: *Bank of Ghana, Accra; CEPA (1999)*

retrenched in 1987. The private sector also retrenched more than 48,000 workers between 1987 and 1995 (see Chapter 10).

The massive cut backs by both government and private sector has reduced the size of the labor force in formal sector employment to less than 10 percent. This has resulted in high levels of unemployment and underemployment especially among educated people, including university graduates, who have traditionally been employed by this sector. Unemployment and underemployment rose from probably about 10 percent in 1980 to 18.5 percent in 1987 and to 21.3 percent in 1993. In 1999, 20.3 percent of the work force was unemployed. Given the poor reporting system, these figures may represent an undercount.

It was envisaged that retrenched public sector workers could be absorbed by the private sector, the engine of development under Structural Adjustment Programs. However, given the limited elasticity, small size and under-capitalization of the private sector, it has not been able to absorb a fraction of the army of retrenched workers. (UNDP, 1997a; Ocran, 1998). Even sectors of the economy like mining, quarrying, and construction which have experienced tremendous growth under the adjustment programs are mainly capital intensive or require highly skilled labor, and therefore have not made any meaningful impact in reducing unemployment. In any case, these sectors account for only 2 percent of total formal sector employment (Ocran, 1998). The 12-16 percent growth in tourism and related services during the Structural Adjustment Programs era may have resulted in 17,000 direct and 45,000 indirect jobs. These jobs, however, tend to be short term and low paid, and provide little or no with little employment protection (Ocran, 1998). Available evidence points out that the only sector that has been able to expand to any appreciable extent is the informal sector. The high involution associated with this sector has enabled it to absorb thousands of retrenched workers from the formal sector, particularly women, as well as the underemployed and new entrants to the labor market. While the rate of employment creation in the sector was 6.5 percent (compared to only 1 percent in the formal sector), the employment generated by this sector has not kept pace with the rate of growth of the labor market or the increasing numbers of unemployed and underemployed. Moreover, the expansion of this sector with its attendant problems of low income, absence of employment protection etc., raises questions about the quality and character of employment. Furthermore, increasing emphasis on informal sector work is likely to create more unskilled, unprotected work with low levels of pay. More importantly, it is unlikely to reduce inequalities in employment based on gender, class, disability and geographical location (Ocran, 1998).

2:4.4 Drastic Cutback in Social Services

Another salient feature of the adjustment programs is the requirement that the state in adjusting countries makes drastic cuts in government expenditure on social services like education, health and welfare. In an effort to meet these conditions, the government of Ghana made an onslaught on these services and shifted a significant portion of the cost to the beneficiaries of these services by introducing user-pay system for health and education. Although state expenditure on health care may have fallen from 8.2 percent of public expenditure in 1974 to 4.3 percent in 1983 prior to the introduction of the restructuring programs, the Structural Adjustment Programs era has worsened the situation. While available data may well indicate that public expenditures on health actually increased from an average 0.8 percent of GDP in 1981-86 to 1.3 percent of GDP in 1987-90, and to in 1.7 percent 1999, these funding levels are way below the 1970s level when expenditure on health averaged around 3-4 percent of GDP. Under current spending levels, the government healthcare expenditure amounts to $6 per capita. This is woefully inadequate and is far lower than the funding in most developing countries. The inadequate funding of the public health sector has engendered a number of problems that include dilapidated, outmoded and insufficient infrastructure, understaffed facilities, low pay and low morale that has resulted in massive brain drain of health professionals out of the public sector health system and also out of the country. For instance, while in 1965 the physician population ratio was 1:13,740, it was 1:22,000 in 1998 (World Bank, 1990; 1999). The number of nurses, pharmacists and paramedic has also dwindled along similar lines (see Chapter 16).

One observable feature of healthcare delivery in Ghana is spatial disparity and unequal access. The rural areas do not receive a fair share of the health care budget and one's place of residence determines to a great extent one's access to available health care services. Although only 36 percent of Ghanaians reside in urban areas, they accounted for over 42 percent of the total government health budget and over 50 percent of total outpatient spending in 1989 (GLSS, 1992). By 1992 this urban share had increased to 49 percent and 55 percent respectively. The 1992 GLSS also showed that whereas the top 20 percent of the population, according to income, received 33 percent of government expenditure on health, the poorest 20 percent enjoyed only 12 percent of this expenditure (Hormeku, 1997). It is estimated that 8.36 million people living in 47,000 rural settlements do not have any, or ready, access to the basic government-provided health facilities, which are largely urban based (Batse *et al.*, 1999). While the urban-bias nature of healthcare distribution has long existed, the so-called poor people friendly nature of the World Bank programs have not redressed the rural-urban imbalances, and if anything, seems to have become worse. Indeed, enough evidence can be adduced to support the fact that the introduction and aggressive implementation of user pay-system in healthcare delivery has forced people to flee the system and to pursue alternative treatments that in any case have always been the most accessible and affordable by Ghanaians (Batse *et al.*, 1999:10; GDHS, 1994; Enyimeyew, 1987).

Regional disparities are also apparent in the distribution of and access to health services. The "core" regions like Greater Accra have more healthcare personnel per capita than the "periphery" regions such as the Northern, Upper West and Upper East. While the entire population of the Accra Metropolitan Area had access to health facilities, only 11 percent of the population in the North had access compared with 77 percent in the Central Region and 26 percent in the Western Region (Batse *et al.*, 1999). The adjustment programs have not been able to redirect resources to the undeserved regions.

Acute malnutrition has also been a fact of life in an era of Structural Adjustment Programs. A 1990 report by the United Nations Children's Fund (UNICEF), showed that 30 percent of all Ghanaian children are malnourished to some degree, 28 percent of the 12 to 23-month-olds was considered wasted and 31 percent of 24 to 59-month-olds was stunted (GSS/DHS, 1994). The 1998 De-

Table 2:3 Changing Values of the Cedi *vis a vis* the U.S. dollar
1979-1999

Years	cedis/US$	% of Change
1979	2.75	
1980	2.75	0.00
1981	2.75	0.00
1982	2.75	0.00
1983	8.83	221.00
1984	36	307.70
1985	54	50.00
1986	89	64.81
1987	153	71.91
1988	202	32.00
1989	303	50.00
1990	345	13.86
1991	390	13.00
1992	520	33.33
1993	815	56.73
1994	982	20.49
1995	1200	22.20
1996	1500	25.00
1997	2000	33.33
1998	2350	17.50
1999	3400	44.68
1979-1999		124,000

Sources: Ghana Statistical Service, 1989;1999; www.Africaonline.com

mographic and Health Survey also reports that malnutrition is high in Ghana, with one in four children under five years of age stunted, 10 percent wasted and 25 percent underweight. Eleven percent of all women are also considered malnourished, falling below the cut off point of 18.5 kgm/m2 for the body mass index (GSS/DHS, 1999). Regional variations also exist in the levels of malnutrition. Forty-five percent of all women and 65 percent of pregnant women in northern Ghana are malnourished compared to 30 percent and 45 percent respectively in the south. Compared to the 1960s and 1970, these are the worst of times for infant and maternal nutrition.

Education, like health has also suffered from government cutbacks. Whereas education expenditure represented 4-6 percent of GDP and 21 percent of government expenditure in the immediate pre-Structural Adjustment Programs years, it accounted for only 2-3 percent of GDP throughout the 1990s (compared to an average of more than 6 percent for Sub-Saharan Africa). While considerable efforts have been made to improve access to education considering the fact that primary school enrollment increased by 31 percent between 1987 and 1994 (GNDPC, 1997), the introduction of a number of user fees that include book costs, furniture, and building fees has driven a number of poor children out of school. Surveys by the Ministry of Education and UNICEF show that in addition to the inability to pay fees, students also stay out because there are a large number of schools which either operate in dilapidated buildings with no protection from the natural elements, or are without buildings altogether. It is estimated that 23 percent of all primary schools operate in classrooms with leaking roofs, and in 50 percent of primary schools in 14 districts do not have any classroom facilities at all (MOE/UNICEF, 1993; GSS, 1993). Those who stay in school in the rural areas and deprived regions do not have access to the teaching and learning facilities that are available to their urban counterparts. Since the implementation of the Educational Reform Program in 1987, the chances of poor rural children attending the well-endowed schools and consequently making it to the universities have been diminished. Indeed, Ministry of Education statistics indicates that students who attend private preparatory and international schools do better in the Basic Education Certificate Examination (BECE), and therefore get a disproportionate share of the slots in the elite senior secondary schools. These private schools are, without exception, located in the urban areas and may charge hefty basic fees that can amount to over $2500, effectively two full years gross income of a senior high school teacher. It is those who attend the elite schools who also gain access to the few universities in Ghana (Frimpong, 1999). If this trend continues, then it should be expected that the increasing inequalities that have been perpetuated under the auspices of the adjustment programs would continue to get worse.

2:5 Increasing Poverty

Another controversial aspect of the adjustment program is its impact on poverty alleviation. Indeed, the World Bank has claimed over and over that its primary objective is poverty alleviation. In reality, however, there may be more poor people in Ghana now than there were in the 1970s, and inequalities between the rich and the poor seem to be widening. If living in poverty is defined as subsisting on less than two-thirds of the national average income, then 37 percent of Ghanaians lived below poverty line in 1988. An additional 7 percent who subsisted on less than one-third were classified as "ultra poor". According to the Ghana Living Standards Surveys [GLSS] (GSS, 1988; 1992; 1995), about 80 percent of the poor live in the rural areas. Another dimension of poverty in Ghana is the regional variations. For instance, whereas the three savannah regions of the country–Northern, Upper West and Upper East–are home to only 12 percent of the population, they have 18 percent of the country's poor and 35 percent of the extremely poor. The benefits of the restructuring programs have not yet trickled down to those who live in the natural resource poor savanna, and who produce food for the local market and do not participate much in the export market (UNDP, 1997b).

Although the incidence of poverty in the nation as a whole dropped to 32 percent in 1992, the face of poverty changed from being a rural phenomenon to encompass urban areas as well. Poverty in Accra, the capital city increased from 8 percent to 23 percent, a situation attributable to Structural Adjustment Programs induced cut backs in formal sector jobs, increased urban migration and the introduction of cost sharing strategies. In addition, the wealth gap between the richest and the poorest in society is much wider than ever before with the lowest 40 percent of the population sharing less than 20 percent of the national income while the top 20 percent have 44 percent.

Contrary to what the program sponsors and the ruling government have been telling the world about Structural Adjustment Programs in Ghana, life has not improved for a majority of Ghanaians. Indeed, the nation has been sliding down on the UNDP's HDI scale. Ghana ranked 129th among 174 countries in 1995 and 133rd in 1999. The IMF has indicated that at the recent rate and pattern of economic growth, it will take the average poor in Ghana at least 30 years to cross the poverty line.

2:6 Conclusions

We conclude the chapter by revisiting the two questions we raised at the beginning. What is development, and what should its ultimate goals be? If development is conceptualized as leading to improvements in quality of life measured not only in terms of per capita income growth, but more important, access to life sustaining goods such as shelter, food health and protection; access to more jobs, better education, freedom from servitude, ignorance and human misery, then there are nagging questions about Ghana's accomplishments under Structural Adjustment Programs. The economic restructuring program in Ghana may have succeeded in effecting essential macro-level structural changes. However, it has not succeeded in improving the quality of life at the micro level. Unemployment, poverty, and socio-economic disparities have not disappeared, and if anything appear to be worse. Basic indicators of well being such as malnutrition and under five mortality rates have not improved very much. Primary school enrollment has dropped, and the percentage of GDP spent on education and health is still below the level reached in the 1970s. While Ghana has managed to achieve growth rates of 4-6 percent throughout most of the adjustment era, it's been projected that the average poor person will have to wait for 30 years to raise their incomes above the poverty line. The ultra-poor will have to wait 40 years more. Among other things, the slow rate of poverty alleviation has been attributed to the fact that a large segment of the population, who live in the resource poor and historically deprived areas as well as in many rural communities, have been excluded from the growth process. Moreover, the unprecedented retrenchment of formal sector workers and high unemployment levels has created a new generation of urban poor (Oxfam, 1994). As aptly summed up by Boachie-Danquah:

> The broad mass of working people, including the middle class, wage laborers as well as lower-level employees in both the public and private sectors, have been seriously affected by retrenchment and redeployment policies, which are critical components of Structural Adjustment Programs. Even for those who remain, the decline in real earnings of both the top and lower echelons has led to mass impoverishment. The escalating prices of all basic goods and the astronomical increases in the cost of public services, brought about by the adherence to the Thatcherite principle of value for money and cost-recovery introduced in tandem with Structural Adjustment Programs, have added to the catalogue of woes. The demands of the Structural Adjustment Program have eroded the standard of living of most working people (1992:246).

Ghana's is not an isolated case. Wherever Structural Adjustment Programs have been the main staples of economic management, unemployment, poverty and inequalities tend to increase (Stewart, 1995; Development GAP, 1995). In the words of an Oxfam Report, structural adjustment programs are undermining recovery prospects, compounding inequalities, undermining the position of women, and failing to protect access to health and education services. Project interventions often continue to cause unacceptable and unacceptably violent human displacement and environmental damage (Oxfam, 1994). Data from the 19 highly intensively adjusting countries support this conclusion (See Table 2:4).

While the Bretton Woods institutions may see development as being synonymous with economic growth measured in terms of Gross National Product per capita, the problems associated with this view are well documented (Arrighi and Saul, 1986; Streeten, 1995). National economic performance is commonly measured through the GNP. However, "economic success is not the same thing as welfare, and measures of economic success are not necessarily measures or indices of welfare. There are a number of reasons for this. In the first place, economic assets do not solely determine welfare. Health, individual well being, quality of life, environmental quality, individual and collective security, all make contributions to welfare which are not reflected in GNP" (Streeten, 1995:32). Since the GNP per capita "does not reflect the distribution of wealth in society, nor does economic growth over time indicate changes in the distribution of wealth, economic growth may well take place without either reducing the poverty of a majority of the population or developing structures necessary to permit self-supporting progress" (Samoff and Samoff, 1976:71). More important, it should be emphasized that economic success does not necessarily translate into welfare and well being for the citizens. Indeed, the UNDP has introduced the Human Development Index (HDI), which consists of the logarithm of GDP per head, calculated at the real purchasing power, not at the exchange rates, up to the international poverty line; literacy rates and life expectancy, as better criteria for gauging development (Streeten, 1995). Dudley Seers captured the essence of this argument succinctly three decades ago, when he wrote:

> The questions to ask about a country's development are therefore: What has been happening to poverty? What has been happening to unemployment? What has been happening to inequality? If all three of these have declined from higher levels, then beyond doubt this has been a period of development. If one or two of these central problems have been growing less, especially if all three have, then it will be strange to call the result development, even if per capita income doubled (1969: p.3).

While nobody disputes the need for the national economy to be efficient, the methods that are being used seem to sacrifice the total well being of the weak and vulnerable, who cannot survive in an environment where unbridled market forces determine access to resources and services. Development should be re-

Table 2:4 The Performance of the 19 Highly Intensively Adjusting Countries

Country	Annual per capita GDP	Annual GDP growth rate (%)	Trend in poverty/ inequality	
	1980-90	1965-80	1980-90	
Argentina	-1.7	3.4	-0.4	increased rural poverty
Bolivia	-2.6	4.4	-0.1	increasing poverty
Cote D'Ivoire	-3.2	6.8	0.5	increasing poverty
Ghana	-0.3	1.3	3.0	increasing rural poverty
Jamaica	0.3	1.4	1.6	not known
Kenya	0.4	6.8	4.2	increasing rural poverty
Malawi	-0.5	5.5	2.9	increasing rural poverty
Mauritania	-1.0	2.1	1.4	not known
Mexico	-1.0	6.5	1.0	increasing rural poverty
Morocco	1.4	5.7	4.0	declining rural poverty
Pakistan	3.1	5.2	6.3	rising inequality
Philippines	-1.5	5.7	0.9	increasing poverty
Senegal	0.0	2.3	3.0	not known
Togo	-1.8	4.3	1.6	not known
Tunisia	1.3	6.5	3.6	declining rural poverty
Turkey	2.6	6.2	5.1	not known
Uganda	0.3	0.6	2.8	declining poverty
Venezuela	-1.7	3.7	1.0	increasing poverty
Zambia	-2.8	2.0	0.8	increasing poverty

Sources: *Khan, 1993*

conceptualized to involve more than generating higher per capita incomes. It should encompass as ends in themselves better education, higher standards of health and nutrition, less poverty, a cleaner environment, more equality of opportunity and greater individual freedom (World Bank, 1991:4). It must be conceived as a multi-dimensional process involving major changes in social structures, popular attitudes, and national institutions as well as the acceleration of economic growth, the reduction of inequality, and the eradication of poverty. Development, in its essence must represent the whole gamut of change which an entire social system, tuned to the diverse basic needs and desires of individuals and social groups within the system, moves away from a condition perceived as unsatisfactory towards a situation or condition regarded as materially and spiritually better (Todaro, 1996:16). Since the Structural Adjustment Programs in Ghana have failed to address these concerns, it's our opinion of that the programs will have to be redesigned and given a more human face (Cornia *et al.*,1988).

Bibliography

Adedeji, A. (1995). "An Alternative for Africa" in: Larry Diamond and Marc F. Plattner, eds. *Economic Reform and Democracy*. Baltimore: Johns Hopkins University Press.

Adepoju, Aderanti (1993). "Introduction" in: Aderanti Adepoju, ed. *The Impact of Structural Adjustment on the Population of Africa: The Implications for Education*. Health and Employment. Portsmouth: UNFPA: 1-6.

Anyinam, Charles (1994). "Spatial Implications of Structural Adjustment Programs in Ghana". *TESG 85* (5): 446-450.

Armstrong, R. P. (1996). *Ghana Country Assistance Review*. Washington, DC: World Bank.

Batse, Z.K.M., Botschie, G. and Agyemang Mensah, M. (1999). *Integrating Capacity Building Within the Context of Social Policies for Poverty Reduction in Ghana*. Dakar: IDRC.

Boachie-Danquah, Yaw (1992). "Structural Adjustment Programs and Welfare Interventions: The Case of Ghana". *Africa Insight* 22 (4): 244-248.

Bradshaw, Y. and Michael Wallace (1996). *Global Inequalities*. Thousand Oaks, California: Pine Forge Press.

CEPA (1999). *Ghana Macroeconomic Review and Program*. Accra: CEPA.

Cornia, Giovanni, Andrea, Richard, Jolly and Frances Stewart (1987). *Adjustment with a Human Face*, Vol. I. Oxford: Clarendon.

Department of International Development (1999). *Ghana Country Strategy Paper*. London: DIP.

Development GAP (1995). "Structural Adjustment Programs at the Root of Global Crisis". *Case Studies from Latin America*. Paper Prepared for the Social Summit, Copenhagen.

Dordunoo, C.K. and Nyanteng, V.K. (1997). "Overview of Ghana's Economic Development" in: V. K. Nyanteng, ed. *Policies and Options for Ghanain Economic Development*. Legon, Accra: ISSER.

Duncan, Alex and John Howell (eds.) (1991). *Structural Adjustment and the African Farmer*. London: Zed for ODI.

Enyimayew, Kwesi (1988). *Cost and Financing of Drugs Supplied in Ghana: The Ashanti-Akim Experience*. Paper presented at WHO Conference, Geneva, June.

Frimpong, J. (1999). *Funding Education in Ghana*. <www.graphic.com.gh/features/af4.html>.

Ghana Government (1998). *1998 Budget Statement*. Accra: Government Printer.

Gilpin, Raymond (1994). "Pill or Poison?". *West Africa*. 28th March - 3 April.

GNDPC (1997). *Ghana-Vision 2002*. Accra: Government of Ghana.

Gogue, T.A. (1996). "Impact of Structural Adjustments on school attendance: The case of Togo". *Development Studies*, Vol. 17: (2) 221-239.

Goulet, D. (1971). "The Cruel Choice". *A New concept in the Theory of Development*. New York: Atheneum.

GSS/GDHS (1994). *Ghana Demographic and Health Survey*. Accra: Ghana Statistical Service and Macro International.

GSS (1988). *Ghana Living Standard Survey*. Accra: Ghana Statistical Service.

GSS (1989). *Quarterly Digest of Statistics, June.* Accra: Ghana Statistical Service.

GSS (1989). *Ghana Living Standard Survey.* Accra: Ghana Statistical Service.

GSS (1992). *Ghana Living Standard Survey.* Accra: Ghana Statistical Service.

GSS (1993). *Quarterly Digest of Statistics, June.* Accra: Ghana Statistical Service.

GSS (1995). *The Pattern of Poverty in Ghana.* Accra: Ghana Statistical Services.

Hettne, B. (1990). *Development Theory and the Three Worlds.* Harlow, Essex: Longman.

Higgot, R. (1984). "Export Oriented Industrialization, the NIDL, the Corporate State and the Third World". *Australian Geographical Studies,* 27(1) 58-71.

Hittle, Alex (1992). *World Bank: New Items for Central and Eastern Europe.* Vol. 1. Friends of the Earth.

Hormeku, T. (1997). *Ghana: The Numbers in the Last Five Years.* Montevideo, Uruguay: Social Watch.

IMF (1999a). *Ghana – Enhanced Structural Adjustment Facility: Economic and Financial Framework Paper,* 1998-2000. Washington, DC: IMF.

IMF (1999b). "Ghana", Selected Issues. *Staff Country Report No. 99/3.* Washington, DC: IMF.

Institute of Statistical, Social and Economic Research (ISSER) (1993). *The State of Ghana's Economy in 1992.* Legon: ISSER.

Institute of Statistical, Social and Economic Research (ISSER) (1995). *The State of Ghana's Economy in 1994.* Legon: ISSER.

Jonah, K (1989). "Social Impacts of Ghana's Adjustments Program 1983-1986" in B. Onimode ed. *The IMF, the and the African Debt,* Vol. 2. London: Zed Press.

Jubilee 2000 (1998). *Structural Adjustment Programs.* Washington, DC: Jubilee 2000.

Kapur, I., M.T. Hadjimichael, P. Hilbers, J.Schiff, P. Szymczak (1991). *Ghana: Adjustment and Growth,* 1983-91. IMF.

Karnik, A. (1996). "Why Structural Adjustment Programs Go Awry?" Political Economy Perspective. *Cybernetics and Systems,* 27: (1) 93-103.

Kendie, S.B. (1995). "The Environmental Dimensions of Structural Adjustment Programs: Missing Link to Sustaining Development". *Singapore Journal of Tropical Geography,* 16: (1) 45-57.

Khan, Rhaman A. (1993). *Structural Adjustment and Income Distribution: Issues and Experience.* ILO.

Konadu-Agyemang, Kwadwo (1988). "IMF Sponsored Structural Adjustment Programs and the Perpetuating of Poverty in Africa: Ghana's Experience Revisited". *Scandinavian Journal of Development Alternatives,* Vol. 17: 3-4 127-144.

Kraus, J. (1991). "The Political Economy of Stabilization and Structural Adjustment in Ghana" in Rothchild, D. ed. *The Political Economy of Recovery,* Boulder: Rienner.

Mikell, Gwendolyn (1991). "Equity Issues in Ghana's Rural Development" in *Ghana: The Political Economy of Recovery.* ed. Donald Rothchild. Boulder, CO: Lynne Rienner Publishers.

Ministry of Information (1991). *An Official Handbook of Ghana.* Information Services Department. Medical Services Educational Statistics.

Ocran, Sarah (1998). *Ghana: Dignity and Sustenance.* Montevideo, Uruguay: Social Watch.

Ould-Mey, Mohammed (1996). *Global Restructuring and Peripheral States: The Carrot and the Stick in Mauritania.* Lanham, MD: Rowman and Littlefield.

Panford, Kwamina (1994). "Structural Adjustment, the State and Workers in Ghana". *Africa Development,* XIX (2) 71-95.

Republic of Ghana (1996). *Policy Focus for Poverty Reduction.* Accra: Government Printer September.

Rimmer, Douglas (1992). *Staying Poor: Ghana's Political Economy 1950-1990.* Oxford: Pergamon Press.

Roe, Alan and Hartmut Schneider (1992). *Adjustment and Equity in Ghana*. Paris: OECD.

Rojas, R. (1997). *Notes on Structural Adjustment Programs*.

Rostow, W. W. (1962). *The Process of Economic Growth*. New York: Norton.

Rothchild, Donald (1991). *Ghana: The Political Economy of Recovery*. Boulder CO: Lynne Rienner Publishers.

Sahn, David E. (1994). *Adjusting to Policy Failure in Africa*. Ithaca: Cornell University Press.

Samoff, J. and Samoff, R. (1976). "The Local Politics of Underdevelopment". *The African Review*, Vol. 6, No. 1.

Seers, D. (1969). "The Meaning of Development". *International Development Review*, Vol. II, No. 4.

Seers, D. (1972). "What Are We Trying to Measure?". *Annals of Development Studies*, 8, 21-36.

Seers, D. (1977). "The New Meaning of Development". *International Development Review*, Vol. 19, 2-7.

Slater, D. (1986). "Capitalism and Urbanization at the Periphery: Problems of Interpretation and analysis" in D. Drakakis-Smith, ed. *Urbanization in Developing Countries*. London: Croom Helm.

Sowa, Nii K. (1993). "A Decade of Structural Adjustments". *West Africa*. 11-17th October.

Stewart, Frances (1995). *Adjustment and Poverty*. London: Routledge.

Streeten, P. (1995). "Human Development: The Debate About the Index". *International Social Science Journal*, Special Issue. March, pp 25-37.

The Hunger Project-Ghana (1996). *Agenda for the Sustainable End of Hunger in Ghana*, Draft. Accra.

Thirlwall, A.P. (1995). *Growth and development: With Special Reference to Developing Economies*. Boulder, CO: Lynne Rienner Publishers.

Todaro, M. (1996). *Economic Development*. New York: Addison-Wesley Publishing Co.

Toye, J. (1987). *Dilemmas of Development: Reflections on the Counter-revolution in Development Theory and Policy*. Oxford [Oxfordshire], UK: New York, NY, USA: Blackwell.

UNDP (1994). *Human Development Report*. New York: Oxford University Press.

UNDP (1997a). *Ghana National Human Development Report*. UNDP.

UNDP (1997b). *Country Co-operation Framework and Related Matters*. UNDP.

UNICEF/OXFAM (1999). *Debt Relief and Poverty Reduction: Meeting the Challenge*. Position Paper. Oxfam.

Wallerstein, E. (1974). *The Modern World System I*. New York: Academic Press.

Watkin, Kevin (1995). *The Oxfam Poverty Report*. Oxford: Oxfam.

Werlin, Herbert H. (1994)."Ghana and South Korea: Explaining Development Disparities". *Journal of African and Asian Studies*. 3-4: 205-225.

World Bank (1990). *World Development Report*. New York: Oxford University Press.

World Bank (1993). *Ghana: Poverty–Past, Present and Future*, Report No. 14504-GH. Washington, DC: Country Operations Department, World Bank.

World Bank (1994). *Adjustment in Africa: Reforms, Results and the Road Ahead*. New York: Oxford University Press.

World Bank (1995a). *Ghana: Growth, Private Sector and Poverty Reduction – A Country Report*. Report No. 14111-GH. Washington, DC: Country Operations Department, World Bank.

World Bank (1995b). *Country Briefs*. Washington, DC: World Bank.

World Bank (1996). *World Development Report*. New York: Oxford University Press.

World Bank (1997). *World Development Report*. New York: Oxford University Press.

World Bank (1999). *World Development Report*. New York: Oxford University Press.

World Health Organization Annual Health Reports on Ghana, 1991-1994.

3 The Growth of Public Debt in a Reforming Economy[1]

JOE AMOAKO-TUFFOUR

> National debt is like a toothache; it is best not to have one, but if you
> have got one it is next best to get rid of it as soon as you can (E.H.
> Young, 1915).

3:1 Introduction

From the perspective of those concerned with public finances, the immediate
post independence Ghanaian economy, in many respects, worked reasonably
well. The period 1955-59 was one of "consolidation" and the overall public
finance was kept approximately in a balance. Only about 5 percent of the nation's
(annual) gross domestic product (GDP) was owed to debt holders and nearly 70
percent of the debt was internally held (Cox-George, 1961). "Independence is
expensive" (Ahmed, 1967:21), because it brings its own financial commitments.
The demands of nationhood and the stimulus to develop spurred growth in
public expenditures in the 1960s. Imbalance between expenditures and revenues began to put strains on public finances, the task of mobilizing resources
for development would become urgent, and falling into debt had begun. Three
features characterized the early history of the public debt: limited reliance on
concessional external resources, greater reliance on non-concessional short-
term and medium term supplier's credit, and special placement of domestic debt
with state-owned financial and non-financial institutions (Agama, 1968). The
debt reached 50 percent of GDP by 1964 and around 60 percent by mid 1970.
Two stabilization attempts in 1966-68 and 1978-79 failed to save the fiscal drift,
deficits widened, and recourse to large scale borrowing from the banking sector
became routine.[2] With the onset of economic reform in 1983, there has been a
marked change in domestic borrowing and inflow of foreign resources has in-

creased becoming larger since the mid 1980s. The net result is that the public debt reached 100 percent of GDP by 1992, nearly 120 percent in 1998, raising concerns about whether the economy can both borrow and grow without risking unsustainable debt buildup. This chapter looks at how we have come to where we are, the impact debt accumulation may have had, and the possible remedies.

Noticeable about Ghanaian debt is the rising debt occurred in peacetime, concurrently with unprecedented privatization, liberalization and a transformation of the economy. Commenting on Italian debt growth, Dornbusch (1988) noted that while transitory and moderate growth of debt could be defended any time, steady growth of debt in peacetime raises "questions about what happens at the end of the rainbow" (p. 25). Many economists share the view that massive debt and the accompanying debt service burden can have a crippling effect on economic restructuring effort, especially when debt growth becomes self-generating because of mounting debt service. As far as the effect of debt burden on growth is concerned, Tanzi and Blejer (1988) observed that for many countries, the level of debt and the rates of economic growth are highly negatively correlated. Even in developing countries, private investment may decline if real interest rates rise, or if there is a repression of domestic interest rates, in response to higher domestic borrowing (Easterly, *et al.* 1994). Dornbusch (1993:350) remarked that the "problem of debt is that the burden of debt service, in the budget and in the external balance, is a source of inflation, overly low standards of living and much too low investment". The financial instability associated with strained budgets engenders capital flight. Attempts to halt capital flight by high interest rates create budget problems and undermines private sector growth. And maintaining undervalued exchange rate to stem capital flight also risks further balance of payment problems and depresses real wages. Ricardian vision apart, most economists concur that debt growth can have deleterious effects on every economy hoping to develop sooner or later.[3]

The intriguing questions are these. If reform requires austere policies which respect budget constraints (Rodrik, 1996), then that debt growth should accompany reform requires explaining. Was growing indebtedness necessary for implementing economic reform? Has the growth in debt enabled or impaired the efficacy of reform policies? And is public finance anywhere to facing the kinds of strains and stresses that all too often accompany accumulation of debt? This chapter considers issues of both historical and prospective nature raised by the rise in debt since the 1980s. Section 2 outlines the history of the debt, the dimensions of the debt, and the cause of the growth in debt. Section 3 looks at the debt burden, explores why debt servicing now appears to be a major problem and the distributional effects on public spending. Section 4 explores the risks and policy challenges of the debt and provides a brief overview of how countries have dealt with large debts and the lessons for Ghana. Section 5 concludes the chapter.

3:2 The Dimensions and History of the Public Debt

3:2.1 Dimensions of the Public Debt

What is so special about Ghana's debt? Figure 3:1 provides the simple answer: the extraordinary leap in the domestic debt, the almost uninterrupted rise in total debt since 1981, the steady speed of its rise between 1986 and 1994, and the high level of the total debt/GDP ratio.

Figure 3:1 Debt/GDP Ratios

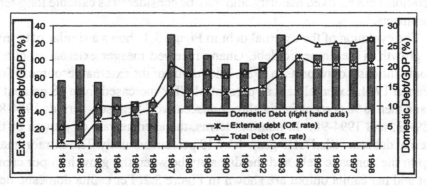

Data for this study are compiled by the author from various sources including World Bank Tables, International Monetary Fund, Bank of Ghana, Ghana Statistical Services, and Centre for Policy Anaysis (CEPA), Ghana. External debt are converted to domestic currency units at the official exchange rate.

The domestic debt/GDP ratio declined from about 18 percent in 1982 to 10 percent in 1985/86, largely on account of growth in GDP rather than on account of a ecline in the absolute level of the debt. The debt rose sharply to 27 percent in 1987 at the height of reform with the advent of government securities. The annual rate of growth of the domestic debt increased from 43 percent between 1989-93 to 55 percent between 1994-98 and exceeded nominal GDP growth by 16 percent for the period 1989-98.

The overall picture of the domestic debt can be characterized by the following elements: rather short average maturity, a domination of treasury bills in the hands of the public and banks, and the placement of medium and long term bonds with non-bank public institutions. In an environment beset by high and variable inflation and with fixed-rate government debt instruments, the average life of debt has shortened sharply from 10 years in 1984 to 2.2 years in 1992 and has since 1994 fallen sharply. Weekly auctions have since 1996 been confined to only 91 day T-bills as placing new issues of over 1 year securities prove arduous. T-bill issues soared in 1996 by 225 percent, more than tripling the year-end 1995 stock, with the result that its share jumped from 11 percent in 1995 to nearly 49 percent in 1998 of total outstanding domestic debt. Initially, T-bills were sold mainly to the non-bank public because of the constraint imposed to avoid crowding out private sector borrowing. This changed in 1997. Opening T-bill auctions to the commercial banks has also meant increasing amount of bills on the market and thus increasing direct borrowing from the banks. The 31 percent of the debt held by the Bank of Ghana, partly as a result of open market operations, has no fixed maturity and may be considered as callable long-term instruments.

The evolution of the external debt in Figure 3:1 shows a similar pattern to the growth in the domestic debt. Ghana received meager external resources prior to the onset of reforms in 1983. The run-up of the external resource inflow began in 1984. External debt as percent of GDP increased from 48 percent to 69 and then to 96 percent for the three consecutive 5-year sub-periods 1984-88, 1989-93, and 1994-98.[4] In nominal terms, the average rate of growth of the external debt exceeded GDP growth by 13 percent annually. The extraordinary leap in the domestic debt and the disparity between the growth in per capita debt and per capita output are shown in Figure 3:2. Per capita domestic debt increased by 3740 percent between 1990 and 1998.

For the same period, total debt per capital debt increased nearly 1200 percent, whereas GDP per capita lagged behind at 642 percent. Except at the height of reform in 1988-92, per capita total debt exceeded per capital GDP on average by 8 percent annually since 1982.

Is Ghana's debt too high? Economic theory does not provide us with any easy answer to this question. Spaventa (1987:357) remarked poignantly from the facts of different debt episodes "that it is meaningless to look for a critical value of the ratio of debt to GDP beyond which the system breaks down and traumatic solutions become necessary". In other words, creditors and debtor may not know precisely when debt crisis will come judging by debt/GDP ratio alone. Far better to bear in mind that debt problems roll in quietly with rising debt/GDP ratio. The question cannot be answered apart from the more funda-

Figure 3:2 Index of Growth of Per capita debt & GDP

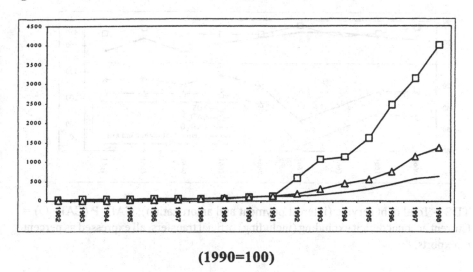

(1990=100)

mental question of debt service capacity: Can the economy conveniently service its debt without trapping itself in a debt spiral? This basic question is explored later in section 3. For now, it is tempting to make some international
comparisons. For example, the average domestic debt/GDP ratio of the heavily
indebted G-15 countries identified in the 1985 Baker Plan,[5] increased from 10
percent in 1981-82 to 16 percent in 1987-88 (Guidotti and Kumar, 1991). In
comparison, Ghana's domestic debt increased from 16 percent in 1981 to 24
percent in 1988. For the external debt, there is a view that foreign debt in
excess of 50 percent of GDP is dangerous (Caprio, 1997). Moore (1990) provides three conceptual ratios used in determining whether a country can be
regarded as having over-borrowed. By his criteria, "danger points" are reached
when, as percent of exports of goods and services, gross external debt exceeds
200 percent; the total debt service–interest and amortization payments–exceed
20 percent; and when the current account deficit reaches 20 percent. Granted
the limitations of such debt ratios, Figure 3:3 shows that the debt service and
gross debt ratios exceeded their respective bounds of 20 percent and 200 percent throughout the 1990s. The current account deficit has since 1995 exceeded
the 20 percent benchmark if official transfers are excluded. Even if these ratios
are not sufficiently high, the associated debt service relative to the capacity of
the economy may be worrisome. The extent of the debt service burden is addressed later in section 3.[6]

Figure 3:3 External Debt (Percent of Exports)

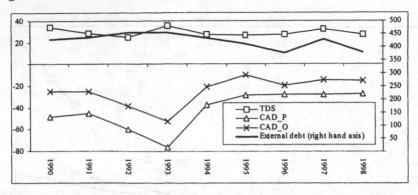

TDS = Total debt service (interest payment and amortization); CAD_P (CAD_O) = Current account deficit excluding (including) official transfers, all expressed as percent of exports.

3:2.2 Sources of Growth of the Public Debt

What explains the growth of the public debt? The centrality of fiscal deficits is emphasized in the analysis of Latin America, Turkey and Phillipines debt crisis of the early 1980s (Sachs, 1985). Governments in these countries did run chronically large budget deficits in the years leading up to the debt crisis. The evidence is that the demands of war finance, persistent trade and current account deficits, deterioration in terms of trade, mismanagement of exchange rates, rising interest rates, assumption of private sector debt, and availability of cheap finance (Buchanan, 1957; Tanzi and Blejer, 1988; Dornbusch, 1993) may all intertwine to push economies towards larger deficits and greater borrowing. Some of these factors, certainly not the demands for war finance, may explain the emergence of public debt in Ghana.

The exploratory regression results in Table 3:1 throw some light on the sources of growth of debt. The sources of growth of domestic and external debt may be different. Fiscal deficits appear more important in the determination of the domestic debt. The external debt is contracted to close the balance of trade gaps. The regressions (2)-(4) have a terms of trade variable which appears to have a "wrong" sign, in that it suggests favorable terms of trade has been accompanied by increase in external indebtedness.[7] Such a relationship, however, is all the more likely if increasing terms of trade is perceived as increasing prospects of export revenues and hence the ability to service the debt. Although Ghana's terms of trade declined somewhat between 1990 and 1996, by an annual average of 0.6 percent, increased borrowings during this period may have been based on the premise that increased export revenues also mean decreased chances of debt default. The estimated effect of domestic interest rates is weak but positive. The effect of foreign interest rates on the external debt is negative as expected, but quantitatively less important for the simple reason that about 80 percent of Ghana's external debt is multilateral and bilateral loans contracted at concessional rates.

Table 3:1 The Sources of Growth of Debt

	FDEF/ GDP	R_d	R_f	BOT/ GDP	TT	R2	D_h
Dependent Variable							
(1)Δ(DD/GDP)	1.66	0.04				0.52	1.74
	(4.2)	(0.4)					
Standardized	*0.74*	*0.07*					
Coefficients							
(2) Δ(XD/GDP)	0.37 (0.3)		-0.55 (0.3)	3.55 (2.9)	0.48 (3.2)	0.36	2.54
(4) Δ(XD/GDP)	0.42 (0.4)			3.57 (3.0)	0.49 (3.5)	0.41	
(5) Δ(TD/GDP)	2.08 (2.1)			3.04 (2.6)	0.38 (2.7)	0.49	
Standardized Coefficients	*0.43*			*0.95*	*0.92*		

The simple regression is estimated over the period 1981-1998 and has as the dependent variables changes in the domestic debt/GDP ratio (DD/GDP), external debt/GDP ratio (XD/GDP) and total debt/GDP ratio (TD/GDP). The explanatory variables are the non-interest fiscal deficit (FDEF/GDP), lagged domestic interest rates on Treasury bill (R_d), lagged interest rate on new external loan commitments (R_f), the balance of trade deficit (BOT/GDP), and the terms of trade (TT); t-values in parentheses; R^2 is the adjusted R-square; D_h is the Durbin's h- statistics. The standardizes coefficients measure the change in the dependent variable, other things being equal, for a unit change in each of the dependent variables.

Since the units of measurement of the explanatory variables are different, the standardized coefficients measure the changes in the debt/GDP ratio, all things being equal, for unit change in the explanatory variable. The influence of fiscal deficits on total debt is nearly half the influence of the trade payment gaps and variations in the terms of trade. The results confirm the centrality of the twin deficits—fiscal and trade—in the evolution of the debt, mutually reinforcing the demand for domestic and foreign borrowing to close domestic spending and foreign payment gaps.

Figure 3:4 History of Fiscal Deficits

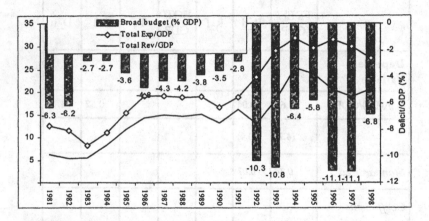

Source: Data based on Government of Ghana budget reports, International Monetary Fund and Centre for Policy Analysis Ghana Macroeconomic Review, various issues.

Figure 3:4 shows a self-evident path of public finances. The deficit, broadly defined to include the domestic counterpart expenditures on foreign financed development projects, fell to nearly 3 percent of GDP in 1983-85 and increased moderately in the late 1980s. There was a large increase in the deficit as a percentage of GDP not because of a fall in GDP, but because of a widening fiscal gap. The reasons for the widening fiscal gap are many, but we will not dig deeper here. It is sufficient to note that the highest deficits recorded since 1980 were in the election and immediate post election years of 1992-93 and 1996-97. Further, although tax receipts increased noticeably since 1990, it has been extremely difficult to control expenditures despite the attempts at fiscal restructuring. It is difficult to give exact magnitudes of how much of the deficit was financed by borrowing on year to year basis. In recent years for which reliable data are available, domestic borrowing averaged 55 and 49 percent of total deficit for the periods 1990-93 and 1996-98, respectively. And for the latter period, total borrowing amounted to nearly 82 percent of the cumulative deficit of 4050.8 billion cedis.

Just how important has the trade deficit been in the rise in public external liabilities? The balance of payment and national accounts provide us a simple framework for analyzing the link. National income accounts show that current account balance is related to domestic saving and investment spending as follows:

Current account balance = Domestic saving − Domestic investment (1)

Current account is the sum of the balance of trade in merchandise, services and net factor income. Domestic saving is the sum of private saving and government saving. Domestic investment is the sum of private investment and government infrastructure spending. A further relation in national income accounting reveals that a country's current account indicates capital flows in and out of the country and this is related to the domestic saving–investment gap as follows:

Net capital inflows = Domestic saving − Domestic investment (2)

Simply put, a country must be receiving foreign capital whenever its domestic saving is insufficient to finance domestic investment plans. A country is a net exporter of capital if domestic saving exceeds domestic investment plans. The fact that the right hand side of (1) and (2) are identical implies that current account surplus is matched by net capital outflows and a deficit by net capital inflows equal to the domestic saving-investment gap. Following Dornbusch (1993), a further decomposition of the current account into non-interest current account (NICA) and interest payment (I*) gives the relationship:

NICA + Interest payments (I*) = "New money" + Other net capital inflows (3)

The right hand side of (3) is net capital inflows defined by the two components: "New money" and other net capital inflows. New money represents borrowing from abroad. Other net capital inflows represent reserve draw down, direct foreign investment inflows and long-term portfolio flows. The non-interest current account is often called the net resource transfer because it measures the net imports of goods and non-interest services. According to Dornbusch, for countries with low domestic saving relative to investment, NICA deficits are the normal channel through which resources are transferred from rich to poor countries to support capital formation and growth.

The data in Figure 3:5 and Table 3:2 bring out strikingly that current account has been in a deficit during the past 15 years and has turned to even greater deficits since 1989. Key driver of the widening deficit is the NICA which in turn reflects primarily the widening deficit on the merchandise trade balance. We see in Table 3:2 that the cumulative trade balance accounts for 95 percent of the NICA deficit and 68 percent of the current account deficit over the period 1984-98. Figure 3:5 also shows that the deterioration in the current account has almost been matched by inflow of new money. It is by no means coincidental that both the saving-investment gap and the current account reached their greatest single year deficit yet in 1993 followed in 1994 by the greatest single year increase in the external debt of about US$1 billion. The cumulative current account deficit of US$6232 million breaks down into a cumulative rise in external debt or "new money" of US$5560 million and a rise in other net capital inflows of only US$672 million. The fact that nearly 90 percent of the cumulative current account deficit for the period 1984-1998 has been financed by debt capital can also be read to say that new private capital inflow in response to reform has been extremely low. Table 3:2 also shows the savings-investment gap as a counterpart of the current account deficit.

Figure 3:5 Current Account/Changes in External Debt (US$, millions)

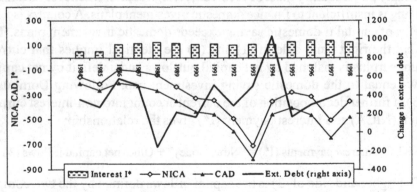

3:2.3 Reform and the Domestic Debt

The exploratory regressions in Table 3:2 explained no more than half of the changes in the debt/GDP ratio, suggesting that the twin deficits, even if the most important in recent years, have not been the only contributory factors. Ghana, like Czechoslovakia, Romania, Hungary and Poland (Allen, 1992) provides a remarkable example of transition economies where domestic debt either did not

Table 3:2 Trade, Current Account Balance and External Debt

		5-year average (US $,m)			Cumulative Balance (US $,m)
		1984-88	1989-93	1994-98	1984-1998
Balance of Trade		-58	-397	-399	-4271
Net factor income (excl I*)		-82	-22	40	-315
NICA		-140	-419	-358	-4586
Interest payments (I*)		92	109	129	1646
Current Account Deficit (CAD)		-232	-528	-487	-6232
Changes in external debt		310	334	468	5560
		1986-95	1991-95	1996-98	
	Domestic savings (% of GDP)	7.2	6.3	13.1	
	Domestic Investment	13.1	14.6	20.3	
	Savings-Investment gap	-5.9	-8.3	-7.2	

Data on domestic savings and domestic investment are taken from CEPA: Macroeconomic Review and Outlook, various issues. I* is interest payments on the external debt and NICA is non-interest current account balance.

exist or were very small before reform, and yet have had to deal with debt accumulation as part of the needed restructuring. For Ghana, reforms induced a growth in domestic debt in 4 ways: (1) the restructuring of the banking system led to the take over of the non-performing loans of the state-owned commercial banks in readiness for privatization; (2) the issue of equivalent government bonds to replace accumulated Bank of Ghana foreign exchange losses; (3) the ex-post validation of government indebtedness to state enterprises; (4) and the de-regulation of domestic interest rates.

The first three channels of debt growth may be viewed as needed stock adjustments to redraw the boundaries between the public sector and the rest of the economy. The financial sector reform that began in 1987 involved the restructuring of the banking sector and deregulation of interest rates. It became apparent that banks were encumbered with non-performing loans which will have made privatization difficult and perhaps unattainable for years. To ensure the soundness of banks for successful privatization, re-capitalization was done through an injection of public funds, in essence replacing bank claims on enterprises with claims on the government in the form of Non-Performing Asset Recovery Trust (NPART) bonds (Chand, 1993). The burden of this debt on public finance depend on the loan recovery rate by the government from debtor enterprises. According to anecdotal evidence gathered through interviews, loan recovery rate has since 1992 fallen short of target by at least 10 percent annually, partly because some private enterprises are now defunct, and partly because some of the bad loans were themselves unrecognized public sector debt contracted by state enterprises.[8]

The Revaluation Account (RA) bond also has its origins in the desire to restructure the balance sheet of the Bank of Ghana. By the mid 1980s, the Bank had become technically insolvent because of exchange rate loses on its foreign liabilities (Sowa, 1996). The difference between the Bank's assets and liabilities is its net worth. Net worth is positive if assets exceed liabilities and the Bank will be considered as solvent. On the other hand if net worth is negative, then the Bank is insolvent and without an increase in its assets, it is not able to meet its contractual obligations. Since the Bank's net foreign assets position has since the 1980s remained negative, currency depreciation meant a widening gap between its assets and liabilities in domestic currency units. The Bank incurred foreign exchange losses and a corresponding decline in its net worth. For the Bank's net worth to be maintained, an increase in foreign exchange losses had to be matched by an increase in assets or a decrease in contractual foreign obligations. The issue of the long-term government bonds to replace the accumulated foreign exchange losses was to increase the Bank's assets if solvency was to be maintained. From a modest 32 billion cedis (or 6 percent of GDP) in 1986, the RA bonds increased to 940 billion cedis in 1995 (or 13 percent of GDP). The growth in RA losses is an increasing function of nominal currency depreciation if net foreign assets is negative, and a decreasing function if net foreign assets is positive. As long as net foreign assets remain negative as it has been for some time now, downward pressure on the external value of the cedi translates into valuation losses, and adds to government indebtedness to the Bank.

While the RA bonds and part of the NPART bonds were not the direct results of accumulated fiscal imbalances, the third stock adjustment is exclusively an ex-post validation of financing past deficits. Typically, though not always, financial resources of state-owned institutions were potentially sources of government funding. In an organic state, the concept of integrity of the balance sheets of state enterprises–banks and non-banks alike–did not matter much.

Debts were placed with state-owned institutions and at an interest rate that did not reflect the cost of attracting and rationing funds. The weakening of the balance sheet of these institutions was indirectly equivalent to the government acquiring liabilities off the budget, largely in the form of IOUs. In many instances, this meant that state institutions lent to the government at negative real interest rates. Examples include the 2.5 percent, 7.4 billion cedis, 10-year government stock issued to the Cocoa Marketing Board (CMB) to replace accumulated debt in 1983. This was rolled over in 1993 and at 2.5 percent, the negative real interest rate was nearly 22 percent in 1993. The 16.5 percent, 16 year, 0.5 billion stock issued to Social Security and National Insurance Trust (SSNIT) carried a negative real interest rate of 13 percent at the time of issue in 1984.

The transition to a market economy and the demands for privatization required clarity as to the assets and liabilities of various state enterprises. This meant the issue of interest bearing debt in many cases to recognize hitherto unrecognized debt. As in the case of the CMB's non-negotiable bonds, subordination of old debt to new debt, in the sense that new debt is offered at higher interest rates than old debt, has remained a feature of this reconciliation. Perhaps the most extensive cleaning of balance sheet has occurred with the nation's pension institution, SSNIT. Between 1983 and 1993, SSNIT holdings of government debt increased from 2 billion in 1983 to 18 billion in 1990 and sharply to 94 billion cedis in 1993, confirming that unfunded pension liabilities are themselves a form of hidden debt accumulation (Easterly, 1999). It is difficult to know the full size of liabilities inherited from the past, let alone the extent of current practice. Centre for Policy Analysis (CEPA), a non-governmental research think tank, observed that the problem of arrears and hidden liability accumulation is neither a thing of the past, nor is it confined to a particular sector. CEPA (1999: 47) remarked on this fact:

> The increase in domestic interest payments by almost 37 percent in the face of reduction in the Treasury bill rate from 47.5 percent in 1997 to 28.7 percent year-end 1998 confirms.. that interest payments may have been payable on some stock of debt not shown in the official debt register. The report by the *Controller and Accountant General* that 1997 arrears of Social Security contributions of 20 billion cedis were converted into a 10-year bond with a 12 percent interest provides further independent confirmation of both the existence of payment arrears and the payment of interest on that debt.

CEPA might have added that it has not been easy for the government to make a clean break with the past as a result of budget deficits. The existence of "arrears" highlights the issue of the proper size of the debt stock and its true fiscal burden. For example, in 1998, the official data underestimated total domestic debt by as much as 11.6 percent (or the equivalent of 3.3 percent of GDP) because of the accumulation of arrears and hidden liability (CEPA, 1998: 46). Had such arrears been counted, the domestic debt/GDP ratio could have been nearly 30 percent.

The Ghanaian experience illustrates poignantly many of the problems of economies in transition. As far as public sector balance sheet is concerned, economic transformation process brings its own challenges in flow and stock adjustments in order to establish the appropriate initial conditions. To this end, debt finance of balance sheet reconciliations may be seen as a rational choice, perhaps, the only choice, even if the magnitude of the needed adjustment may have been underestimated. Beyond these, the government is also incurring new debt, most rapidly in the 1990s, as a result of continuing fiscal deficits and deterioration of current account. The new public sector borrowings, the private sector demand for credit and limited domestic savings all combine with domestic inflation to create upward pressures on interest rates, feeding back into increased cost of borrowing.

Figure 3:6 Gross and Net Capital Investments (percent of GDP)

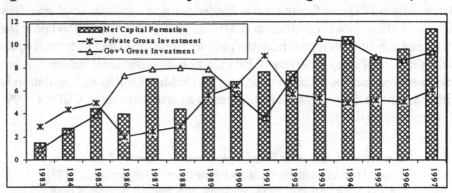

Source: *Ministry of Finance, Central Bureau of Statistics, Accra, Ghana*

As for the external debt whether the "new money" represents a policy concern depends on whether the trade balance associated with it is driven by an investment surge or a consumption boom. An investment surge in areas that would enhance exports or replace imports can generate in the medium to long-term sufficient growth in income and foreign exchange to service the debt and perhaps, eventually, to eliminate it. If this "growth-cum-debt" strategy sounds familiar, it is because many developing country debt problems begin this way. And with great regularity—as documented in Sachs (1987) and Dornbusch (1993) for Brazil, Bolivia, Argentina, Venezuela, Turkey and Ivory Coast, to name a few—countries have done well for short periods, only for a myriad of internal and external factors to intertwine to hinder growth; some eventually to decline as debt service obligations mount. It is fair to say that Ghana's economic restructuring could not help being shaped by this "growth-cum-debt" strategy. Great emphasis was placed on attracting external financing and, as Chand (1993:367) pointed out, the fiscal budget became the principal instrument to this end.

Figure 3:6 shows the sporadic trends in net capital formation as a percent of GDP and the counter movements in public and private investments. Granted that years of policy neglect allowed stocks of public infrastructure capital to decline, we see a rising public-to-private capital growth since 1986. Even if the government used these resources to develop an enabling infrastructure for growth, that private investment has fallen markedly since 1991, well below government investment, is indicative of a "crowding out" and the slow growth of the economy's productive capacity. Expanding public infrastructure investment apart, Aryeetey (1994) pointed out that the poor growth in private investment may be attributed to the perception of uncertainty in the political and economic environment. He observed through interviews that a segment of the private sector chose importation of finished goods over long-term physical investment. With trade and price liberalization on one hand and the perceived government antipathy toward private ownership of capital on the other, trade in finished goods was deemed safer and provided high turnover. This no doubt undermines the expectations that public investment in infrastructure is a precondition for private sector growth and improved ability to service debt. Many would argue that credibility in political behavior and consistency in policy-making remain lacking, and that the system may not have matured sufficiently yet to appreciate these equally essential ingredients to successful reform. The limited growth in private capital formation and the flourishing of consumption goods' trade are warning signals of an undesirable trend in so far as enhancing the productive capacity of the economy and debt servicing capacity are concerned.

3:3 The Debt Service Burden

It cannot be denied that the issue of government debt instruments has contributed to the development of the money market, even if far from what it could have been. It was clear from the outset of financial sector reform that the privatization of commercial banks will not always lead to the growth and functioning of financial markets. Adequate and continuous supply of debt instruments are essential, especially if the central bank is to run more market-based monetary policy. In the absence of reliable and adequate private debt instruments, the issue of treasury securities improved the menu of financial instruments and permitted the central bank to perform open market operations. Treasury securities have been sold since 1987 in maturity tranches ranging from one to five years. In this role, the government became a financial innovator, providing securities that allow the money market to function in ways otherwise impossible. Even if the advent of government securities may have led to some substitution away from bank deposits, the danger of portfolio crowding out in the early years, arguably, seemed rather minimal because of the underdeveloped nature of savings mobilization in the economy.

Useful as treasury securities have been in this regard, and indeed moderate borrowing need not be a bad thing, the focus here is on the unsustainable levels of borrowing. The basic fiscal and monetary problems can be daunting if indebtedness is relatively high and debt management falters. Rising debt service restricts fiscal freedom and preempts other areas of spending in the allocation of revenue. The maturity structure of the domestic debt and the associated interest cost can be matters of monetary concern. They may limit the central bank's ability to use interest rates as an instrument of monetary control since rising rates feed back into the budget if frequent refinancing becomes a feature of debt management. An additional concern has to do with the linkages between the domestic and external debt if the latter's debt-servicing becomes onerous. In the desire to maintain international credit worthiness, governments may choose to add to the domestic debt in order to meet external debt obligations.

This is all the more likely when export revenue falls due to adverse shocks. Guidotti and Kumar (1991) in their study of 15 heavily indebted countries observed that for most of them growth in domestic debt occurred at a time when external debt was also increasing, which is by no means coincidental. How close is Ghana to any of these concerns?

3:3.1 Domestic Debt

Table 3:4 presents the trends in debt service capacity during the 1990s. Rising interest cost on domestic debt reflects overall rising interest rates on account of higher inflation rate and higher demand for domestic credit. The average interest cost (the ratio of interest paid to total domestic or external debt) increased nearly 8-fold between 1990 and 1998. With inflation rate remaining in double

Table 3:4   Debt Service Burden (Percent)

1. Domestic Debt (DD)	1985	1990	1991	1992	1993	1994	1995	1996	1997	1998
Average Interest Cost	12	2.2	6.7	5.9	9.0	14.9	13.9	16.5	18.8	17.6
Interest on DD/Total Interest	89	33	79	56	68	72	71	75	76	72
Percent of total expenditures										
Total (domestic and foreign) interest	9.6	8.1	9.4	9.3	12.6	14.6	14.9	17.3	19.9	24.1
Wages & Salaries	29.6	24.4	23.1	26.1	21.3	18.9	16.8	18.3	17.6	21.8
Other purchases of goods and services	24.6	13.3	13.2	10.6	9.6	8.7	7.3	7.2	6.7	8.4
Subsidies and Transfers	9.1	12.9	12.0	11.0	13.1	10.7	10.6	5.6	4.7	4.6
Total interest/Tax revenue	15.9	12.4	13.4	20.2	26.8	27.8	28.9	33.4	40.7	40.4
Total interest/GDP	1.5	1.3	1.8	2.2	3.7	4.7	5.4	6.0	6.6	
Domestic Debt by maturity										
T-Bills (under one-year)	12	5	4	7	6	8	11	26	45	49
Government bonds	88	20	31	37	41	43	33	33	17	20
RA Bonds		75	65	56	54	49	56	42	38	31
2. External debt burden										
Average Interest Cost	.4	1.5	.6	1.4	1.4	1.2	1.4	1.4	1.5	1.9
Interest due/Exports		11.0	10.0	11.0	10.0	9.0	9.0	8.8	9.2	6.7
Budget Interest paid/Exports	1.6	6.3	2.5	6.2	6.2	5.4	5.6	5.4	6.4	7.3
External debt service/GDP	NA	NA	NA	4	7	7	9	7	8	7
External debt service/Exports	NA	34	29	25	36	27	27	27	32	27
External debt Exports	377	429	436	454	455	435	411	376	430	380
Maturity Composition										
Long term	55.1	62.5	64.0	68.8	69.0	70.5	75.4	73.1	81.3	83.6
Medium term	36.5	29.2	27.0	22.0	21.3	18.9	14.0	16.2	7.4	5.1
Short term	8.4	8.3	9.0	9.2	9.7	10.6	10.6	10.7	11.3	11.3
Composition of Creditors										
Concessional	91.5	91.5	92.3	92.5	92.2	91.8	90.6	88.2		
(Non-Concessional)	8.5	8.5	8.7	7.5	7.8	8.2	9.4	11.8		

Notes: IMF Financial Statistics, World bank Debt Tables, Government of Ghana Budget Reports, Ghana Central Bureau of Statistics and Ministry of Finance.

figures (20-30 percent) and the uncertainty surrounding near-term economic conditions, it is unlikely that market interest rates will fall. And with little or no prospect of debt reduction, the outlook is one of continuing growth of interest payment on the domestic debt.

The ratio of total interest payments to expenditures is a measure of how much of government outlays are non-discretionary and, therefore, the potential trade-offs between expenditure categories. As percent of government expenditure, interest payments have since 1997 ranked as quantitatively the most important, increasing three fold from 8.1 in 1990 to 24.1 percent in 1998. Between 1989-93 interest payments grew by 10 percent annually and rose sharply to 17 percent annually between 1993-1998. As in 1997/98, interest payment is projected to be the largest single item in government expenditures in the years 1999-2001. The optimistic forecast is that interest payments will average 15 percent of total expenditures and 4.7 percent of GDP in 1999/2001. Most significant is that the interest payments will be about twice the government's estimate of total budget deficit of 2.6 percent of GDP in 1999-2001.[9]

It is also significant that interest payments averaged 40 percent of tax.revenue in 1997/98 compared to 13 percent in 1990/91 and projected to remain about 28 percent of tax revenue for 1999/2001. Many will agree that even with prudent fiscal planning and management, it will be many years before the debt service charge as a fraction of revenue and spending is brought down significantly. The numbers cited above suggest that there must have been important changes in spending in order to accommodate interest payments. The broad changes between 1985 and 1998 are evident in Table 3:4. As interest outlays increased by nearly 14 full percentage points of total spending between 1990/91 and 1997/98, subsidies and transfers fell by 7 percent and outlays on other public services by 4 percent. The result has been the introduction of a wide range of user fees in health care delivery and cost recovery in education.

As mentioned earlier, in an environment beset by high and variable inflation and where government debt instruments carry fixed coupon, the average life of debt has shortened significantly, especially since 1994 with weekly auctions confined to only 91-day Treasury bills. Shortening maturity implies shifting towards large maturing issues which must be frequently redeemed or refinanced. Neither option is without burden. Redemption or debt retirement will depend on the health of public finances in the near term 1999-2001. And since the overall budgetary cash balance for these years is projected to be in a deficit of about 2.6 percent of GDP annually, reducing the size of the domestic debt seems unlikely. Despite the popularity of government debt, short-term refinancing makes nearly 50 percent of the domestic debt vulnerable to capital flight and short-term interest rate increases. This inevitably limits the recourse of the central bank to upward adjustments in interest rates as an anti-inflation tool in the conduct of monetary policy. In fact, when the market is dominated by T-bills, the Bank cannot vary the size of offerings tactically in response to short run changes because every issue is more likely to be determined by the amount maturing than by the needs of monetary control.

3:3.2 External Debt

With regard to the external debt, rising average interest cost reflects the increasing recourse to non-concessionsal loans which typically carries higher interest cost. It is worth noting that the data reported here may understate the average interest cost because of arrears in debt service. For example, as reported in Table 3:4, the interest due on the external debt exceeded the interest paid (as a share of exports) by nearly 4 percent annually between 1990 and 1997.

The external debt service (interest and amortization) payments as a share of exports and GDP increased from the mid 1980s and peaked around 1993 and showed no signs of stabilizing or falling. The reason for this trend is that export growth exceeded debt service growth by 15 percent between 1992/93 and 1997/98. The rise in the volume of the external debt service from about $279 million in 1992 to $533 million in 1998 reflects the combined effects of the increase in the size of the debt, the change in its composition (with some share of obligations contracted on commercial terms) and the hardening of the financial terms of the loan. For example, non-concessional loans amounted to 15 percent of total external financing in the period 1992-94 (World Bank, 1995). This increased to 35.5 percent of total new loans of $233 billion contracted in 1997.[10]

With declining average maturity, coupled with debt service charges that are not likely to fall appreciably in the near term, the external debt service burden would pose great difficulties if falling commodity prices as observed in the late 1999 fail to recover in the medium term. Since these charges are, for the most part, immune to direct government intervention, the only recourse will be to seek external refinancing, or borrowing at home to service the external debt. Both options are not without precedents. In 1996 the government mobilized domestic resources to finance the net external repayments because maturing foreign loan repayments exceeded new loan disbursements (CEPA, 1997: 5). The data in Table 3:4 also suggest that in fact, external debt service paid has consistently fallen short of debt service due with 1998 as the exception. The World Debt Tables report that of the accumulated interest arrears of $33 million in 1989-90, $27 million was capitalized in 1992. And $14 million of the principal was also rescheduled in the same year. At the end of 1994, external payment arrears amounted to $93 million (World Bank, 1995). We are informed by Latin American experience that arrears presage a debt crisis. Crisis occur when the resources for continued, quiet refinancing are lacking in the budget or in the foreign balance. Crisis can be precipitated by tightening foreign credit markets, deterioration in export revenues and domestic mismanagement (Dornbusch, 1993).

3:3.3 Sustainability of the External Debt

The interesting question, perhaps, is not whether the debt burden will rise relative to GDP but whether the burden will be sustainable. A comparative analysis of Ghana's external debt burden with similar low income and other heavily indebted countries are presented in Table 3:5. Of the heavily indebted low income countries, Ghana's sustainability index (ratio of the net present value of external debt to exports) of 236 falls close to the midrange of the World Bank's threshold of debt sustainability of 200-250 percent. Ghana's index is exceeded by only Cameroon, Cote d'Ivoire and Uganda. Ghana's debt service burden, however, is the highest in the group. In comparison with the low middle income and heavily indebted middle income countries, only Argentina's debt sustainability index of 408 exceeds Ghana's index of 236. And only Argentina, Mexico and Venezuela's debt service burdens exceed Ghana's burden of 30.5. Even if the debt statistics by themselves are not terrifying, it is noteworthy that Ghana's GDP growth of about 4 percent between 1994-98 is much lower than most of the countries, and is only about one-half that of the heavily indebted middle income countries.

Notwithstanding the difficulties of international comparisons, we see that Ghana's position is anomalous. Although its external debt index falls within the range of sustainability, the debt service burden is not, and this is how debt service difficulties begin, leading slowly but surely to strained budgets.

In its 1997 Article 4 consultation, the IMF reported that in spite of Ghana's high external debt indicators, the debt burden is considered manageable and consistent with projected rates of economic growth over a 20-year period.[11] For that reason, Ghana was judged not to require assistance under the Highly Indebted Poor Countries (HIPC) Initiative.[12] Figure 3:7 highlights the sensitivity of Ghana's external debt indicators to projections of export and GDP growth. The baseline scenario represents the fund's assessment of Ghana's debt sustainability, with average growth of exports set between 7-9 percent for the projection period. The other lines are derived from a more modest annual growth of exports of 5 percent. The baseline scenario clearly requires a strong export per-

Table 3:5 Comparisons of the Burden of External Debt for Selected Low Income, Low Middle Income, Upper Middle Income Countries, 1998

	Real GDP growth	Debt Sustainability Index (200-500%)	Debt Service Burden Index (20-25%)
Low Income			
Ghana	4.0	236.5	30.5
Cameroon	5.1	324.5	20.9
Cote d'Ivoire*	6.0	278.6	28.5
Honduras	4.5	227.3	30.2
Kenya	2.1	162.7	21.6
Nigeria*	3.9	165.6	8.6
Uganda	5.4	249.3	23.1
Zimbabwe	3.2	129.6	20.9
Lower middle income			
Algeria	1.3	200	30.1
Philippines*	5.2	107.5	11.3
Sri Lanka	6.4	95.5	7.7
Upper middle income			
Argentina*	8.6	408.1	68.1
Chile*	7.1	147.1	21.3
Malaysia*	7.8	147.1	21.3
Mexico*	7.0	118.8	34.8
Poland	6.9	100.7	7.2
Venezuela	5.1	136.1	33.5

Source: *Data from World Bank Economic Indicators, 1999. Data as of end of 1998.*

It is assumed that the definition of external debt as reported by the World Bank is based on a data consistent series across countries. Debt Sustainability Index is the ratio of the sum of short term debt and the present value of medium and long term debt to exports. Debt service burden is the ratio of the sum of interest payments and principal repayments to exports. The numbers in parenthesis represent the World Bank threshold of debt sustainability.

* Countries identified under the Baker Plan as problem debtors.

Figure 3:7 Simulation Path of External Debt Indicators

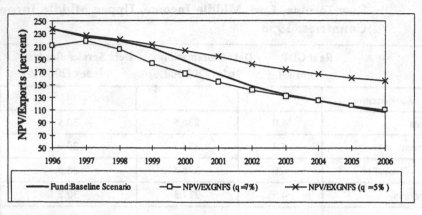

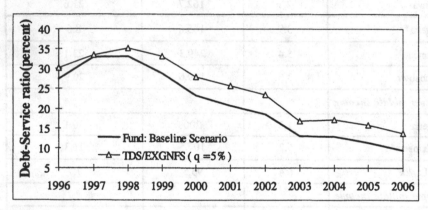

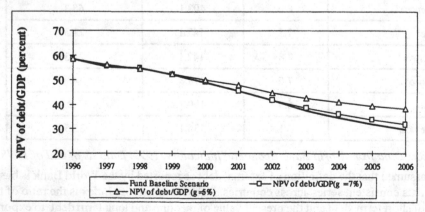

Reproduced with permission of CEPA from Amoako-Tuffour (1998) in Centre for Policy Analysis –CEPA (1998). TDS = Total debt service, EXGNFS = Exports of goods and non-factor services

formance, nearly a cumulative improvement of 86 percent between 1997-2006, and quickly in the near term, if debt indicators are to fall rapidly as predicted. At 5 percent export growth, the indicators follow a gradual decline and diverges from the baseline. By this scenario, debt indicators will remain above their threshold levels of 200 percent for debt-to-export ratio and 20 percent for debt service-to-export ratio in the years 2000/2002.

Table 3:6 provides approximate indication of what it might take, in terms of sources of export growth, to achieve export growth sufficient to put debt indicators on a sustainable path. The results suggest that a continuation of the 1990-98 average growth paths for traditional and non-traditional exports will be just enough to guarantee an overall rate of growth of exports of about 8 percent for the projection period. Are these growth rates attainable in the near term? Given the downward trends in commodity prices, it is unlikely that there will be major price gains to sustain the kind of performance witnessed in the early 1990s. Moreover, the major sources of growth of traditional exports–the opening of new gold mines and the maturing of new bearing cocoa trees–witnessed in the early 1990s have stabilized, if not already in decline. And after falling by 51

Table 3:6 Different Scenarios of Non-Traditional Export Growth of Traditional Exports

		Growth of Traditional Exports						
		0.02	0.03	0.05	0.06	0.07	0.08	0.09
	0.05	0.34	0.25	0.05	-0.05	-0.15	-0.24	-0.34
Desired	0.06	0.45	0.35	0.16	0.06	-0.04	-0.14	-0.23
Export	0.07	0.56	0.46	0.27	0.17	0.07	-0.03	-0.13
Growth	0.08	0.67	0.57	0.37	0.28	0.18	0.08	-0.02
	0.09	0.77	0.68	0.48	0.38	0.29	0.19	0.09
	0.1	0.88	0.78	0.59	0.49	0.39	0.30	0.20

Export growth is decomposed into growth of traditional and non-traditional exports. The relative weights of 0.9 and 0.1 for traditional and non-traditional exports respectively are based on 1991-97 averages. Between 1991-1997, average annual growth of traditional exports is about 7 percent and for non-traditional exports 23% (or 17% if the jump in 1995 by nearly 70% is excluded).

percent between 1994-97, growth in the volume of timber is expected to be fairly modest, if any at all. This leaves the non-traditional export sector as the main engine of growth if Ghana is to achieve spectacular export performance above the 5 percent mark. We see from the table that even with a combination of 5 percent growth of traditional exports and desired overall export growth of 7-9 percent, non-traditional exports must grow in the range of 27-48 percent. The policy challenge is how to promote the non-traditional sector to fill the export growth gap, against the background of a ten-year annual growth performance of about 18 percent between 1989-98 (or 23 percent between 1991-97). The risks and policy challenges of the public debt are explored in the next section.

3:4 The Risks and Policy Challenges

The above evidence would suggest that even if Ghana has not yet faced a true debt problem in the likes of Brazil, Mexico or Argentina up to now, it has built up all the conditions and vulnerability for a debt crisis to emerge with the slightest adverse shock to its balance of payment and the least weakening of its future fiscal position. Greater instability of export revenues, higher domestic inflation, persistent fiscal deficits and lower output growth will make the system more fragile and compel awkward fiscal and monetary choices. The currently depressed commodity prices exemplify the fragility of the system. The Deputy Minister of Finance, after disclosing to parliament in November 1999 that revenues have fallen significantly following deterioration in commodity prices, sought either increases in the value added tax, or more external and domestic borrowing. The irony here, and policy-makers may seize upon it, is that the trend of falling commodity prices and export revenues in 1999 may have been just what Ghana needed to increase its prospects of debt relief under HIPC sooner than later.

The recent proposals to reform debt relief, especially shortening the time frame for the implementation of the HIPC initiative, applying less restrictive eligibility criteria, setting a ceiling for the share of fiscal revenue allocated to external debt service (UNCTAD, 1999) are encouraging. Assuming that debt relief is forthcoming, that the generosity will be significant enough to ameliorate debt service burden where it hurts most, and that there will be safeguards to ensure that spending of any money freed is closely monitored, two questions remain: Can Ghana pursue sufficient stabilization so as to avoid a recurrence of the accumulation of debt? And how does Ghana deal with its large domestic debt? I address these questions in turn.

Successful stabilization involves a multi-pronged attack: fiscal and trade deficits must be brought under control coupled with growth in output. Improvements in non-interest current account, real exchange rate and fiscal policies are at the center of discussion for two reasons. First, unlike commodity price trends, these measures are within the reach of domestic macro management. Second, it is the mismanagement of fiscal policy along with trade policy that often cause countries to succumb to debt crisis (Sachs, 1985).

3:4.1 External Payments Outlook

Granted that successful stabilization depends not only on past balance sheet relationships. Tables 3:7 and 3:8 highlight just how the prospects of improvements may be. The first four rows of Table 3:7 highlight the composition of trade balance. The large increase in exports as percent of GDP, especially since 1994, would lead one to expect that Ghana should have done well. But even with sharply increased exports, trade deficit deteriorated from 1 percent in 1984-89 to about 6 percent in the 1990s. Export growth beyond its historic trends and in excess of 25 percent of GDP will be most important for two reasons. It will reflect the modest improvements in the productive capacity of the economy because of capital investments and also improvements in external payments capacity. But as noted earlier, export growth will depend as much on improvements in domestic supply constraints, on improvements in the terms of trade, as on gains in competitiveness.

Key to the latter is the exchange rate management. The real exchange rate changes showed substantial to moderate real depreciation in the mid 1980s to 1993. This was followed by four successive years of real exchange rate appreciation. The 1994-98 average of 4.3 percent conceals the 19 percent and 7 percent appreciation in 1995 and 1998, respectively, and the cumulative real exchange appreciation of 32 percent in the period 1995-98. Before the exchange rate trend started to unravel in 1999, over-valuation had been a contributory factor to the import growth (CEPA, 1999). Although exports increased by nearly 34 percent between 1994-98, it was more the result of improvements in volume and, later in 1998, in terms of trade, than to any measurable gains in competitiveness (CEPA, 1999).

Growth in exports is not the only solution to reduce the trade imbalance, a contraction in imports in a manner compatible with growth of the economy may accelerate improvements in trade balance. Even if export growth can maintain its thrust of the late 1980s, we have seen that import growth exceeded export growth by $31.5 million annually for the period 1984-98. The import growth of $122.7 million annually splits into non-oil imports of $118.4 million and oil imports

Table 3:7 Trade and Current Account of the Balance of Payments (pe cent of GDP)

	Percent of GDP			
	5-year Averages			Average Annual Growth
	1984-88	1989-93	1994-98	(1984-1998) US$, millions
Rate of Real exchange depreciation (+) and appreciation (-)	53.8	5.6	-4.3	
Merchandise Exports	13.5	16.1	24.4	91.2*
Merchandise Imports	14.6	22.7	30.7	122.7
Trade Deficit	-1.1	-6.7	-6.3	-31.5
Services (net)	-5.2	-6.0	-6.2	-9.6
Private Transfers (net)	2.1	3.8	4.7	21.7
Current Account Balance (CAB)	-4.2	-8.9	-7.8	-19.4
Official Transfers (net)	2.4	3.8	3.4	7.2
CAB (including official transfers)	-1.8	-5.1	-4.4	-12.2

Data Sources: *Bank of Ghana – Annual Reports (various issues), CEPA Macroeconomic Review, 1999*

* Numbers in this column represent the average annual changes in the corresponding variables between 1984-1998. Thus merchandise exports increased US $91.2 million against the average increase in imports of US $122.7 million, resulting in average annual worsening of trade gap by US $31.5 million.

of $4.3 million annually. Estimates of the annual rate of growth for the period 1984-98 suggests that on average, non-oil imports accounted for nearly 95 percent of total imports, suggesting that although oil imports had witnessed an increasing trend, it cannot "explain" the increase in the non-interest current account (NICA) deficit and hence the external debt. It is difficult to determine the exact proportion of capital goods imports in total non-oil imports because detailed data on the commodity composition is not publicly available. Commenting on import composition, CEPA (1999:116) noted that the import of consumer

Table 3:8 Measures of Budgetary Performance, GDP growth

Percent of GDP			
	5-year Averages		
	1984-88	1989-93	1994-98
Operating Fiscal Balance	1.55	2.03	4.00
Primary Fiscal Balance	-1.24	-2.29	-1.54
Interest Expenditures	1.55	2.06	5.43
Structural Fiscal Balance	-2.79	-4.35	-6.97
Target real GDP growth	5.3	5.0	5.2
Actual real GDP growth	4.8	4.5	4.0
Nominal average interest rate	19.1	26.0	37.8
Inflation rate	29.1	23.1	35.0
Implied real rate on financial instruments	-10.0	2.8	2.8

Operating fiscal balance is total tax revenue less current non-interest expenditures. *Primary fiscal balance* is total tax revenue less current non-interest expenditures, domestic capital expenditures. *Structural fiscal balance* is total tax revenue less current non-interest expenditures, domestic capital expenditures, interest payments.

goods may have grown faster than GDP. In fact, criticisms of import liberalization policy surfaced as early as in 1989. Local manufacturers feared that the importation of consumption goods–textile, leather products, cosmetics, plastics– which flooded the market undermined local productive capacity.[13]

Policy measures to reduce demand for non-oil imports and to moderate real appreciation of the currency, if not to reverse it, should take precedence in the effort to reduce trade deficit and to improve the NICA deficit. There are two reasons for this refocusing. First, it is reasonable to suggest that Ghana may have significantly liberalized its import regime way ahead of export growth. If that is the case then import growth as witnessed in the past must be curtailed especially since the prospects of astounding export growth remain low. Second, experience elsewhere, mostly in Latin American countries, demonstrate that attempts to gain export competitiveness through devaluation risks greater adverse social consequences. It erodes the purchasing power of all who derive income from domestic sources and whose income are quoted in fixed nominal units of the local currency. The losers in this case are mostly civil servants, teachers, factory workers, and the growing service sector. From their standpoint, export promotion via currency devaluation represents a levy on their real wages and income. The winners are the mining and cocoa sectors, timber merchants and non-traditional exporters. In an economy already beset by rising cost of fuel and petroleum products–the catalyst for many domestic price increases– a depreciation induced price increases will accelerate the loss of real income and fuel social protest.[14] While real wage loss resistance may be countered by government repression of labor unions and with pro-government union leaders, it will be difficult to sustain wage repression over long periods of time, particularly in a fragile democracy. Moreover, because the bulk of employment remains in the non-exportable sector, it is doubtful whether real wage cutting can translate into export competitiveness. Surely, continued transformation of the economy arguably will require increases in import of capital goods. But import growth if driven by consumption boom rather than by direct investment surge poses a threat to economic growth because of the force majeure of debt servicing obligations.

3:4.2 Fiscal Outlook

The fiscal budgetary performance in Table 3:8 highlights the daunting task ahead in fiscal adjustments. Attention here focuses on the operating budget, primary budget and the interest payments because this is where much of the fiscal crowding out occurs, and this is where much of the new domestic debt is coming from. The operating budget surplus has improved markedly since 1989 from 2 to 4 percent of GDP, but this has come more from improvements in tax revenue than from overall cut in operating expenditures despite the changes in the composition of the latter, as we saw earlier in Table 3:6. Most striking is that primary budget balance continues to fall short of covering interest payments. How long can this continue? Others have thought about these fiscal matters and come up with an answer. For example, in their discussion of issues of public debt policies, Velthoven, Verbon and van Winden (1993) show that the ability of governments to service debt depends on there being future primary surpluses.

Solvency requires that the existing public debt equals the present discounted value of future primary surplus. And in steady state, solvency condition requires that the government must have a surplus on its primary budget that equals its domestic and foreign interest payments (Buiter, 1985).

The high debt/GDP ratio has led to high interest payments which reached 6 percent of GDP in 1998. Whether the situation will improve and how quickly it does will depend on the rate of growth of the economy, the trends in domestic interest rates and the commitment of the government to meet inter-temporal budget constraint. Substantial growth in output in excess of 5 percent would be beneficial in two ways: it will reduce the debt-GDP ratio and provide additional financial resources through tax revenues (assuming a buoyant tax system). But as we see in Table 3:8, the reality is that actual output growth has consistently fallen short of the 5 percent target.

With economic growth that in all probability will not exceed 5 percent, the bulk of stabilization rest on fiscal management. After years in which the tax structure was little changed from its design deficiency and associated distortions and omissions, the decade of the 1980s and 1990s saw a redoubling of tax reform efforts to encourage compliance, to broaden the tax base, and to stimulate private sector growth (Public Expenditure Review, 1993). Significant progress in the area of taxation and revenue system suggest that the task of fiscal stabilization must turn to (1) rationalization of spending and (2) rehabilitation of budgeting practices and procedures to ensure more efficient expenditure use and effective expenditure control and monitoring. Generating primary budget surplus over the near term is not without social and economic costs, but, it will provide the credible evidence that accumulated debt will not be serviced by further borrowing. Moreover, it is the only realistic way of achieving needed fiscal adjustment if even harsher social and economic costs are to be avoided in the immediate future.

To this end, first, an expanded role of the Comptroller and the Accountant General in the management and conduct of government operating (non-interest current) spending functions would be beneficial. Second, more effective mechanisms in the implementation, evaluation and monitoring of capital or development projects will also be helpful. Such a task can be performed by a coordinating agency with the responsibility to oversee, coordinate, monitor and evaluate capital projects across line ministries and departments. When this is done, borrowing can be justified for capital expenditures since these add to public assets and the government can allocate the costs to both current and future beneficiaries.

3:4.3 Domestic Debt Outlook

Finally, we turn to the issue of dealing with large domestic debt, which increased by 93 percent between the first quarter of 1997 and first quarter of 1999, and is not likely to decline in view of the health of public finances in the near term. As mentioned earlier, the domestic debt/GDP ratio of 27 percent in 1998 may not be exceptionally high by international standards nor necessarily "worrisome". But Guidotti and Kumar (1991:11) remarked that, domestic debt problems may still occur if, at the same time, "either the ratio of external debt to GDP is high or the current or expected future fiscal position is weak". The preceding evidence show that both conditions prevail in Ghana. We also know that when the out-standing debt is all short term with interest rate insufficient to compensate for inflation, and when the debt growth becomes self-generating, owing to mounting interest payments, there is always the temptation to take short-cut to get rid of at least some of it.

In his survey of episodes of large public debts, Alesina (1988) observed that solving large public debt historically turns on the income and wealth redistribution that accompanies the alternative remedies. And since a redistributive problem is intrinsically a political problem, there has been no single optimal remedy. Remedies pursued by different countries and their degree of success depend on the political context. Alesina conjectured that the less cohesive and unstable the political situation, the greater the likelihood of fiscal deadlocks, the greater the likelihood that public debt will grow and eventually induce an inflationary spiral because the government will resort to overt or covert monetization of deficits.

Faced with a fiscal deadlock, Germany in 1919-23 issued new debt and monetized its deficits. The ensuing inflation process led to near complete default of outstanding debt. By the fall of 1922, the real value of outstanding debt was only about 5 percent of its value in 1919. The French solution to the debt problem was a combination of moderate inflation, default of the debt, successful stabilization, prudent debt management, and a strong and independent central bank. The ability of the latter to resist the pressure of monetization and its role in stabilizing the franc added credibility to the French's stabilization effort of 1926-28. With successful stabilization, France redeemed its short term debt and issued long term debt, twisting the maturity structure of the debt, and in so doing reducing the debt service burden. Unlike Germany and to some extent France, Great Britain resorted to tight fiscal policies to reduce public sector borrowing requirements, and to tight monetary policy to control inflation and capital flight which it feared may happen. Like the British, the U.S. adopted a non-inflationary, non-defaulting approach to solving its debt problems. Key was the rapid growth of the U.S. economy and budget surpluses obtained largely through drastic expenditure cuts rather than tax increases. In recent times, Canada in the 1990s, with debt service charges almost one-third of total expenditures, solved its public debt problem by strong fiscal stabilization, tight monetary policy to curb inflationary expectations and strong economic growth.

Will a depreciation of the nominal debt with a burst of inflation work in Ghana? And what are the dangers of such a policy move? The inflation tax on the domestic debt averaged 30 percent for the period 1984-98. The tax peaked at 50 percent in 1995/96 and fell to 18 percent in 1998. Spaventa (1988) notes that debt repudiation is permanently beneficial if and only if primary budget deficits are eliminated and credible fiscal responsibility is established. Otherwise, debt repudiation can be an expensive proposition not worth contemplating. Besides the loss of reputation, it has perverse wealth redistribution effects and may generate financial instability (Buiter, 1985). Moreover, with under one year T-bills representing about 49 percent of the debt, it will be difficult to depreciate short maturity debt without seriously risking refinancing problems and capital flight out of the domestic financial market. The latter can put pressure on the exchange rate causing further depreciation, inflation and financial instabilities. Confidence in a turnaround, therefore, must rest on the belief that fiscal policy will at some stage sooner generate surpluses to service and to reduce the accumulated debt in order to avoid debt pathology. Meanwhile, the government must be willing to experiment with lengthening the maturity structure of the debt by issuing two to five year instruments with reasonable premium. This will be in line with encouraging the development of the secondary market and increasing the depth of the market and the scope of debt management.

3:5 Conclusion

This chapter has provided an overview of the growth of public debt, the causes, the possible impact and the risks and challenges the debt represents. The overall picture of the debt since 1980 can be characterized by the following elements: the extraordinary leap in the domestic debt, the uninterrupted rise in external debt, the rather short average maturity of the domestic debt, a domination of treasury bills in the hands of the public and banks, and the growing debt service burden on the fiscal budget and current account balance.

Debt growth has been partly the result of economic restructuring, partly the result of a growth strategy, and partly the result of debt overhang from a succession of past quasi fiscal activities. But there is also the new debt and key to this are the fiscal and trade deficits. The Ghanaian experience illustrates some of the problems of economies in transition. As far as public sector balance sheet is concerned, establishing the appropriate initial conditions for economic transformation is a challenge too often taken for granted. To this end, debt finance of balance sheet reconciliations may be seen as a rational choice, perhaps, the only choice, even if the magnitude of the needed adjustment was underestimated. Moreover, since the benefits of reform extend to future generation, debt financing of infrastructure investment finds support if public investments enhance the productivity of complementary private sector inputs, improving the revenue generating capacity to service and to amortize the debt. However, in Ghana's case, the slow growth in long-term private capital formation and the flourishing of trade in consumption goods cast doubt on the viability of the "debt-cum-growth" strategy.

Does Ghana have a debt problem? Arguably, the answer must be yes. The evidence suggests that even if Ghana has not yet faced debt problem in the likes of Brazil, Mexico or Argentina up to now, it has built up all the conditions and vulnerability for a debt crisis to emerge with the slightest adverse shock to its balance of payment and the least weakening of its future fiscal position. Can the current course of deficits be sustained without imploding debt? The answer is no. It is doubtful whether Ghana can both borrow and grow without risking the buildup of unsustainable debt. As far as the external debt is concerned, debt relief will be welcome. But relief should not be mistaken as a spell for success in debt management. As the experiences of Brazil, Argentina, Mexico and elsewhere demonstrate, the stresses and strains on public finance will ultimately re-emerge if fiscal and trade deficits are not rectified.

But can Ghana pursue successful stabilization so as to ensure needed fiscal and trade balance? The large increases in exports during the past decade would lead one to expect that Ghana should have done well. But even with sharply increased revenues, imports have surged way ahead of export growth, driven in part by consumption boom and public investment surge. The present reality is that diversification of and growth in exports could not be the only solutions, even if external and internal conditions are favorable, even if export growth can maintain its thrust of the late 1980s. A contraction in non-oil imports in a manner compatible with long term growth, and measures to moderate real appreciation of the local currency, if not to reverse recent trends, should therefore take precedence in the effort to improve non-interest current account balance. As far as fiscal adjustment is concerned, confidence in a turnaround must rest on the belief that fiscal policy will at some stage soon generate sufficient surpluses to service and to reduce the accumulated debt in order to avoid debt pathology.

Significant progress in the area of taxation and revenue system suggest that attention must turn to one of rehabilitating budgeting practices and procedures to ensure more effective expenditure use, control and monitoring and greater accountability.

Finally, Ghana's experience is a poignant example of how one can go further in orthodox reforms–liberalization, privatization, free trade, undistorted market prices, competitive exchange rates–and yet face a potential crisis if fiscal policy is not managed well, if import liberalization runs way ahead of export growth, and if dependence on external resources is carried too far. In this sense, Ghana is yet another interesting laboratory of reform, underscoring the view that what constitutes appropriate stabilization and structural reform–approaches and sequencing–remain on a shaky grounds than we may be willing to admit.

Notes

1   This paper has benefited from my association with the Centre for Policy Analysis (CEPA), Accra, Ghana since 1994. The support of the research fellows at CEPA and the research assistance of Lawrence Apaloo also of CEPA are gratefully acknowledged. The comments by Dr. Leornard Pluta, and Dr Melville McMillan are also gratefully acknowledged.

2   Ghana Commercial Bank, *Quarterly Economic Review*, Oct-Dec. 1978.

3   The Ricardian vision brought to the fore by Barro (1974) states that increased government borrowing may induce increased saving by forward-looking taxpayers. Substituting borrowing for taxation therefore makes no difference for the real state and development of the economy. The empirical support ranges from mixed to weak.

4   External debt reported by the Bank of Ghana tend to vary from IMF data. For example, Bank of Ghana reported 1998 external debt of US$5918.8 million. The IMF reported a much larger debt stock of US$7011 million. The higher debt reported by the fund is attributed to large non-concessional borrowing by the government in the mid 1990s (CEPA, 1999, p 117). This paper uses IMF data whenever there is a discrepancy between official Government of Ghana data and IMF data.

5   The 15 countries (G-15) identified in 1985 under the Baker Plan for dealing with heavily indebted countries include Argentina, Bolivia, Brazil, Chile, Columbia, Cote d'Ivoire, Ecuador, Mexico, Morocco, Nigeria, Peru, Phillipines.

6   It is tempting to compare Ghana's debt/GDP ratio to some historical precedents. For long periods of her history, Britain exhibited ratio in excess of 100 percent. This happened in the 1920s and 1930 in the Napoleonic wars. In France the ratio peaked above 180 percent in 1922 and remained above 100 until 1928. It remained above 100

in the USA in the postwar period until 1950. In Italy debt levels were well above 80 percent in the 1980s and in Belgium and Ireland in the 130 and 120 percent, respectively by the mid 1980s. But except for Italy, Belgium and Ireland, these debt episodes were all due to previous war efforts or of an uncommonly severe recession and swiftly dealt with in the immediate post war years. Like Italian debt of the 1980s, Ghana's debt grew in peace time, but unlike Italy, only one ruling party presided over the past 18 years of the growth in debt.

7   According to Thirwall (1991) declining terms of trade may be highly correlated with indebtedness. The higher the debt/GDP ratio the faster the percentage decline in unit value of exports and thus the terms of trade. The direction of causation here is that indebtedness leads to falling terms of trade not the other way around.

8   NPART bonds peaked at C41.5 billion (or 1.7 percent of GDP) in 1991 and fell to C34.9 billion (or 0.4 percent of GDP) in 1995. State-owned enterprises accounted for 37.5 percent of the bad loans in 1991. Data and anecdotal evidence on NPART bonds were gathered by the author through interviews with the government agency charged with the debt recovery (FINSAP Secretariat) and also from Bank of Ghana Monetary Survey.

9   The 1999/2001 budget figures are taken from Government of Ghana Budget Statement 1999 and CEPA Macroeconomic Review, 1999.

10   Comptroller and Accountant General: Public Accounts, 1997.

11   This section of the chapter draws on previous work by Amoako-Tuffour (1998).

12   The HIPCs include 41 countries (Ghana included) considered to be heavily indebted to foreigners. In all of these countries, the external debt stock was at least twice the value of exports on average during 1992-94. Adjusted for concessionality, the median debt-to-exports ratio was 340 percent. Information is taken from the work of Stijn Claessens and others (1997).

13   Ghana Commercial Bank, *Quarterly Economic Review*, July-Sept. 1989.

14   There were five upward revisions in fuel prices in 1999–twice in June, then in September, October, and December. The price of fuel increased by approximately 56.6 percent between June and December 1999.

References

Agama, G.K. (1968). "The Growth of Money and Public Debt in Ghana: 1957-66". *Economic Bulletin of Ghana*, vol. 2: 8-33.

Ahmed, Nassem (1967). "Deficit Financing, Inflation and Capital Formation". *The Ghanaian Experience: 1960-65*. Weltforum-Verlag, Munchen: Germany.

Alesina, A. (1988). "The end of large public debts" in Guivazzi and Spaventa: 34-79.

Allen, M. (1992). "Government Debt Management" in Vito Tanzi (1992): 67-79.

Amoako-Tuffour, J. (1996). "Public Sector Corruption, Embezzlement and Economic Reform". *Review of Human Factor Studies*, 2(1): 27-48.

Amoako-Tuffour, J. (1998). "Sustainability of the External Debt" in Centre for Policy Analysis, *Ghana: Macroeconomic Review and Performance*, 106-113.

Aryeetey, E. (1994). "Private Investment Under Uncertainty in Ghana". *World Development*, vol. 22 (8):1211-1221.

Bank of Ghana. *Quarterly Economic Bulletin*. Various Issues. Accra: Government of Ghana Printing Office.

Bank of Ghana. *Annual Report*. Various Issues. Accra: Government of Ghana Printing Office.

Barro, R. (1974). "Are government bonds net wealth?". *Journal of Political Economy*, 82:1105-1117.

Buchanan, J.M. (1957). *Public Principles of Public Debt*. Richard D. Irwin, Inc.: USA.

Buiter, W. (1985). "A Guide to Public Sector debt and Deficits". *Economic Policy* (1): 14-79.

Caprio, G. (1999). Cited in *Journal of Economic Literature* (December: 1725).

Centre for Policy Analysis (1997). *Macroeconomic Review and Outlook*. Accra: Ghana.

Centre for Policy Analysis (1998). *Ghana: Macroeconomic Review and Performance*. Accra: Ghana.

Centre for Policy Analysis (1999). *Ghana: Macroeconomic Review and Performance*. Accra: Ghana.

Chand, S. K. (1993). "From Controls to Sustainable Liberalization: Ghanaian Lessons" in Vito Tanzi :353-373.

Claessens, S., Detragiache, E., Kanbur, R. and Wickham, P. (1997). "HIPC's Debt Review of Issues". *Journal of African Economies*. vol. 6 (2): 231-54.

Cox-George, N.A. (1961). "Studies in Finance and Development: The Gold Coast (Ghana)". *Experience*. Dennis Dobson: London.

Dornbusch, R. (1988). "Comments on Spaventa" in Guivazzi and Spaventa (eds.), *High Public Debt: The Italian Experience*. Cambridge University Press.

Dornbusch, R. (1993). *Stabilization, Debt, and Reform: Policy Analysis for Developing Countries*. Prentice Hall: NJ.

Easterly, W. (1999). "When is fiscal adjustment an illusion". *Economic Policy*. April, 57-86.

Easterly, W., Alfredo Rodriguez, C., Schmidt-Hebbel, K. (1994). *Public Sector Deficits and Macroeconomic Performance*. Oxford University Press.

Ghana, Republic of, "Quarterly Digest of Statistics". Various Issues. Central Bureau of Statistics Accra: Government of Ghana Printing Office.

Ghana Commercial Bank (1978). *Quarterly Economic Review*. Oct-December, vol. 1(4), Accra: Ghana Commercial Bank Press.

Ghana Commercial Bank (1989). *Quarterly Economic Review*. July-Sept., vol. 12 (3), Accra: Ghana Commercial Bank Press.

Giavazzi, F. and Spaventa, L. (eds. 1987). *High Public Debt: The Italian Experience*. Cambridge University Press.

Guidotti, P. E. and Kumar, M. S. (1991). *Domestic Public Debt of Externally Indebted Countries*. IMF Washington. Occasional Paper 80.

Ministry of Finance (1994). *Public Expenditure Review 1993*. Accra: Ghana.

Moore, D. (1990). "Debt—Is it Still a Problem?". *Australian Economic Review* 87. pp.

17-32.

Pagano, M. (1988). "The management public debt and financial markets" in Giavazzi and Speventa: 135-176.

Rodrik, D. (1996). "Understanding Economic Policy Reform". *Journal of Economic Literature*, 34:9-41.

Sachs, J. (1985). "External debt and Macroeconomic Performance in Latin America and East Asia". *Brookings Papers on Economic Activity*, vol. 2: 523-564.

Sachs, J. (1987). *Developing Country Debt and Economic Performance Country Studies*. University of Chicago Press.

Spaventa, L. (1987). "The Growth of Public Debt". *IMF Staff Papers*, vol. 34 (2): 374-399.

Spaventa, L. (1988)."Is there a public debt problem in Italy" in Giavazzi and Spaventa (1988): 1-24.

Sowa, Nii K. (1996). "Policy Consistency and Inflation in Ghana". *African Economic Research Consortium* Research Paper (43).

Tanzi, V. (1992). *Fiscal Policies in Economies in Transition*. IMF Washington.

Tanzi, V. (1993). *Transition to Market: Studies in Fiscal Reform*. IMF Washington.

Tanzi, V. and Blejer, M. L. (1988). "Public Debt and Fiscal Policy in Developing Countries" in Arrow, K. J. and Boskin, M. J. (eds.) *The Economics of Public Debt*. MacMillan Press.

Thirlwall, A.P. (1991). "The terms of trade of primary commodity prices, debt and development" in Davidson, P. and Kregel, J.A. (eds.) *Economic Problems of the 1990s: Europe and Developing Countries*. Worcester: U.S.A.

United Nations Conference on Trade and Development (1999). *The Least Developed Countries: 1999 Report*. United Nations: Geneva.

Velthoven, B., Verbon, H., and van Wilden, F. (1993). "The Political Economy of Government Debt: A Survey" in H. Verbon and A.A.M. van Winden, (eds.), *The Political Economy of Government Debt*. Amsterdam: North-Holland.

World Bank (1995). *Ghana: Growth, Private Sector, and Poverty Reduction* Report No. 14111-GH. Washington.

4 Fiscal Impacts of Structural Adjustment

KOJO APPIAH-KUBI

4:1 Introduction

In 1983, the government of Ghana launched a comprehensive but gradualist economic recovery program (ERP) on the principles of structural adjustment (Armstrong, 1996). This was done against the background of massive macroeconomic distortions in the production and exchange systems. One policy measure that has been at the core of Ghana's financial and structural adjustment has been fiscal management.

Fiscal adjustment has successfully contributed in turning the economy around. It has contributed successfully to raise the annual average real GDP growth rate of the country from -2 percent in 1970-1982 to about 5 percent in 1983-1998. Other economic achievements include an increase in the per capita income growth rate from about -5 percent in the previous decade to a positive 2 percent per annum and a reduction in inflation from 123 percent in 1983 to about 19 percent (average) in 1998 (G.S.S., 1986, 1989, see also World Bank, 1995a). Notwithstanding these positive achievements, the country still ranks among the poorest in the world with a per capita income of about US $400 as at the end of 1998. Even after more than one and a half decades of reform the country has yet to attain the investment and savings levels of the early sixties and remains far from the levels observed in the rapidly growing Asian countries. These anxieties have called forth new economic strategies to transform the reform into a consolidation phase of a new economic regime of sustained accelerated development—on the pillars of openness, market orientation, macroeconomic stability, and poverty alleviation. Here, too, fiscal policy is expected to play the leading role of not only ensuring a macroeconomic balance in the economy, but also stimulating considerable increases in the levels of savings and investment.

This chapter evaluates the success of adjustment in Ghana with particular reference to fiscal policies: tax and expenditure reforms and their relationship to other macroeconomic indicators such as savings and investment. It also attempts to assess Ghana's past fiscal policy in the context of efficacy of mobilization of resources for development.

4:2 Fiscal Adjustment

By late 1982, fiscal finances in Ghana were completely out of hand, with government revenue having fallen from 20.5 percent of GDP in 1970 to 5.4 percent in 1982. Under the ERP, fiscal adjustment was, therefore, assigned the following objectives (Kapur, 1991):
• to correct the imbalances in government finances through:
  i) tax reform aimed at improving the efficiency and effectiveness of the tax system and tax administration, and increases in prices of public services to cover cost;
  ii) reform of the government expenditure management aimed at reducing total government expenditure and restructuring of government expenditure in favor of capital spending;
• disengagement of the state from direct production and divestiture of loss making state owned enterprises (SOEs) and privatization of social services provision;
• removal of fiscal controls that hinder trade, savings, investment, price and foreign exchange stability, to allow allocation of resources to be guided by market forces;
• restoration of fiscal and monetary discipline, coupled with the rehabilitation of the basic social and economic infrastructure.

In the following sections, we turn to key measures and events that were adopted or contributed to the achievement of the above-mentioned objectives.

4:2.1 Tax Reform

As part of the efforts to boost resource mobilization, a conscious tax reform was embarked upon in 1986 within the framework of the Structural Adjustment Programs. The main thrust of tax reform included: (1) simplifying the whole system of indirect taxes; (2) reducing the level of both direct and indirect taxes; (3) rationalizing the tax structure and (4) strengthening the revenue institutions to improve tax compliance and administration (Terkper, 1994). During the period under study the tax system witnessed successive discretionary changes in the tax structure and tax rates. The most significant of these changes included the consolidation of all incomes and allowances and the reduction in the highest marginal tax rates on personal and company incomes from 65 percent in 1983 to 35 percent in 1995. These reductions became necessary since past attempts to increase revenue without any corresponding change in the tax structure had led to high marginal tax rates on a shrinking base. Final taxes on dividends were equally reduced from 30 percent in 1989 to only 10 percent since 1992 after being converted from withholding tax. Exemptions were also widened, particularly, to attract foreign investments and personal tax relief also witnessed successive upward adjustments over the years for various economic and social reasons. Besides all personal allowances including transport, housing, rent and other in-kind incomes and allowances were consolidated with income and made taxable.

Indirect taxes similarly witnessed numerous frequent discretionary changes with the view to rationalizing the tax regime. To this end the sales tax was identified as a major tax for expanding the relative base and was thus made neutral in 1987 with the unification of rates (17.5 percent) applicable to domestic output and imports. Export taxes, with the exception of port jet and vessel fuel sales and cocoa, were also abolished. This was done with the view to eradicate the discrimination of Ghanaian exports and increase their competitiveness against exports from other countries, which are taxed. A new super sales tax on luxury goods was introduced in 1990 with tax rates ranging from 50 percent to 500 percent and modified on several occasions. Excise duties on all locally produced goods except tobacco, alcoholic and non-alcoholic beverages, for instance, were merged into the sales tax, and their rates reduced. Tariffs were also simplified to conform to the Brussels' tariff nomenclature and substantially reduced to standard rates of 0, 10 and 25 percent. The reform of the indirect tax also paved the way for petroleum tax to assume new significance in the tax structure. During this period its relative share of the total tax revenue rose from about 5 percent in 1987 to about 30 percent in 1993 before declining to about 17.4 percent in 1996. This massive increase in revenue had been the result of persistent upward shifts in tax rates. Export taxes also experienced major changes with the coverage being reduced to only cocoa exports and jet fuel sales at the ports.

Another event worth mentioning was also the attempt in 1995 to convert the sales tax into a value-added tax (VAT) at a flat rate of 17.5 percent for both imported and domestic produced goods, with luxury consumer goods attracting a further 17.5 percent excise duty. However, implementation problems arising from poor planning and public opposition led to its withdrawal in June 1995, barely 3 months after its introduction. Conscientious efforts to enhance administrative efficiency witnessed the reorganization of Customs, Excise and Preventive Service (CEPS) and Internal Revenue Service (IRS); organizations assigned to handle the indirect and direct taxes respectively. The National Revenue Secretariat of the Ministry of Finance was also reconstituted to oversee and co-ordinate the activities of the other two revenue administrative institutions and efforts are advanced to convert this secretariat into an agency board for all the revenue administration services with more powers.

4:2.2 Expenditure Reform

Under the SAP fiscal policy was directed not only at resource mobilization, but also focused on better control, monitoring and reporting of expenditure to achieve greater efficiency in public resources management. Reform measures (Tsikata, et al., 1997; Nashashibi and Gupta, 1992) undertaken to achieve these goals included the following:

• Reform of the government wage structures through increases in wages and salaries for civil servants, expansion of the wage differences among the various skilled grades and reduction of the wide disparities within the public sector, with the objective to boost working morale and productivity in the public service as well as to attract skilled workers for the public sector.

• Realignment of government expenditure in favor of capital spending.

• Ensuring the timely preparation, presentation and publication of a broad-based budget.

• Ensuring proper monitoring and execution of the budget.

Attempts to achieve the above objectives were made with the help of a series of reform programs. Shortly after the introduction of the ERP, an Integrated Personnel and Payroll Program was launched not only to reduce the size of the government wage bill, which had accounted for over 90 percent of the total recurrent budget in the early eighties, but also to restructure the civil service remuneration system with the view to reversing the severe compression in differentials between the top and bottom grades. To this end a Civil Service Reform was also embarked upon within the framework of a National Institutional Reform Program to enhance efficiency in the civil service. In order to realign government spending towards capital expenditure so as to facilitate appropriate allocations for rehabilitation of the country's economic and social infrastructure, a three-year rolling Public Investment Program (PIP) was launched from 1986-1996. The PIP had covered about 97 percent of total government development expenditure, out of which 60 percent had been to rehabilitate and develop economic infrastructure including roads, highways, and ports and 30 percent for agricultural services, cocoa and mining sectors. In 1986, an SOE reform program was also launched to privatize some SOEs as well as improve efficiency and profitability of remaining SOEs and thereby reduce the managerial and financial burden that they had placed on public resources.

These efforts were complemented with the establishment of close expenditure monitoring systems in the various ministries, departments and agencies to avoid overspending and to achieve better expenditure control. One such program is the Expenditure Tracking System (EXTRACON), which was intended to monitor expenditure in MDAs. The latest major expenditure management program, dubbed PUFMARP, was launched in July 1995 with the main objective to develop an integrated computerized financial management system that would assist in addressing issues of the weak budgetary framework.

The preparation, monitoring and review of the budgeting process has undergone varied changes under the SAP, all with the objective to improve upon government expenditure management. Beginning in 1998, the government for the first time presented its budget in the "broad format". In this broad-based budget all government expenditures and domestic revenue as well as foreign inflows in the form of project grants disbursed directly to specific projects and programs were captured in one budget for the first time. Furthermore, the budget of 1998 introduced the concept of "domestic primary balance", defined as domestic revenue less domestic expenditure excluding interest payments, to facilitate a better and accurate assessment of national revenue mobilization performance.

As part of the efforts to strengthen government machinery in the preparation and execution of the national budget, the 1999 budget also ushered in a path-breaking innovation in the preparation and presentation of government budgeting. Government expenditure was prepared against the background of a Medium Term Expenditure Framework (MTEF) and presented in the form of a three-year rolling expenditure program, which began in 1999 and covers the period 1999-2001. Whereas the budget allocations for the financial year 1999 represent government commitments for that year, those of 2000 and 2001 are just indicative budget levels, which would be subject to revisions in their respective fiscal years.

The MTEF as a component of the PUFMARP sought to correct the observed weaknesses in the preparation and execution of the national budget. To this end it abolished the traditional split between recurrent and development expenditures and reclassified expenditures into discretionary and non-discretionary spending, in order to take account of the budgetary implications of the capital budget on recurrent expenditure.

4:3 Fiscal Trends under SAP

The structural adjustment program has recorded many fiscal changes. The most significant of these changes occurred with revenue, expenditure, saving and investment indicators. The following sections discuss the changing nature of the trends in the fiscal indicators under the SAP.

4:3.1 Trends in Government Revenue

Under this SAP, Ghana's fiscal situation has improved substantially through better mobilization of revenue, improved tax collection and rationalization of consumption and user charges (World Bank, 1995). Table 4:1 gives a simple outline of the main features of government finances in the Ghana between 1983 and 1998. It can be seen from the table that government revenues have increased tremendously from 5.38 percent of total GDP in 1982 to about 19.89 percent of total GDP in 1998. Tax revenues, which in 1998 accounted for about 80 percent of total government revenue, have also witnessed tremendous increases relative to the GDP; i.e. from an equivalent of about 4.56 percent in 1982 to about 15.7 percent 1998. The increase in the tax effort has been the result of upturns in almost all the major tax revenue components of the Ghanaian tax system. The relative share of taxes on income and property to the GDP, for instance, increased from 1.7 percent in 1982 to 4.4 percent in 1998. Similarly, export taxes have, despite the reduction in coverage to include only cocoa exports and jet fuel sales at the ports, recorded substantial increases in revenue receipts, rising from 0.25 percent of GDP 1982 to 2.3 percent in 1998. Import duty has also witnessed substantial increases, from 0.58 percent of GDP in 1982 to 2.2 percent in 1998. Petroleum tax had, until 1986, generated virtually nothing rose to account for over a fifth of total government revenue and 4.1 percent of GDP in 1994, before declining to 2.22 percent in 1998. As can be seen in Table 4:1, the SAP period also witnessed an upsurge in non-tax revenue from about 0.82 percent of GDP in 1982 to its highest peak of 8.78 percent in 1994 before declining to 3.53 percent in 1998. This increase reflects the government's revenue mobilization efforts and has been particularly more pronounced with respect to the surge in revenue from income and fees from 0.06 percent of GDP in 1982 to a high peak of 2.64 percent in 1996, before falling again to 1.26 percent in 1998. Another corollary of the reform has also been the change in the relative composition of the tax revenue and the gradual shift from direct taxes to a reliance on indirect taxation, even though the ratio of tax revenue to total revenue has remained almost the same (about 80 percent) over the years. This shift has also been forged by the rationalization of consumption taxes, culminating in the replacement of the single stage manufacturer's sales tax with a multistage value added tax in 1998.

Table 4:1 Source of Government Revenue in Ghana, 1982-1998 (percent of GDP)

	1982	1983	1984	1985	1986	1987	1988	1989	1990	1991	1992	1993	1994	1995	1996	1997	1998
Income & Property	1.70	0.89	1.52	2.36	2.95	3.23	3.85	3.24	2.73	2.57	2.22	3.00	3.44	3.10	3.94	4.81	4.40
P.A.Y.E.	0.51	0.29	0.34	0.55	0.68	0.61	0.57	0.51	0.53	0.48	0.56	1.01	1.05	0.88	1.10	1.39	1.31
Self-Employed	0.36	0.19	0.27	0.35	0.35	0.48	0.48	0.36	0.34	0.29	0.19	0.20	0.23	0.16	0.21	0.25	0.23
Companies	0.77	0.41	0.89	0.97	0.75	0.85	1.83	2.21	1.67	1.64	1.23	1.48	1.80	1.71	1.22	1.47	1.42
State ent.	0.00	0.00	0.00	0.38	1.07	1.08	0.80	0.00	0.00	0.00	0.00	0.00	0.00	0.00	0.82	1.05	0.77
Rent	0.01	0.01	0.00	0.01	0.01	0.01	0.00	0.00	0.00	0.00	0.00	0.00	0.00	0.00	0.00	0.00	0.00
Others	0.05	0.00	0.02	0.10	0.09	0.20	0.17	0.16	0.18	0.16	0.23	0.32	0.37	0.36	0.60	0.66	0.67
Domestic Goods	1.19	0.95	2.06	2.44	3.41	3.49	3.41	3.67	3.53	5.08	4.60	6.06	6.24	5.39	5.20	5.42	4.76
Excise Duty	1.65	0.83	1.89	2.10	1.93	1.77	1.43	1.35	1.13	1.08	1.10	1.08	1.16	0.95	1.02	1.09	0.92
Sales Tax	0.26	0.13	0.16	0.34	0.62	1.07	1.18	1.26	1.15	0.93	0.98	0.85	1.01	1.00	1.56	1.77	1.62
Petroleum Tax	0.00	0.00	0.00	0.00	0.86	0.65	0.81	1.06	1.25	3.07	2.51	4.13	4.07	3.44	2.62	2.55	2.22
International Trade Trans.	0.95	2.69	3.05	4.90	5.55	6.53	5.03	5.40	4.57	4.91	4.06	4.79	7.01	6.60	5.90	6.47	6.62
IMPORTS	0.69	1.04	1.21	2.22	2.78	2.91	2.70	3.18	3.23	3.42	3.34	3.82	4.32	4.52	2.95	4.36	4.30
Import Duty	0.58	0.76	0.92	1.65	2.03	2.09	1.46	1.71	1.70	1.79	1.77	1.96	2.12	2.31	1.41	2.15	2.22
Sales Tax	0.09	0.24	0.26	0.46	0.58	0.70	1.06	1.31	1.37	1.43	1.39	1.49	1.74	1.86	1.25	1.61	1.59
Special Tax	0.00	0.00	0.00	0.05	0.10	0.08	0.14	0.15	0.12	0.15	0.14	0.21	0.10	0.07	0.07	0.13	0.11
Super Sales Tax	0.00	0.00	0.00	0.00	0.00	0.00	0.00	0.00	0.00	0.00	0.00	0.00	0.00	0.00	0.00	0.47	0.00
Purchase Tax	0.03	0.03	0.03	0.06	0.08	0.03	0.04	0.00	0.00	0.00	0.00	0.00	0.00	0.00	0.00	0.00	0.00
Other Taxes	0.00	0.00	0.00	0.00	0.00	0.00	0.00	0.02	0.04	0.04	0.03	0.16	0.35	0.29	0.22	0.00	0.38
EXPORTS	0.25	1.65	1.84	2.67	2.77	3.62	2.33	2.22	1.34	1.49	0.73	0.97	2.69	2.08	2.95	2.11	2.32
Cocoa Export Tax	0.00	1.52	1.67	2.58	2.72	3.62	2.33	2.22	1.34	1.49	0.73	0.97	2.69	2.08	2.95	2.11	2.32
Others	0.25	0.13	0.17	0.09	0.05	0.00	0.00	0.00	0.00	0.00	0.00	0.00	0.00	0.00	0.00	0.00	0.00
Tax Revenue	4.56	4.54	6.63	9.69	11.91	13.25	12.29	12.31	10.83	12.52	11.32	13.93	16.61	15.70	15.04	16.70	15.57
Non-Tax	0.82	1.03	1.74	2.39	2.59	2.07	2.34	2.82	2.35	2.93	2.30	5.85	8.78	5.08	5.01	3.88	3.53
Grants	0.76	1.00	1.40	1.91	1.84	0.68	1.24	1.51	1.37	1.49	1.17	1.81	0.80	1.81	0.72	0.99	0.94
Income & Fees	0.06	0.03	0.34	0.47	0.76	1.39	1.10	1.32	0.96	1.24	0.92	1.71	2.47	2.17	2.62	0.84	1.26
Divestiture Rev.	0.00	0.00	0.00	0.00	0.00	0.00	0.00	0.00	0.01	0.20	0.21	2.33	5.52	1.11	1.67	2.05	1.33
Grand Total	5.38	5.56	8.37	12.08	14.50	15.32	14.63	15.14	13.17	15.49	13.17	19.71	25.48	20.17	20.06	20.58	19.31

4:3.2 Trends in Government Expenditure

In contrast to structural adjustment programs of other African countries, whose implementation were characterized by massive reductions in government spending (Killick, 1993), the ERP of Ghana recorded substantial increases in government spending. This occurred both in nominal and real terms. In nominal terms government spending increased from 15.2 billion cedis in 1983 to 3,725 billion cedis in 1998, representing an increase by over 245 times over the same period.

Contributing to this substantial increase in total government spending has been large increases in recurrent and capital spending from 13 billion cedis and 1.6 billion cedis in 1983 to 3,053 billion cedis and 582 billion cedis in 1998 respectively. The introduction of additional new expenditure items including special efficiency funds, transfers to ECOMOG peacekeeping operations and the creation of Non-Performing Asset Recovery Trust (NPART) and district assemblies common fund for the 110 districts, inflation and population pressures (Table 4:2) have all led to increases in government expenditure. As a proportion of the GDP, total government spending increased more than two and half fold from 8.2 percent in 1983 to 21.5 percent in 1998, implying a substantial increase in the government's claim on real resources available in the country. The highest increase of over 520 percent relative to the GDP during the period was registered in the capital expenditure, which reflects the extent of the government's efforts to re-align resources in favor of capital expenditure. The growth of recurrent expenditure relative to GDP was equivalent to about 139 percent between 1983 and 1998.

The picture depicting these increases, however, characterizes varying phases of government expenditure management (Figure 4:1). Whereas the initial three years of the ERP show rapid increases in government expenditure and thus higher budget deficits, the second half of the eighties depicts an era in which the government achieved its best fiscal performance. Not only did the government succeed in maintaining almost a constant level of expenditure relative to GDP between 1986-1991, it also managed to achieve successive (narrow) budget surpluses during that period equivalent to an average of 0.8 percent of GDP.

This period was followed by massive increases in government expenditure (1992-1996), which led to the re-emergence of macroeconomic imbalances.

Figure 4:1

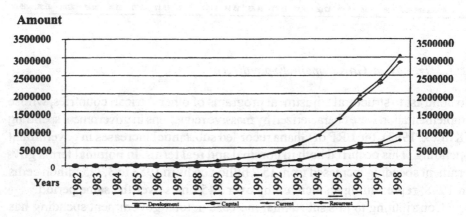

Table 4:2 Source of Government Revenue in Ghana, 1982-1998 (percent of GDP)

	1982	1983	1984	1985	1986	1987	1988	1989	1990	1991	1992	1993	1994	1995	1996	1997	1998
Current Expenditure	10.32	7.36	8.62	11.59	12.17	10.99	11.04	10.82	9.90	11.11	14.59	16.76	17.10	17.49	18.10	18.90	17.66
Personal Emoluments	2.66	2.03	1.95	4.36	5.46	5.14	4.68	4.79	4.34	4.41	6.21	6.82	6.61	6.46	6.12	6.42	5.91
Other Good & Services	2.43	1.86	3.75	3.79	2.38	2.24	2.78	2.58	2.07	2.28	3.47	2.74	2.64	2.84	3.12	2.67	2.83
Military Exp. on Mach. & Equip.	0.02	0.01	0.04	0.08	0.14	0.00	0.01	0.02	0.01	0.00	0.00	0.01	0.01	0.13	0.11	0.08	0.14
Rent	0.05	0.04	0.03	0.06	0.06	0.06	0.05	0.06	0.05	0.08	0.08	0.11	0.06	0.11	0.10	0.10	0.11
Interest Payments (Public Debt)	2.51	1.20	1.27	1.48	2.22	1.42	1.14	1.32	1.34	1.76	2.18	3.70	4.65	4.43	5.38	6.70	6.23
Transfers: Domestic	2.65	2.22	1.57	1.82	1.92	2.13	2.37	2.05	2.09	2.36	2.58	3.32	3.02	3.46	3.15	2.83	2.36
Transfers: Abroad	0.00	0.00	0.00	0.00	0.00	0.00	0.00	0.00	0.00	0.21	0.07	0.06	0.11	0.06	0.12	0.10	0.08
Capital Expenditure	0.98	0.65	1.24	1.74	1.64	2.70	2.65	3.04	2.62	2.91	4.28	3.95	5.95	5.64	4.63	4.20	4.04
Acquisition of new Capital Assets	0.83	0.54	1.08	1.47	1.32	2.04	2.10	2.16	1.99	2.44	3.16	2.62	3.05	3.69	3.53	3.14	3.37
Purchase of existing capital goods	0.04	0.03	0.03	0.02	0.01	0.00	0.01	0.01	0.02	0.01	0.01	0.01	0.01	0.03	0.03	0.10	0.06
Cap transfers to domestic sectors	0.12	0.07	0.14	0.25	0.32	0.25	0.26	0.20	0.22	0.07	0.14	0.15	1.87	1.23	0.82	0.93	0.09
Transfers: Public Ent	0.00	0.00	0.00	0.00	0.00	0.00	0.00	0.00	0.00	0.00	0.00	0.00	0.00	0.00	0.00	0.00	0.00
Transfers: Local Govt.	0.00	0.00	0.00	0.00	0.00	0.00	0.00	0.00	0.00	0.00	0.00	0.00	0.00	0.00	0.00	0.00	0.52
Cap transfers to abroad	0.00	0.00	0.00	0.00	0.00	0.00	0.00	0.00	0.00	0.00	0.00	0.00	0.00	0.00	0.00	0.00	0.00
Special Efficiency Fund	0.00	0.00	0.00	0.00	0.00	0.40	0.29	0.68	0.39	0.38	0.97	1.17	1.03	0.69	0.25	0.03	0.00
Financial Claims	0.01	0.01	0.00	0.01	0.00	0.00	0.00	0.00	0.00	0.00	0.00	0.00	0.00	0.00	0.00	0.00	0.00
Net Lending	0.41	0.23	0.29	0.62	0.52	0.65	0.57	0.54	0.47	0.47	0.43	0.60	0.17	0.21	1.01	0.04	-0.15
Other Items																	
Capital (Foreign financed)	0.00	0.00	0.00	0.00	0.00	0.00	0.00	0.00	0.00	0.00	0.00	0.00	0.00	0.00	0.00	7.42	4.40
Total Exp.	11.72	8.25	10.16	13.96	14.34	14.34	14.26	14.41	12.99	14.48	19.29	21.30	23.22	23.35	23.74	23.14	21.55

Source: *Statistical Service, Quarterly Digest of Statistics, various issues*

The corrective measures, which followed, however, seem to bear positive results. The fiscal situation in 1997 and 1998, for instance, is characterized by not only a falling trend in relative increases in annual government spending, but also by an equilibrium of spending out-turns and budget projections.

4:3.3 Savings and Investment Mobilization

Similar to the role envisaged for the public sector in previous development plans of Ghana, fiscal policy under the ERP was assigned the vital role to assist in augmenting the level of savings of both public and private sectors, so as to increase the leeway for higher capital formation and investment. To this end, fiscal policy was directed to impact upon the supply (savings) side and the demand (investment) side in the economy.

On the savings side efforts were intensified under the SAP to increase total savings by ensuring that a significant share of the national income was directed toward the financial system. In pursuit of this objective several policy recommendations were followed. These included a restructuring of the tax system of Ghana, with a view to broadening the potential coverage and increasing the returns to future consumption (savings), laying more emphasis on taxes of personal consumption expenditure. These efforts culminated in the conversion of the single stage manufacturer's sales tax to a multistage general VAT in 1995 and again in 1998 after its withdrawal on account of poor planning and implementation. Additionally, the marginal tax rates on various personal and corporate incomes were drastically reduced so as to make more funds available to potential savers.

A rural banking system was also introduced in 1984 to help mobilize rural savings. This facilitated the introduction of the "Akuafo" check system, whereby cash crop purchases from farmers were paid with check drawn on (rural) banks instead of physical cash. By so doing farmers became used to keeping cash in excess of daily needs in bank accounts, thus making it available for investment instead of keeping it in their homes.

On the demand side, emphases were placed on increasing the efficiency and effectiveness of financial sector management. This involved a financial sector reform, the establishment of a stock exchange and the removal of administered interest rates, credit ceilings and government directed bank lending.

4:3.3.1 Trends in Savings

Under the ERP, domestic savings have risen from its low level of 3.62 percent of GDP in 1983 to a peak of about 13.38 percent in 1997 (see Figure 4:2 and Table 4:5). The improvements in savings can be attributed partly to improvements in public finances, which enabled the government to change from negative savings, which had characterized the early eighties, to rising positive savings. For instance, government savings increased from its low level -1.72 percent of GDP in 1983 to about 8.11 percent in 1994 before falling once again to 2.34 percent in 1997. The initial stage of the reform recorded the highest increases in government savings. The nineties have, however, apart from the steep rises in 1994 and 1995 (Table 4:3), shown stagnation in government savings, as a result of massive increases in government wage and interest pay-

Figure 4:2 Selected Savings and Investment Indicators as Percentage of GDP, 1983-1999

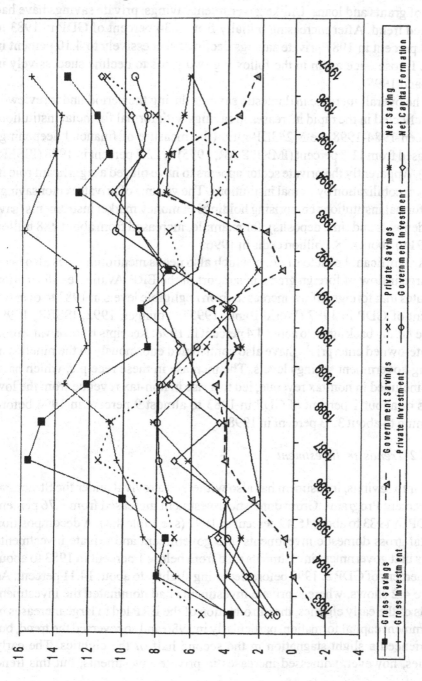

ments on the national debt. It must, however, be mentioned that a substantial proportion (80 percent) of the savings of Ghana had come in from abroad in the form of grants and loans. Unlike government savings, private savings have had a mixed trend. After increasing initially from 5.34 percent of GDP in 1983 to 11.48 percent in 1987 private savings declined successively to 4.10 percent in 1991. It rose once again in the following two years to decline successively in 1994 and 1995.

The overall increase in domestic savings during the period under review is also reflected in the rapid increases in savings with formal financial institution. Between 1984-1998 the M2/GDP ratio (the measure of financial deepening) increased from 11.8 percent (IMF IFS AR, 1995) to 22.7 percent in 1998 (ISSER, 1999). Apparently the private sector appears to have played a significant role in savings mobilization by formal institutions. The volume of private sector savings with formal institutions comprising holdings of money market instruments (savings deposits and time deposits), for example, increased from about ¢88 billion in 1991 to about 478.1 billion cedis in 1996.

A significant boost to savings, which also needs mentioning, has also been the large inflow of foreign grants in support of the ERP. Available information indicates that foreign inflow increased from negligible levels in 1983 to over 6.6 percent of GDP in 1997 (World Bank, 1995a; Younger, 1992; ISSER, 1999) before falling back again to about 4.4 percent in 1998. Receipts from privatization of state-owned enterprises, have also contributed enormously in the nineties in raising government savings levels. The increase in these receipts, which have been included in non-tax revenue, led to a rise in non-tax revenue from the low levels of about 1 percent of GDP in 1983 to almost 9 percent in 1994 before declining to about 3.53 percent in 1998.

4:3.3.2 Trends in Investment

Similar to savings, investment has also witnessed a revival under the Structural Adjustment Programs. Gross domestic investment increased from 3.76 percent of GDP in 1983 to about 15.49 percent in 1997 (see Table 4:3). A decomposition of total gross domestic investments into government and private investments shows that government investments rose from below 1 percent in 1983 to about 9.43 percent of GDP in 1993 before declining slightly to about 14.11 percent. As Figure 4:2 shows, whereas private investments had dominated the investment efforts of the early eighties, the introduction of the ERP led to large increases in government capital formation, particularly in 1986 and so reversed the trend, but experienced a slight stagnation in the second half of the eighties. The early nineties, however, witnessed increases in private investments, but this trend

Table 4:3 Composition Gross Capital Formation by Type of Goods at Current Prices (percent)

Year	Building	Other Construction	Land Improvement	Transport Equipment	Machine & Equipment	Changes in Stocks	Gross Capital Formation
1983	68.83	5.06	0.46	12.62	13.33	-0.30	100.00
1984	56.40	5.52	0.37	18.62	18.74	0.35	100.00
1985	52.41	9.12	0.91	12.98	24.17	0.42	100.00
1986	43.30	9.56	0.73	16.14	29.56	0.71	100.00
1987	43.34	9.85	1.09	16.27	28.74	0.71	100.00
1988	45.26	1.34	0.07	19.21	33.37	0.77	100.00
1989	40.76	11.28	0.17	18.58	28.70	0.52	100.00
1990	41.65	11.47	0.12	18.16	28.05	0.55	100.00
1991	38.84	12.69	0.12	18.74	29.12	0.48	100.00
1992	39.51	12.11	0.09	18.27	29.54	0.47	100.00
1993	31.90	10.90	0.13	23.71	32.77	0.59	100.00
1994	30.59	15.20	0.17	16.86	36.59	0.58	100.00
1995	30.21	13.74	0.16	16.76	38.61	0.52	100.00
1996	33.59	14.34	0.15	18.99	32.51	0.42	100.00
1997	42.44	10.53	0.10	17.68	25.51	3.74	100.00

Source: *Statistical Service*. Quarterly Digest of Statistics, *Accra, various issues: Author's own computations.*

was short lived. In 1992, due to the re-emergence of economic uncertainties, private investments experienced substantial declines from which it has not fully recovered.

Generally, it can be said that investment activity of the private sector has been low. Whereas the low response of the private sector during the early adjustment period, for instance, can be attributed to the utilization of idle capacity, which had existed prior to ERP, the stagnation since 1992 can be ascribed to the re-emergence of economic imbalances, which has been observed since the return to civilian rule in 1992. This situation can also be attributed to the harsh realities of the trade liberalization under ERP, which exposed the economy, probably, to a too high degree of world competition and thereby maimed investment efforts of existing firms or frightened new potential ones away (Armstrong, 1996; Lall and Stewart, 1996). The slow pace of the privatization of SOEs can also be another reason.

The marked increase in domestic investment has also contributed to changes in its composition (Table 4:3). In 1983, for instance, investments in buildings accounted for almost 70 percent of the total gross capital formation, whereas investment in machines and equipment accounted for only 13 percent. The improvements in business prospects, under the ERP, have shifted the composition of capital formation in favor of machines and equipment. The relative share of machines and equipment to total gross capital formation increased successively from 13 percent in 1983 to a peak level of 38.6 percent in 1995 before falling to 25.5 percent. Investments in transport equipment have also recorded a similar upward trend, rising from less than 10 percent of total gross capital formation in 1983 to about 24 percent in 1993 before declining to about 17.68 percent in 1997. There have also been large increases in investments in construction between 1983-1997, whose proportionate share of total gross investments rose from 5 percent to about 10.53 percent. Even though a marked change in emphasis towards investments in business machinery and transport under the Structural Adjustment Programs can generally be observed, the nineties appear to depict a reversal of the trends in favor of investment in real assets.

It is also noteworthy that the formation of new capital–as reflected in the growth volume of machines and equipment–has been very much concentrated. Most of these investments have gone in favor of a few sectors, notably the mining and the cocoa sectors, where successive increases in producer prices have led to increased mining activities and the planting of saplings, respectively.

4:4 Assessment of the Fiscal Impact

The contribution of fiscal policy to the turn-around can be judged by the positive changes, which most macroeconomic indicators have experienced since 1983. A sample of these indicators has been presented in Table 4:4. It can be observed that the annual growth rate of the real GDP, for instance, has changed from -2.53 percent in 1981-1983 to an average of 4.9 percent between 1984-1998. Government revenue relative to GDP has risen from its low level of 5.7 percent in 1981-1983 to 19.8 percent in 1995-1998. Gross fixed capital formation (investment) and savings have also increased from 3.7 percent (1981-1983) and 0.7 percent to 16.98 percent and 21.64 percent in 1995-1997 respectively. During the same period the export sector recovered tremendously from its low level of 4.7 percent of GDP in 1981-1983 to 22.64 percent in 1995-1997. The economy has also experienced a substantial reduction in inflation and black market exchange rate premium. Fiscal deficits were also reduced from -4.2 percent of GDP in 1981-1983 to an annual average surplus of 0.8 percent in 1986-1991 before deteriorating to annual average of -2.6 percent in 1995-1998.

Table 4:4   Selected Macroeconomic Indicators, 1977-1998

	1977 - 1980	1981 - 1983	1984 - 1987	1988 - 1991	1992 - 1994	1995 - 1998
Real Growth 1975 Prices						
GDP (1975 prices)	5,433.93	5,022.00	5,564.00	6,753.75	7,846.33	9,214.63
Growth (%)		-2.53	5.60	4.80	4.32	4.83
Agriculture	2820.65	2713.00	2839.68	3094.78	3274.23	3728.54
Growth (%)		-1.27	1.56	2.48	1.60	4.20
Manufacture./Industry	946.60	642.00	726.53	974.63	1138.80	1309.89
Growth (%)		-10.73	4.39	8.54	4.80	3.93
Services	1817.43	1793.00	2143.58	2874.90	3628.23	4392.65
Growth (%)		-0.45	6.52	8.53	6.43	5.85
Fiscal Performance (% GDP)						
Total Revenue	9.40	5.70	12.60	14.60	19.45	19.83
Tax Revenue	8.10	4.90	10.40	12.00	13.96	15.75
Total Expenditure			12.41	13.32	19.39	22.36
Current Expenditure	12.90	9.60	10.60	10.40	16.15	17.53
Capital Expenditure (without Net Lending)	3.50	1.30	2.10	2.84	4.73	4.51
Overall Surplus/Deficit (annual average)	-1.44	-4.20	-1.20	0.80	-2.00	-2.65
External Trade Balance						
Trade Balance billion cedis	159	-95	-33	-234	-325	
Current Account Bal. billion cedis	6.4	-235	-78	-161	-178.53	292.04
Composition of Output (% GDP)						
Total Consumption	93.63	96.32	92.47	93.68	92.89	88.87*
Private Consumption	82.28	89.27	83.04	82.86	79.83	83.20*
Government Consumption	11.35	7.05	9.43	10.82	13.06	5.67*
Gross Capital Formation	6.68	3.65	9.05	12.56	14.58	14.43*
Private Investment	3.60	2.69	4.09	7.53	5.32	5.42*
Government Investment	3.08	0.96	4.96	5.03	9.26	9.00*
Exports	9.65	4.73	13.60	18.06	18.43	22.64*
Imports	10.43	5.87	15.10	19.15	26.38	29.11*
Prices & Exchange Rates						
Inflation.	73.6	87.22	28.60	27.08	19.95	38.31
Official Exchange ¢/$	2.23	4.27	75.82	291.63	680.93	1800.52

Source: *Statistical Service*, Quarterly Digest of Statistics, *various issues, Accra. Author's own computation.*
*refers to 1997 economic indicators over time and relate these ex-post changes to ex ante (anticipated)
projections of the government as represented in its annual budgets.*

The positive changes have also led to a massive decline in macro-economic instability from its peak of 4.4 in 1982 to 0.6 in 1992 (Schmidt-Hebel, 1994). Indeed the contribution of fiscal policy variables to this turn-around has been very dominant and is assumed to account for an average of about 91 percent of the structural correction of, for instance, government deficits (Schmidt-Hebel, 1996). Macro-economic indicators, however, appear to have deteriorated since 1992. Not only has the growth rate of the GDP fallen since 1992, but also a re-emergence of macro economic imbalances can be observed in the economy as mirrored in the rising budget deficits and general price levels. The fiscal prudence of the 1986-91 era has given way to rising budget deficits equivalent to about 2.65 percent of GDP in 1995-1998. This is underlined by stagnating government revenue and rising government expenditure.

In the following section, we offer an appraisal of the impact of fiscal adjustment as reflected in changes in major.

4:4.1 Fiscal Performance According to Tax and Revenue Effort

As can be seen in Table 4:4, Ghana's fiscal situation has made substantial improvements under the ERP. Total revenue and tax receipts as a percentage of GDP has tripled from its low level of about 4.5 percent and 5.6 percent in 1983 to 15.6 percent and 19.3 percent in 1998 respectively. This strong inter-temporal revenue performance is reflected in almost all the major revenue instruments. The increase in tax receipts is also reflected in substantial increases in income elasticity of tax revenue from 0.69 in 1969 to 1.08 in 1996 (Appiah-Kubi, 1997).

Between 1983 and 1998 sales and excise taxes increased from 0.95 percent of GDP to 4.8 percent. This was followed by taxes on personal income and property, which increased from 0.9 percent relative to GDP in 1983 to about 4.4 percent in 1998. Export taxes, despite the reduction of the coverage to include largely cocoa exports recorded substantial increases in receipts, rising from 0.25 percent of GDP 1982 to 2.3 percent in 1998. From Figure 4:3 it can be observed that the positive impact of the adjustment has resulted in upward trends in overall revenue indicators. It is also possible to discern 3 phases in these trends. The first phase, which characterized the high growth phase (1983-1987), began with the introduction of the ERP in 1983. During this period government revenue registered strong annual average growth rates of over 96.4 percent, with the highest growth rate over 120 percent and lowest about 52 percent being achieved in 1983 and 1987 respectively. These high growth rates were largely facilitated by the "revolutionary drive" at the beginning of the Rawlings' regime, backed by threats of reprisals for attempts to evade or delay in paying

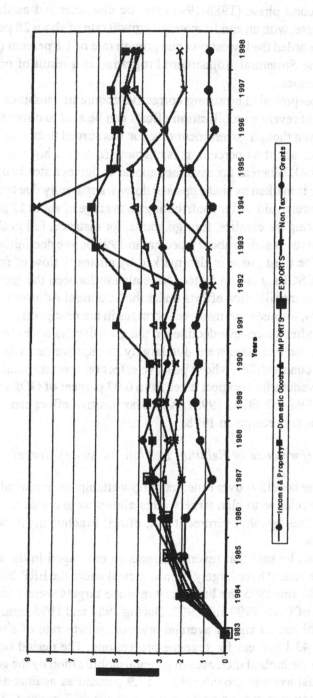

Figure 4:3 Selected Government Expenditure Indicators, 1983-1998 (percent of GDP)

taxes due. The second phase (1988-1992) can be characterized as the low revenue growth phase, with an annual average growth rate of about 28 percent. This period also recorded the lowest revenue growth rate of 1.8 percent (1992) in the history of the Structural Adjustment Programs, as a result of massive declines in tax revenues.

The third phase–preceded by strong corrective economic measures partly to boost government revenue mobilization efforts can be said to cover the period 1993-1998. Even though government revenue has surged to an annual average growth rate of about 47 percent, it is characterized by a high degree of volatility (Figure 4:4). Whereas tax revenues appear to depict a steady upward trend, total revenue is marked by sharp ups and downs, apparently due to inconsistent inflows in foreign aid. After contributing an average of about 12 percent of total revenue during the eighties, foreign grants, for instance, fell to about 3 percent in 1994, but rose again to about 9 percent in 1995 before declining to 4.8 percent in 1998. The third phase is also marked by unsteady flow of receipts from divestiture of SOEs. Foreign factors have also influenced the growth of government revenue mobilization efforts under the Structural Adjustment Programs. During 1994, for instance, massive increases in tax receipts from cocoa exports were seen after successive declines in the growth rates over the previous two-period. These increases were due largely to improvements in world prices for primary commodities, whose positive effect on revenue resulted directly in raising revenues from export taxes from 0.97 percent of GDP in 1993 to 2.7 percent in 1994 (UNECA, 1995). A similar positive effect can also be inferred from cocoa tax receipts in 1998.

4:4.2 Revenue Performance in Relation to Plan Budgetary Targets

Having analyzed the tax effort over time, let us now attempt an appraisal of the fiscal performance relative to plan targets. This allows us to consider the degree of success realized by the government in its efforts to achieving its budgetary revenue targets.

In general it can be said that revenue targets as envisaged in the annual budgets of the government have largely been achieved under the ERP. For most years between 1984 and 1995 the budgetary revenue targets were exceeded, with the exception of 1986, 1990 and 1992. During 1984 and 1995, actual revenue collection realized an annual average revenue growth rate of about 51 percent as against 43.1 percent for revenue projections. The period between 1984 and 1989 saw the highest increases in revenue mobilization by the government with an annual average growth rate of 57.9 percent as against average annual budgetary revenue growth projections of about 46.7 percent. The respective rates fell to 45.2 percent and 40.1 percent during 1990-1995, with the actual revenue out-turns exceeding the budget estimates. In contrast the picture of revenue mobilization efforts of the government in the second half of the nineties is characterized, not only by dwindling revenue growth rates, but also by difficulty of the government in achieving its budget projections. Between 1996 and 1998, for instance, the annual average revenue growth has been only 27.8 percent against average annual growth estimates of 36.8 percent.

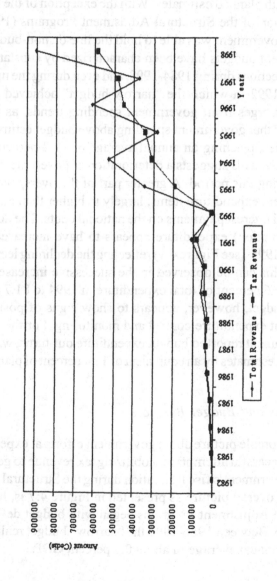

Figure 4:4 Comparison of Government Total Revenue and Tax Revenue, 1982-1998

4:4.3 Expenditure Management

Apparently a basic fundamental weakness of fiscal management in Ghana in the past has been the lack of effective budget planning and expenditure monitoring systems to anticipate and forecast accurately all expected major expenditure developments of the coming fiscal year to curb any imprudent fiscal adventures. This deficiency in expenditure management had also persisted during the reform period and can readily be observed in Figure 4:6 that compares actual government expenditure with planned estimates. With the exception of the first four years of the introduction of the Structural Adjustment Programs (1984-1987), through which the government was able to hold the line on non-budgetary expenditures, government budgets have been characterized by substantial spending above budget projections during 1984-1998, and even during the much heralded periods of 1986-1992 in which the "narrow budget" achieved surpluses. The picture that emerges from government spending trends, as presented in Figure 4:6, shows that government spending above budget estimates increased during the nineties, reaching an annual average of 19.2 percent of planned targets in 1992-1997. This suggests a deterioration of government expenditure management during this period. A greater part of the overspending occurred among the recurrent expenditure items, largely in higher than anticipated government wage and interest payments on the national debts. The dominance of recurrent (consumption) expenditure appears to have increased in intensity, particularly since 1995 (see Figure 4:1) reflecting the declining leeway for government savings. This can be observed in the successive increases in recurrent spending from 73.6 percent of total expenditure in 1994 to 81.7 percent in 1997. The 1998 budget, however, appears to show signs of positive development in government expenditure control and monitoring. For the first time since 1987, the government reported budget expenditure out-turns, which lie below the original budget estimates by an equivalent of 12.4 percent of planned estimates.

4:4.4 Fiscal Developments and Budget Balances

Let us now relate this unfavorable picture about government efforts at expenditure control to its modest successful attempts at mobilizing tax revenue to get an overall picture about the government's fiscal situation during the Structural Adjustment Program era. The overall picture as presented in Figure 4:7 is, however, mixed. The Structural Adjustment Programs began with budget deficits for the period of 1984-1986. Between 1986-1991 the "narrow" budget realized surpluses equivalent to an annual average of about 0.8 percent GDP.

Figure 4:5 Difference between Total Budget Revenue Outturn and Estimates, 1984-1998

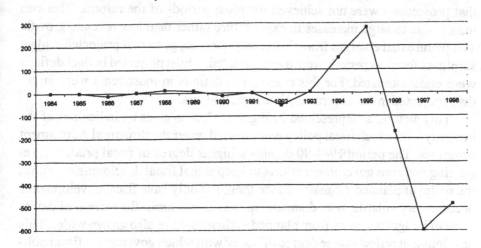

Figure 4:6 Comparison of Planned and Actual Government Expenditure, 1984-1998

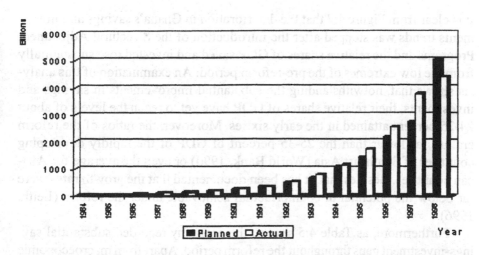

These positive budget balances turned into huge deficits in the period between 1992-1993, which were related to massive expenditure increases and revenue declines, especially in 1992, as a result of the political change-over from military to civilian rule. The strong corrective measures adopted returned the budget back to positive balances in the following two years only to return to deficits in 1996. In 1997, the budget deficits increased from 1.4 percent of GDP in 1996 to 8.6 percent and leveled off in 1998 at 6.3 percent (Figure 4:7). A comparison of the planned budget estimates with the budget out-turns suggests that projections were not achieved for most periods of the reform. This was mainly due to large increases in expenditure rather than low revenue growth. The picture that emerges from this comparison suggests that projected budget surpluses for most periods were overestimated, whilst projected budget deficits were underestimated. For this reason, the deficits in most years were larger than projected in the government's annual budgets.

This picture, as represented in Figure 4:7, also gives an indication of the intensity with which fiscal policy was pursued under the Structural Adjustment Programs. The period 1984-90 depicts a higher degree of fiscal prudence, reflecting vigorous government efforts to keep actual fiscal developments within the realm of planned estimates. Since then, not only have fiscal developments become very volatile, with dramatic up- and down-turns—the margin of deviations of budget out-turns from planned estimates—have also grown wider. This development reflects some degree of laxity with which government fiscal policies of the nineties were pursued as compared to the eighties.

4:4.5 Judging Government Savings and Investment Performance

It is clear from Figure 4:7 that the deterioration in Ghana's savings and investments trends was stopped after the introduction of the Structural Adjustment Programs and the relative shares of GDP saved and invested rose substantially from the low extremes of the pre-reform period. An examination of this analysis shows that, notwithstanding the substantial improvements in savings and investments, their relative shares of GDP have yet to reach the levels of about 24-25 percent attained in the early sixties. Moreover, the ratios of the reform era are far lower than the 25-35 percent of GDP of the rapidly developing countries of Southeast Asia (World Bank, 1990) or even the average for African countries. Besides that, it has been documented that the growth rates were far below the potential or counterfactual achievable under the reform (Leith, 1996).

Furthermore, as Table 4:5 shows, the economy recorded substantial savings-investment gaps throughout the reform period. Apart from macroeconomic imbalances, the main cause for the sizeable excess of domestic investment over domestic savings (finance) and the modest growth in domestic investment lies in the lagging growth of domestic savings to finance the investment. This is attributed partly to the irresolute effort on the part of the government to mobilize domestic savings for financing investment. This is reflected in the fact that net current transfers from abroad have more than financed the net domestic sav-

Figure 4:7

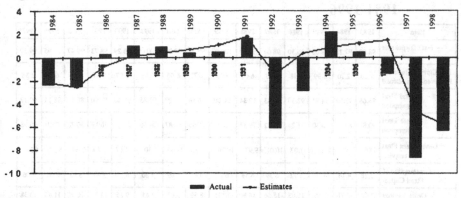

ings. However, these foreign transfers, half private transfers from expatriate Ghanaians and half foreign aid from abroad, averaging 6 percent of GDP annually, have increasingly been used to finance consumption (Leith, 1996) and less additional investment. On the other hand the growth of net capital formation has also been low because net domestic savings have financed it almost solely with foreign debt or equity capital playing only a marginal role. This can be seen in the fact that the net current account balances and hence its contribution towards savings has been almost zero. Another factor, which appears to have impeded the recovery of domestic investment since the introduction of the reforms, has been the high rate of consumption of fixed assets. The implication of the high rate of consumption of fixed assets, which until 1993 had exceeded net capital formation, is that a substantial proportion of the gross fixed capital formation had just simply replaced depreciated and not augmented capital stock.

4:5 Savings Performance in Relation to Planned Savings

An assessment of the savings performance of the Ghanaian economy between 1984 and 1998, comparing planned to actual savings indicates that public savings performance fell short of the target for most periods of the Structural Adjustment Programs implementation. According to Leith (1996) actual savings should have been much higher given the potential of the country. Between 1983 and 1998, for instance, actual savings performance of the government exceeded its target only on four occasions, i.e. in the years 1987, 1988, 1994 and 1995. The picture of government savings performance looks even worse when analyzed in relation with government's tax revenue mobilization effort. Despite the enormous increase in tax revenue relative to the GDP, the level of total tax

Table 4:5 Consumption, Savings and Investment as percent of GDP, 1984-1996

Item	1983	1984	1985	1986	1987	1988	1989	1990	1991	1992	1993	1994	1995	1996	1997*
Nat'l Disposable Income	98.02	97.22	88.02	96.30	98.04	97.60	100.07	99.52	99.23	100.09	98.29	98.73	99.47	97.95	99.39
Net Current Transfers Received from Abroad	1.20	2.70	1.99	4.48	7.26	6.66	8.14	6.77	6.03	6.83	8.67	8.67	8.10	6.95	8.36
Total Final Consumption	96.68	93.39	84.94	92.30	91.83	89.34	94.00	96.38	95.29	97.86	93.40	87.40	88.30	88.14	90.17
Private Final Consumption	90.82	86.13	76.30	81.24	81.82	79.37	83.74	85.45	83.87	84.56	78.73	76.19	80.72	82.91	85.98
Government Final Consumption	5.86	7.26	8.64	11.07	10.01	9.97	10.26	10.93	11.43	13.30	14.67	11.20	7.59	5.23	4.20
Consumption of Fixed Capital	2.28	4.14	4.38	5.72	6.39	6.54	6.08	5.45	5.09	5.20	8.50	7.90	6.63	7.06	7.06
Gross Savings	3.62	7.97	7.46	10.03	15.56	14.80	12.15	8.59	9.03	7.43	9.15	12.70	10.92	11.60	13.38
Government Savings	-1.72	-0.02	0.50	2.50	4.08	4.07	4.65	3.40	4.93	0.76	1.75	8.11	7.28	3.40	2.34
Private Savings	5.34	7.99	6.97	7.52	11.48	10.73	7.50	5.19	4.10	6.67	7.40	4.59	3.64	8.21	11.04
Net Savings	1.34	3.83	3.08	4.00	6.20	8.26	6.07	3.14	3.94	2.23	3.26	7.56	7.44	6.54	6.15
Gross Fixed Capital Formation	3.76	6.85	8.76	9.30	10.36	10.86	13.47	12.19	12.68	12.83	15.86	15.04	14.09	13.71	15.49
Private Investment	2.87	4.37	4.93	2.00	2.49	2.89	5.59	6.48	9.07	5.71	5.35	4.91	5.13	5.06	6.08
Government Investment	0.89	2.48	3.83	7.30	7.87	7.97	7.88	5.71	3.61	7.12	10.51	10.13	8.95	8.65	9.41
Net Capital Formation	1.47	2.74	4.42	3.94	7.01	4.39	7.19	6.81	7.65	7.70	9.15	10.70	8.92	9.60	11.38
Net Recurrent Exp. A/c Balance	-1.72	-0.02	0.50	2.50	4.08	4.07	4.65	3.40	4.93	0.76	1.75	8.11	7.28	3.40	2.34
Account Balance (BoP)	-2.84	-3.72	-4.33	-2.00	-2.13	-1.98	-1.78	-3.58	-3.59	-5.47	-9.58	-4.67	2.23	-4.67	5.60
Gap Sav/Inv	-0.14	1.12	-1.30	0.73	5.20	3.94	-1.32	-3.60	-3.65	-5.40	-6.71	-2.34	-3.16	-2.11	-2.11
Government Savings/ Investment Gap	-2.61	-2.50	-3.34	-4.80	-3.79	-3.90	-3.24	-2.31	1.32	-6.36	-8.76	-2.02	-1.67	-5.25	-7.07
Private Savings/ Investment Gap	2.47	3.62	2.04	5.52	8.99	7.84	1.91	-1.29	-4.97	0.96	2.05	-0.32	-1.49	3.15	4.96

Source: *Statistical Service, Quarterly Digest of Statistics, Accra, various issues, own computations*

revenue had nearly always fallen short of the requirement, with the exception of the 3-year consecutive period of 1986–88, needed to finance recurrent expenditure. The shortfall has always been made good largely by foreign grants and loans.

Part of the explanation for the lackluster public savings performance may lie in low growth of real per capita income of less than 2 percent on average during the Structural Adjustment Programs period, which did not allow for any massive increase in tax revenue and thus government savings. A comparison of the second half of the 1980s with the 1990s with regard to government savings performance, however, leads to the conclusion that the decline in actual savings of the government–particularly in the 1990s–is attributable less to any weakness in mobilizing revenue than to a rapid growth in government recurrent expenditure.

The problem that belies the slow growth of savings by the government is the continued high level of current government consumption. Whereas expenditure reform of the 1980s under Structural Adjustment Programs succeeded in reducing current consumption from over 90 percent of total expenditure in the early eighties to about 73.6 percent in 1994, the nineties have witnessed some degree of laxity in the reform resulting in the reversal of the downward trend. As at the end of 1998, for instance, large increases in wage and interest payments of government debts have sent the relative share of the current total expenditure once again beyond the "critical" border of 81 percent.

From this information we can sum up briefly that, under Structural Adjustment Programs, public savings performance has fallen short of the target. There were substantial increases in savings at the beginning of the Structural Adjustment Programs in the second half of the eighties. However, large increases in private and public consumption have given way to stagnating savings level and thus a widening gap between planned and achieved domestic savings.

4:5.1 Investment

Gross domestic investment has increased substantially under the Structural Adjustment Programs. From its low levels in the early 1980s, net investments rose successively from 1.47 percent of GDP in 1983 to about 11.38 percent. An analysis of the decomposition of the gross domestic investment in government and private investments show that, although government investments have undergone tremendous revitalization under the Structural Adjustment Programs, private investment always performed better than government investment.

Although private investment had increased from 5 percent in 1984 to 12 percent in 1993 before falling to 9 percent in 1996, these increases have fallen short of expectation and the level necessary to transform the Ghanaian economy from the adjustment/recovery phase to a sustainable accelerated growth phase. Two main reasons are usually assigned to explain the lukewarm response of the private sector to investment, in spite of the many incentives accruing from macroeconomic and sectored reforms under the Structural Adjustment Programs. While the first one touches on the fact that the reforms have not gone far enough to address substantive institutional issues, the second one dwells on the credibility issue with regard to the sustainability of reforms in the medium-to-longer terms. According to Aryeetey (1994) a major reason for the poor private investment response has been in part a legacy of past policies, embedded in a deep mistrust by the private sector for government, whose earlier policies can be construed to depict an antagonistic attitude toward private investment.

Another reason for the low investment level in Ghana can be seen in the low level of domestic resource mobilization to finance investment. This deficiency can be illustrated by a high-savings, investment gap that accelerated during the Structural Adjustment Programs period from 10.14 percent of GDP in 1983 to 17.71 percent in 1993 before falling to 2.11 percent in 1997. This is due mainly to the high level of consumption as compared to other developing countries (Table 4:5). Moreover the large budget deficits resulting from large credits to public enterprises, particularly in the first half of the nineties, continue to have depressing effects on private investment.

Indeed the widening savings-investments gap, particularly private savings investment gap throws a dark shadow on the success with respect to efforts at mobilizing resources to finance accelerated growth, given the fact that much of the savings of the nation, about one-third (World Bank, 1990), that finances Ghana's current investment comes from abroad.

4:6 Fiscal Policy, Resource Mobilization and the Parallel Black Economy

The size of the parallel economy in Ghana started to assume significance in the mid-1960s, with the growth of cocoa smuggling to neighboring countries, where producer prices were higher (Stryker, 1990). The market received a significant boost in the 1970s with the introduction of strict foreign trade and exchange controls to correct external imbalances in the exchange system, coupled with massive state intervention and other dirisgitic measures meant to bring the economy under state control. The situation was exacerbated when efforts to raise revenue to finance the increasing government spending led to high marginal tax rates and consequently created disincentives to formal economic activity. By 1982, for instance, the size of the parallel economy relative to the formal economy had grown from insignificant in 1965 to 32.41 percent of GDP (Islam, *et al.*, 1994).

Indeed the Structural Adjustment Programs has done a lot to reduce the size of the black economy considerably. Macroeconomic stabilizing, price and foreign exchange deregulation, as well as financial sector reform policies have succeeded in bringing about a substantial reduction in the black market activity. The successive devaluation of the cedi, for instance, has led to a massive reduction of the black market premiums in the foreign exchange market from 22.43 in 1983 to less than 1.2 in 1998. Similarly the successive increases in producer prices of cocoa have made smuggling unattractive. Furthermore the tax reform, which accompanied the Structural Adjustment Programs, has also reduced the effective rates of almost all taxes and hence the disincentive effects of the tax system, making tax evasion or avoidance less economically attractive.

In spite of these successful developments, a lot still remains to be done, because other subtle determinants of the growth of the parallel economy have come to replace old ones. Rent seeking and patronage relations–major determinants of the parallel economy–are still prevalent in the Ghanaian society. If they had formerly been associated with access to scarce resources such as import licenses and foreign exchange, today they take the form of multiple seals of donor-financed salary supplements to public sector employees and by the non-transparent way in which donor-financed benefits (vehicles, external training, study tours abroad, appointment to advisory groups, commissioning of studies, etc.) are allocated through patronage relationships (Armstrong, 1996). Another less obvious manner in which the black market economy continues to frustrate resource mobilization relates to black incomes reaped through "siphoning off" from public programs and projects. All these constitute "leakages" that go to increase government budget expenditure, since they must be provided for in the form of higher project costs. In essence these leakages represent additional (and illegal) transfer payments, which effectively reduce the value of resources mobilized. Since these transfers end up mostly in lavish luxurious consumption it can be plausibly argued that large black market economy imparts a downward bias to the economy's savings propensity (Armstrong, 1996).

On the other hand, it must be mentioned that the benefits of price increases in cocoa have not been fairly distributed, but have rather favored chiefly large-scale and male farmers. This is underlined by the fact that about 94 percent of the gross cocoa income goes to only 32 percent of the cocoa farmers (Commander, *et al.*, 1989), who constitute the large-scale farmers. Moreover the redeployment of public servants, many of whom moved to the rural areas, increased the demand for and hence the price of farmlands to the detriment of the poor rural households. Another aspect, which also affected the rural poor negatively and thus led to a reduction of their gains from the Structural Adjustment Programs, was the phasing out of all government subsidies on agricultural input inputs.

Furthermore, the reform is usually criticized as being limited in scope with regard to the agricultural sector. Because after more than a one and a half decades of fiscal adjustment, marketing and storage facilities as well as vital physical structures, including rural feeder roads and critical links with external markets, continue to be underdeveloped. For this reason, the performance of the dominant sector in the whole Ghanaian economy has not only been far below its potential, but also comparatively less than that of other sectors of the economy.

The rapidly rising debt stock and its servicing represents not only a drag on government adjustment efforts of the government; it poses a fiscal constraint to resource mobilization for development. Rising indebtedness depresses also capital formation-investment and, consequently, lowers economic growth through its negative effect on liquidity and expected profitability. It also induces increasing disincentive effects on current investments and adjustments due to prospective investors' expectations of higher future tax burdens to service rising debts (Claessens, *et al.*,1990).

An indication of an emerging debt crisis in Ghana, as a result of the rising indebtedness, can be seen in the deterioration of its debt indicators over the adjustment years. For instance, the debt service and the total debt to the GNP ratios increased from 13.1 percent and 31.3 percent in 1980 to 24 percent and 96 percent respectively in 1997 (Appiah-Kubi, 1997; World Bank, 1997). This has been compounded by growing accumulation of debt arrears over the years, which had in the case domestic debt risen to about 140 billion cedis (about US $70 million) in 1997 (ROG, 1997).

4:7 Summary and Conclusion

The beginning of the eighties witnessed massive social and economic decline in Ghana. To avert this downturn, the government introduced a comprehensive structural adjustment reform in 1983. A central policy element of this reform has been fiscal adjustment with the objective to ensure macroeconomic stability as well to boost investment and savings. From the above analysis it can be said that fiscal adjustment has succeeded in turning the economy around. The deterioration in the economy has been stopped to pave way for a timid economic growth. Fiscal tax reform has improved tax administration and revenue generation. Rising government revenues from domestic and foreign sources have facilitated higher government expenditure and consequently improved social welfare and equity. Macroeconomic imbalances have been largely reduced.

Notwithstanding these positive achievements, the economy appears, after one and one-half decades of reform, to have not fully recovered from the economic decline of the seventies and early eighties. Per capita income, for instance, has yet to reach the levels of the early eighties. Real growth has stagnated at about 5 percent through out the adjustment period. Present saving and investment levels are still below the modest peak levels of the post-independence era of the sixties and the African average. There exists an apparent divergent saving and investment ratios, which closely mirror the country's growth performance.

Even though revenue targets as envisaged in the annual budgets of the government have largely been achieved under the Structural Adjustment Programs, it must be mentioned that expenditure out-turns have, for most part of the period, exceeded revenue. This has been the result of government's spending above budget projections for most of the reform period. Compounding this problem in recent times is the substantial decline in recent times in the government's revenue. This has resulted in a great deterioration of budgetary performance of late from a surplus of 2.2 percent of GDP in 1994 to a peak deficit of 8.6 percent of GDP in 1997 before falling to 6.3 percent in 1998.

In conclusion, it is noteworthy that more than a decade of fiscal reform has succeeded in exerting a positive impact on the economy of Ghana. Yet it appears that over the past 15 years Ghana has been grappling with repairing the damage to the economy from the poor economic mismanagement prior to the introduction of the economic reform. These efforts, however, holds promising prospects for transforming the economy to a self-reinforcing growth process in the near future. The apparent policy implication should be to give efforts in raising domestic savings for investment a high priority so as to ensure a sustainable path toward achieving higher per capital growth. It is in this respect gratifying to note that the government introduced a Medium-Term Development Plan in 1999 within the framework of Ghana Vision 2020, which, whilst building firmly upon past achievements and existing development programs under the ERP, should ensure continuity and stability in economic policy and acceleration of socioeconomic development.

References

Alderman, H. (1994). "Ghana: Adjustment's Star Pupil?" in *Adjusting to Policy Failure in African Economies*. D. E. Sahn, ed., Ithaca, New York.

Appiah-Kubi, K. (1995). "Das sozioökonomische Gefälle zwischen Stadt und Land in Afrika: Zur politischen Ökonomie der wirtschaftlichen Entwicklung des ländlichen Raums", in *Ghana, Frankfurt am Main*.

Appiah-Kubi, K. (1997). "The Growing African Debt Crisis". *The Case of Ghana*. mimeo.

Armstrong, R.P. (1996). *Ghana Country Assistance Review*. A Study in Development Effectiveness. World Bank: Washington, D. C.

Aryeetey, E. (1994). "Private Investment under Uncertainty in Ghana". *World Development 1994*, vol. 22, no. 8, August: 1211-1221.

Aryeetey, E. (1996). *Structural Adjustment and Aid in Ghana*. Friedrich Ebert Stiftung. Accra, Ghana.

Asenso-Okyere, W. K., *et al.* (1997). "Understanding the Health and Nutritionals Status of Children in Ghana". *Agricultural Economics*, no.17, p. 59-74.

Asenso-Okyere, W. K., *et al.* (1998). "Cost Recovery in Ghana: Are there any Changes in Health Care Seeking Behaviour?". *Health Policy and Planning*, 13(2), p.182-188.

Bamberger, M., A. M. Yahie, G. Matovu, eds. (1996). *The Design and Management of Poverty Reduction Programs and Projects in Anglophone Africa*. Proceedings of a Seminar Sponsored Jointly by the Economic Development Institute of the World Bank and the Ugandan Management Institute. Washington, DC: World Bank.

Boateng, E. O., K. Ewusi, R. Kanbur, A. Mackay (1990). *A Poverty Profile for Ghana, 1987-1988*. Social Dimensions of Adjustment in Sub-Saharan Africa, Working Paper, No. 5. World Bank: Washington, DC.

Center for Policy Analysis (CEPA) (1999). *Ghana Macroeconomic Review and Program*. Accra, Ghana: CEPA.

Claessens, S., E. Detragiache, R. Kanbur and P. Witham (1997). "HIPC's Debt Review of the Issues, World Bank and IMF". *Journal of African Economies*, vol. 6, no. 2: 231-254.

Commander, S., J. Howell and W. Seini (1989). "Ghana: 1983-1987" in *Structural Adjustment and Agriculture Theory and Practice in Africa and Latin America*. London: Overseas Development Institute.

Demery, L. and L. Squire (1996). "Macroeconomic Adjustment and Poverty in Africa: An Emerging Picture" in *The World Bank Research Observer*, vol. 11, no.1, February: 39-59.

Frimpong-Ansah, J. H. (1992). "The Vampire State in Africa". *The Political Economy of Decline in Ghana*. London: Africa World Press.

Ghana Statistical Services (1986).

Haynes, J., T. Parfitt and S. P. Riley, eds. (1989). "Ghana: Indebtedness, Recovery and the IMF, 1977-1987" in *The African Debt Crisis*. London, New York: Routledge.

Hutchfull, E., B. K. Campbell and J. Loxlely, eds. (1989). "From Revolution to Monetarism" in *Structural Adjustment in Africa*. London: St. Martin's Press.

Islam, R. and D. Wetzel, et al. (1994). "Ghana: Adjustment, Reform, and Growth" in *Public Sector Deficits and Macroeconomic Performance*. New York: Oxford University Press.

ISSER, Institute of Statistical, Social and Economic Research (1997). *State of the Ghanaian Economy Report*. Various issues. Legon: ISSER.

ISSER, Institute of Statistical, Social and Economic Research (1999). *The State of Ghana's Economy 1998*. Legon: ISSER.

Kapur, I. *et al.* (1991). *Ghana: Adjustment and Growth, 1983-1991*. International Monetary Fund Occasion Paper 86. Washington, DC: IMF.

Khan, M. S. (1990). "The Macroeconomic Effects of Fund-Supported Adjustment Programs". *IMF Staff Papers*, vol.37, no. 2 (June): 155-179.

Killick, T. (1993). "The Adaptive Economy. Adjustment Policies in Small, Low-Income Countries". The World Bank, EDI Development Studies. Washington, DC: World Bank.

Klein, T. M. (1994). *External Debt Management: An Introduction*. Washington, DC: World Bank.

Krassowski, A. (1974). *Development and the Debt Trap: Economic Planning and External Borrowing in Ghana.* London: Croom Helm.

Kraus, J. (1991). "The Political Economy of Stabilisation and Structural Adjustment in Ghana" in *The Politics of Recovery.* Boulder, London: L. Rienner Publishers.

Lall, S. and F. Stewart, B. Ndulu and N. van de Walle, eds. (1996). "Trade and Industrial Policies in Africa" in *Agenda for Africa's Economic Renewal.* Washington, DC.

Leechor, C., I. Husain, and R. Faruqee, eds. "Ghana: Frontrunner in Adjustment" in *Adjustment in Africa. Lessons from Country Case Studies.* Washington, DC: World Bank.

Leechor, C., Thomas, V. and Chibber, A., *et al.* (1991). "Ghana: Ending Chaos" in *Restructuring Economies in Distress, Policy Reform and the World Bank.* Washington, DC: World Bank.

Leechor, C., I. Husain and R. Faruqee, eds. (1994). "Ghana: Frontrunner in Adjustment" in *Adjustment in Africa. Lessons from Country Case Studies.* I. Washington, DC: World Bank.

Leith, J.C. (1996). "Ghana: Structural Adjustment Experience". *International Center for Economic Growth Publication.* San Francisco, CA.

Mackenzie, G.A., D. W. H. Orsmond and P. R. Gerson (1997). "The Composition of Fiscal Adjustment and Growth: Lessons from Fiscal Reforms in 8 Economies". *IMF Occasional Paper*, 149, Washington, DC: IMF.

Marc, A., C. Graham, M. Schacter, M. Schmidt (1995). *Social Action Programs and Social Funds. A Review of Design and Implementation in Sub-Saharan-Africa.* World Bank Discussion Papers No. 274. Washington, DC: World Bank.

Nashashibi, K., Sanjeev Gupta, Lori Henri, Claire Liuksila, Walter Mahler (1992). *The Fiscal Dimensions of Adjustment in LLC.* IMF Occasional Paper No. 95, April. Washington, DC: IMF.

Republic of Ghana (RoG) (1998). *The Budget Statement and Economic Policy of the Government of Ghana for the 1998 Financial Year.* Accra, Ghana.

Republic of Ghana (RoG) (1999). *The Budget Statement and Economic Policy of the Government of Ghana for the 1999 Financial Year.* Accra, Ghana.

Ribe, H. and S. Carvalho (1990). *How Adjustment Programs can help the Poor: The World Bank Experience.* World Bank Discussion Papers No. 71. Washington, DC: World Bank.

Sandbrook, R. and R. Oelbaum (1997). "Reforming Dysfunctional Institutions through Democratisation? Reflections on Ghana". *The Journal Modern African Studies*, vol. 35, no. 4: 603-646.

Schmidt-Hebel, K. (1994). "The Macroeconomics of Public Sector Deficits: A Synthesis" in W. Easterly, C. Rodriguez and K. Schmidt-Hebel, eds. *Public Sector Deficits and Macroeconomic Performance.* Oxford.

Schmidt-Hebel, K. (1996). "Fiscal Adjustment and Growth: In and Out of Africa". *Journal of African Economies. AERC Supplement*, vol. 5, no. 3: 7-59.

Schmidt-Hebel, K., L. Servén, A. Solimano (1996). "Saving and Investment: Paradigms, Puzzles, Policies". *The World Bank Research Observer*, vol. 1, no.1, February: 87-

118.

Statistical Service, *Quarterly Digest of Statistics*. Accra. Various issues.

Stewart, F., P. Mosley, ed. (1992). "The Many Faces of Adjustment" in *Development and Finance and Policy Reform. Essays in the Theory and Practice of Conditionality in LDCs*. New York: St. Martin's Press.

Stryker, J. D. (1990). *Trade Exchange Rate and Agricultural Pricing Policies in Ghana. The Political Economy of Agricultural Pricing Policy*. World Bank Comparative Studies. Washington, DC: World Bank.

Terkper, S.E. (1994). "Ghana: Trends in Tax Reform (1985-1993)" in *Tax Notes International*, vol. 8, no. 19: 167-1275.

Tsikata, G. K. and G. K.Amuzu., V. K. Nyanteng, ed. (1997). "Fiscal Policies and Options" in *Policies and Options for Ghanaian Economic Development*. Accra: ISSER.

UNECA (1995). *Economic and Social Survey of Africa 1994–95*. Addis Ababa: UNECA.

UNIDO (1985). *External Indebtedness: Another Dimension of the African Debt*, prepared by the *Global and Conceptual Studies Branch*, UNIDO. December 585, 19.

Warren, J. (1977). "Savings and the Financing of Investment in Ghana 1960-69" in W. T. Newlyn, ed. *The Financing of Economic Development*. London: Oxford.

World Bank (1984). *Ghana: Managing the Transition*, vol. II, Statistical Appendix. Washington, DC: World Bank.

World Bank (1986). *Financing Adjustment with Growth in Sub-Saharan Africa, 1986-1990*. Washington, DC: World Bank.

World Bank (1988). *Adjustment Lending: An Evaluation of Ten Years of Experience*. Policy and Research Series. Washington, DC: World Bank.

World Bank (1990). *Ghana 2000 and Beyond: Setting the Stage for Accelerated Growth and Poverty Reduction*. Washington, DC: World Bank.

World Bank (1994). Adjustment in Africa Reforms. *Results and the Road Ahead*. Washington, DC: World Bank.

World Bank (1995). "Trends in Developing Economic Extracts" vol. 3, *Sub-Saharan-Africa*. Washington, DC: World Bank.

World Bank (1995a). "Ghana Growth, Private Sector, and Poverty Reduction". *A country Economic Memorandum*. Country Operations Division, West Central Africa Department, Report no. 14111-GH.

World Bank (1995b). *World Debt Tables, External Finance for Developing Countries* vol. 1. Washington, DC: World Bank.

World Bank (1997). *Africa Development Indicators*. Washington, DC: World Bank.

World Bank (1997a). *Global Development Finance 1997*. Washington, DC: World Bank.

World Bank. *World Debt Tables, First Supplement*. Various issues. Washington, DC: World Bank.

Yahie, A. M., M. Bamberger, A. M. Yahie, G. Matovu, eds. (1996). "Poverty and Approaches to Poverty Alleviation in Ghana, Uganda, Zambia, and Malawi" in *The Design and Management of Poverty Reduction Programs and Projects in Anglophone Africa*. Proceedings of a Seminar Sponsored Jointly by the Economic Development Institute of the World Bank and the Ugandan Management Institute.

Washington, DC: World Bank.

Younger, S.D. (1992). "Aid and the Dutch Disease: Macroeconomic Management When Everybody Loves You" in *World Development*, vol. 20, no. 12 (November): 1587-1597.

5 From a Developmental to a Managerial Paradigm: Ghana's Administrative Reform under Structural Adjustment Programs

Peter Fuseini Haruna

5:1 Introduction

Structural Adjustment Programs (hereafter SAPs) have spread widely among developing nations since they were first introduced in Turkey (Haque, 1998; Salleh, 1996; Mendoza, 1996; Pai, 1994; OECD, 1993; Mahmud, 1992). Structural adjustment refers to a process whereby developing nations reshape their economies to become more market-oriented. Such nations assume that with less government, their economies will become efficient, healthy and productive in the long run if market forces operate, and government does not protect, subsidize, regulate, or directly produce goods and services (World Bank, 1994; Haque, 1997).

Ghana, one of the first to introduce stabilization and structural adjustment programs in sub-Saharan Africa in 1983, still suffers from high unemployment and poverty despite its initial gains (World Bank, 1999; Konadu-Agyemang, 1998). Critiques of SAPs as applied to Ghana confront the issues from an economic perspective with interests in the practical implications for the operations of government (Rothchild, 1991; Gyimah-Boadi, 1991; Mackenzie, 1993; Anyinam, 1994; Konadu-Agyemang, 1998). I take these critiques further and inquire into the implications of SAPs for the broader issues of the roles of the state and public administration.

The question of interest to this chapter is: How do SAPs affect the roles of the state and public administration in Ghana? This is not only a practical but also

111

a theoretical question and it is important for two reasons. First, there is little or no consensus about the role of the state in Africa (Hyden, 1983; Rasheed and Luke, 1995). Second, with its colonial heritage, the state in Africa bears little relationship with much of civil society (Hyden, 1983; Glickman, 1988). This is relevant because the implementation of SAPs implies dismantling the state apparatus, reducing state responsibility, and raising questions of appropriate roles of the state in development. Likewise, implementing SAPs requires de-bureaucratization, divestiture, and privatization that marginalize the role of public administration in Africa. The issue is that such neo-liberal and market-based reforms have consequences that reflect not only in the falling standards of living but also challenge the very legitimacy of the state (Haque, 1998; Pollitt, 1990).

By examining the questions of whether and how SAPs affect the roles of the state and public administration in Ghana, this chapter (1) broadens the basis of critiques of SAPs, (2) highlights opportunities and challenges posed by performance-based administration, and (3) contributes to an understanding of public administration from a comparative perspective. For, despite widespread market liberalization, Ghanaians' expectations of the state in the functions of governance remain high (Chazan, 1991). This is not to deny the transition of the state from the post-colonial state to the competitive state of the 1990s as the Ghanaian government struggles to respond to growing international political and economic influences.

This chapter considers the state to include those institutions designed by society to perform public functions reflecting the will of the people (Hyden, 1983). It defines public administration to include the institutions, processes, and procedures established by public law, as well as the people working in them (Marini, 1971). It treats the concepts of public administration and the state as phenomena arising from historical changes. The roles of the state and public administration in Ghana are expected historical developments of capitalism with the colonial state as a crucial phase. The state and public administration in Ghana today are the products of the dynamic nature of accumulation of capital at the global level. Change and continuity are dialectical characteristics of the development of social and economic systems (Wamsley *et al.*, 1996). The changes beginning in the last quarter of the twentieth century altered the nature of capitalist economies including developing nations and their respective organizations of governance and administration. But the core of the state and of public administration persists in the broader sense of continuity, although their nature and character are changing from the administrative state (Waldo, 1948) to a corporative state (King, 1999) that seems to turn Ghana back to the colonial period.

After summarizing the roles of the state and public administration since colonialism, the chapter discusses how SAPs (1) infringe on Ghana's sovereignty, (2) challenges the legitimacy of the state, and (3) reduces the capacity of public administration to adopt policies to support the underprivileged in society. The final section of the chapter emphasizes that the autonomy of the state in Ghana has diminished because of the strict adherence to "conditionalities" that donor agencies have imposed. The diminished role of the state accentuates the weakening of the Ghanaian economy, expands foreign ownership of our national resources, and erodes the ability of the state to mediate interests of competing segments of Ghanaian society. I conclude that reform should be based on institutional capacity building grounded not only in the market criteria of efficiency and competition but also in our political, social, and cultural experiences.

5:2 Changing Roles of the State and Public Administration

The state and public administration in Ghana will be a century old in the year 2001. British imperial Orders-in-Council (1901 and 1906) created the "Gold Coast" colony and laid the foundation of the modern state and public administration (Kimble, 1963). These landmark pieces of legislation constituted the basis for the creation of the colonial civil and public services (Ayee, 1994; Adu, 1965), giving birth to contemporary Ghanaian state and its public administration. During this period, the state and its public administration underwent different changes, culminating in a transformation of Ghanaian society characterized by rapid population growth, urbanization, and extensive public services. Analyzing these changes reveals three interrelated phases with distinct roles for the state and public administration: (1) colonial state, (2) post-colonial state, and (3) structural adjustment phase. Each of these phases is examined in the next sections, emphasizing the unique characteristics of the state and public administration and how they affect the development of Ghanaian society.

5:2.1 The Colonial State

The purpose of the British imperial Orders-in-Council (1901 and 1906) was to legalize British colonial domination over the Gold Coast colony (Ghana) and its people. The colony initially included the coastal settlements and later Ashanti and the Northern Territories. The British mandated area of Trans-Volta Togoland became a part of the Gold Coast colony in 1919 (Kimble, 1963). British rule sought to administer the people of these four geographical regions under a centralized colonial administration based on a command and control model. The roles of the state and public administration in the Gold Coast colony included: (1) maintaining law and order, (2) regulating individual and collective behavior, and (3) providing basic infrastructure to facilitate smooth implementation of imperial policies (Ayee, 1994; Greenstreet, 1973; Adu, 1965).

To achieve its goal, the colonial government required an administrative framework that guaranteed two sets of values: (1) integrity, impartiality, and loyalty to British imperial power and (2) control, efficiency, and economy. There was not a better-suited model than bureaucratic administration, which was long established in Europe and elsewhere (Albright, 1970). The essence of bureaucracy is a mix of centralization, order, control, and efficiency (Weber, 1958), which served the interest of British colonialism well. In Ghana, bureaucracy created a grid of a hierarchy of offices and officeholders, and provided the political and administrative context "to prosecute the imperial policies" and to bring the "Pax Britannica to all the dependent territories overseas" (Adu, 1965, p. 14-15). The bureaucratic system so ensured law and order that today, ironically, some Ghanaians remember what they perceive as "good old" colonial days while others wish to see the "white man" return.

It is hardly surprising that the early civil and public services in Ghana consisted of highly centralized power concentrated in the offices of the colonial governor, chief, provincial, and district commissioners responsible for keeping the peace, and maintaining law and order. The colonial governor was the ultimate source of power in Ghana, accountable, not to Ghanaians, but to the British Crown. The colonial public administrative structures were composed mainly of British nationals appointed to serve under the colonial administration. Officeholders included military personnel drawn from British colonial regiments and occupation forces (Adu, 1965).

The core of the civil service was the office of colonial secretary, which formed the heart of the colonial public administrative system. This secretariat operated as a technical departmental establishment with divisional heads coordinating the functions of government (Greenstreet, 1971, p. 17). Subordinate categories of personnel comprised of permanent, non-political officeholders including police, treasury officials, sanitary inspectors, to name a few, responsible for providing mundane and more routine services of government. From its inception, the nature of colonial public administration and of the state reflected three tenets of public administration: (1) centralization, (2) bureaucratization, and (3) limited representation. Given British colonial hegemony, it is understandable but unreasonable why this should be so.

For the most part, public administration and the state under British colonial rule played passive roles because they formulated little positive policy toward social and economic development for Ghanaian society. To be sure, they created the conditions for foreign private business and entrepreneurship to flourish. They helped to keep the peace, created order, and in today's terminology provided the "enabling environment" for economic growth.

But the colonial era was largely marked by economic laissez-faire with limited direct governmental participation in production and productivity. Much of the economic activity laid in private businesses, foreign private companies interested in selling manufactured goods, and purchasing local agricultural produce. Industries tended to be concerned with exploiting and exporting minerals, timber, and rubber to Britain and other European metropolitan centers (Abernethy, 1988; Balogun, 1995). In this mercantilist approach, the colonial government paid scant attention to promoting domestic economic services beyond what it required to function. Balogun states:

> In terms of the general orientation of government and the scale of government intervention in economic activities, the colonial public service was a regulatory law-and-order institution (1995, p. 16).

Colonial fiscal administration typically centered on simple accounting of revenue and expenditure and balancing the budget. Revenue collection, mainly customs and export duty, involved raising funds to meet recurrent expenditures while development expenditure was limited to bear necessities–telecommunications, harbors, and railways linking urban centers within the "golden triangle" of Accra, Kumasi, and Sekondi.

Much commentary exists on Ghana's colonial heritage, some of which stresses landmark infrastructure constructed during the colonial period–Achimota College, Takoradi harbor, Korle-Bu Hospital, and railway lines, among others. By and large, most scholars believe British colonialism should take the blame for much of Ghana's current development problems. However, four related themes need reiterating. First, because public administration and the state were rooted in, and structured to protect British interests, where colonial people had little or no representation, the basis of public administration and of the state was questionable. British colonial rule therefore lacked the moral fiber to administer, as well as the legitimacy required of modern democratic representative government.

Second, public institutions that Ghana inherited from the colonial state, designed as they were, to advance the goals of British colonial interests, remain and in many instances are strengthened by existing political and administrative practices. For example, state officials are reluctant or even resisting ceding power required in attempts to decentralize and restructure the Ghanaian public administration system (Ayee, 1994; Mohan, 1996). This has resulted in "centralized decentralization" whereby political and administrative power remains largely concentrated in Accra despite efforts at pursuing decentralized administration policy.

Third, transportation and communication networks designed to satisfy the needs of trade with British colonial interests and other metropolitan centers persist with slow efforts made to balance urban and rural development needs in Ghana (Ayee, 1997; Crook, 1994). The gap between Ghana's urban and rural areas is widening to the disadvantage of the rural communities, where over 60 percent of the Ghanaian population lives and works.

Finally, as colonial economic activities skewed toward primary commodities for export, little effort was made to expand local entrepreneurial capacity. Thus the private sector in Ghana was ill prepared and lacked the ability to play a meaningful role in national development. It may therefore be concluded that the ghost of Ghana's colonial past lurks around even after a century of public administration and nearly a half-century of nationhood. The question is how Ghana has addressed the roles of the state and public administration in the wake of political independence to make them relevant to the social and cultural experiences of the citizens, which is the subject of the next section.

5:2.2 Post-colonial State

Ghana's political independence in 1957 and the transition to nationhood in 1960 presented Ghana with an opportunity to redefine the state and public administration to make them relevant to, and functional in its economic, social, and cultural context. The GDP growth rate of 6 percent in 1960, which was comparable to many developing nations (Hyden, 1983; Huq, 1989; Werlin, 1994) boosted this opportunity. Ghana's independence inspired other African countries and the reform of its public services assigned different roles to the state and public administration in political, social, and economic development. These reforms resulted in active state leadership, where the state became the sole employer and provider of public services as the fruits of independence and self-government. But public administration remained passive, serving as a tool for implementing policies and programs. As we will see, a pro-active state with a passive bureaucratic public administration system led to growth and multiplicity of public services that did not necessarily advance the public interest in Ghana.

The transition of the Gold Coast colony from its colonial status to a constitutional and self-governing state in 1957, and later a democratic republic in 1960 changed the nature and character of the state and public administration in Ghana. First, with political independence, the name "Gold Coast" was changed to Ghana symbolized by "Freedom and Justice". Second, Ghana's new Constitution expressed the aspirations of a free and an independent people and created a unitary system of government consisting of a national assembly, a prime minister, and a judiciary. Finally, independence brought the people of different ethnic and cultural origins under one central government, which required the reorganization of the state apparatus in Ghana to meet the challenges of social and economic development (Ayee, 1994).

To meet the goals of nationhood, the philosophy of Ghana's government, the state, and of public administration changed, focusing on nationalistic development in what Adu (1965) described as a "revolution". By 1960, when Ghana became a self-governing republic, the nature and character of the state and public administration had undergone transformation from law and order to development and provision of public services (Ayee, 1993; Adu, 1973).

One of the most important policy instruments guiding the reorganization of the state and public administration in Ghana centered on "Africanization", the process of replacing expatriate employees with Ghanaians (Table 5:1). Adu (1965, p. 31) states:

> Few people would deny that an indigenous government should be
> served as soon as possible by an indigenous and local Civil Service.

Began in the early 1950s, the process of Africanization led to steady increases in Ghanaian participation in government, especially in the top echelons of the public services. This policy transformed especially the colonial civil service into a predominantly Ghanaian organization, giving public administration a new character by 1960.

Table 5:1   Africanization of the Ghanian Civil Service, 1952-1960

Nationality	1952	1953	1954	1955	1956	1957	1958	1959	1960
Ghanain	620	898	1043	1277	1553	1941	2984	2320	2766
Expatriate	1332	1350	1350	1241	1123	984	880	859	749
Total	1952	2248	2393	2518	2656	2925	3864	3179	3515

Source: *Adopted from Ayee, J.R. 1993, p. 279*

The underlying assumption of reform was to make the state and public administration more representative of, and responsive to Ghanaian society and aspirations of the government. The imperative of development underlying independence resulted in the creation and expansion of public services constructed around governmental ministries, departments, and agencies (hereafter MDAs). MDAs were established as administrative arrangements to pursue specific policy goals and programmatic objectives. The government's policies and programs determined the number and size associated with each. Under British colonial rule, there were relatively fewer MDAs but with independence they increased as part of comprehensive public sector reforms. The number of ministries in the post-colonial state has been influenced by the tasks the government assigned to itself.

The government ministries, throughout the colonial and post-independence eras, constituted pivots of policymaking and implementation that made Accra, and to a large extent the regional capitals, centers of political and administrative power. The numbers of ministries and departments have been changing throughout the years (see Table 5:2).

In 1967, the 31 ministries established under the Nkrumah administration were reorganized and reduced to 17. In the 1980s, many departments including the Central Bureau of Statistics, Customs and Excise, Income Tax, Civil Aviation, and Immigration assumed semi-autonomous statuses with changes in their focus and function. Public administration has also been organized along productive sector lines consisting of programs and sub-programs. Programs defined the framework within which intended expenditure fell, making them synonymous with departments. Government budgetary expenditures tended to be clas-

Table 5:2 Ministries/Departments in Ghana for Selected Years

Year	Ministries	Departments
1966	31	-
1970	17	48
1978	20	53
1982	22	47
1989	20	43
1990	19	41

Source: *Ghana Office of the Head of the Civil Service Records*

sified in terms of sectors: energy, housing, mining, export, health, education, transport, communications, and more. Programs have undergone various transformations, assuming permanency such as the Ghana Cocoa Board or being absorbed (e.g., National Energy Board and Upper Regional Agricultural Development Program).

As part of public services restructuring, the state increased its participation in the economy, leading to the proliferation of governmental agencies in Ghana. In addition to the MDAs, as well as regional and district administrations, the government created public corporations or state-owned enterprises (SOEs) to promote economic and social development (Gyimah-Boadi, 1991; Greenstreet, 1973):

> The 1950s and the early 1960s saw rapid expansion in the number of SOEs. The Nkrumah CPP nationalist government came to rely heavily on SOEs as instruments for the achievement of socialist goals. During this period the government established state enterprises in various sectors of the economy including commerce, agriculture, transport, construction, manufacturing, services, mining, finance, mass communications, and the Volta River Authority, multipurpose agency (Gyimah-Boadi, 1991, p. 194).

The post-colonial state witnessed continued growth and expansion of the public services in Ghana. Table 5:3 indicates growth in civil service employment through the 1970s, reaching the highest level of 143, 143,237 employees in 1986. Throughout the 1970s, uncontrolled public sector growth stifled, rather than promoted, social and economic development. This includes the establishments or the number of employees allowed for all grades approved for ministries and departments based on capacity and projected enlargements. In some instances establishments have been allowed for employees' advancement as well.

Employment in the state-owned enterprises expanded from 11,000 at independence to 240,000 in 1984. About 60 percent of the government's non-debt current expenditure went into wages (Werlin, 1994, p. 210). The distribution of civil servants by Ministry indicates that the largest Ministries include Health, Local Government, Agriculture, Lands and Natural Resources, and Works and Housing. These accounted for 81 percent of the total civil service labor force in 1987. The difference of 19 percent is distributed across the other ministries. Thus Ghana's civil servants tend to be concentrated in a small number of ministries. Higher education also saw expansion and improvement based on the establishment of universities, secondary, and technical institutions to enhance op-

Table 5:3 Ghana's Civil Service Employment: 1970-1989

Year	Size	% Change	Year	Size	% change
1970	97,789	-	1980	124,012	2.34
1971	100,136	2.34	1981	126,988	2.34
1972	102,359	2.17	1982	130,036	2.34
1973	105,000	2.52	1983	133,157	2.34
1974	107,520	2.34	1984	136,353	2.34
1975	110,520	2.71	1985	139,625	2.52
1976	112,743	1.97	1986	143,237	-9.27
1977	115,449	2.33	1987	131,089	-5.59
1978	118,210	2.34	1988	124,148	-5.87
1979	121,105	2.39	1989	117,267	-

Source: *Ghana's Office of the Head of Civil Service Records*

portunities for Ghanaians to qualify for higher public service responsibility. The University College of the Gold Coast (The University of Ghana) attained a degree awarding status to educate and train managerial, professional, and supervisory personnel for Ghana's public services. The government established two new universities in Kumasi (Kwame Nkrumah University of Science and Technology) and Cape Coast (University of Cape Coast) to train professionals for industrial and educational purposes. In addition, the government established the College of Administration (School of Administration at the University of Ghana) for training and research in business and public administration. Later, the Institute of Public Administration (Ghana Institute of Management and Public Administration–GIMPA) was set up to augment existing facilities for public service education and training in Ghana. The establishment of GIMPA was:

Consummation of the Nkrumah government's belief that the success of its plans for social and economic development depended on a good, efficient, and honest administrative system and that such a system could best be developed through giving young graduates newly recruited to the higher echelon of the civil service a thorough grounding in the principles and techniques of administrative planning and management before they take on their duties (Greenstreet, 1978, p. 5).

The cost of running the public services in Ghana, especially the civil service rose steadily over the years. Such expenditure includes wages and salaries of public employees, their living allowances, and government's contribution to the Social Security Fund. Costs relative to goods and services consisting of expenditures on traveling and transport, repairs, and maintenance of fixed capital, assets, stationery, printing of reports, and provision of utilities form a substantial component of the civil service cost (Ghana Government's Annual Estimates, 1980). The MDAs depend on the governments consolidated fund for all their expenditures.

Rising government's wage bill is attributed to increases in wages and salaries of civil servants, the result ultimately of steady inflation over the period 1970-1990. For example, in 1975 labor cost to the government was 395.9 million cedis, indicating a 40 percent increase over the previous year. In 1985, the labor cost to the government rose by 180 percent over the 1984 cost. Government's expenditure on goods and services followed a similar pattern: expenditure in 1980 indicated an increase of 31.3 percent over 1981 while that for 1983 increased 63 percent over the 1982 cost. These increases were the result of the rapid increases in the prices of printing material, utility services, and cost of repairs and maintenance of capital assets in the civil service. The effect of increasing cost of labor is the reduction in the relative cost of goods and services. In other words, funds that should go into goods and services needed for productive work tended to go into labor cost.

Due to inflation, which reached a peak of 123 percent in 1983, the labor cost to the government increased while real labor cost declined. A similar pattern is depicted in the expenditures on goods and services for the civil service. Such expenditure at 1975 constant prices had been declining, reaching the lowest levels in 1984. As a proportion of the government's total expenditure, the cost of labor to government rose steadily since 1970: 27 percent in 1970, 29.2 percent in 1983, and 38.4 percent in 1986. Thus although government's expenditure on the civil service rose between 1970 and 1990, at 1975 prices the labor cost to the government actually declined (ODA, 1988).

This overview indicates that the primary concerns of the state and public administration in post-colonial state included (1) creation of a self-conscious national identity, and (2) pursuit of political, social, and economic development. The strategy for achieving these goals focused on restructuring and expanding the political and administrative framework to meet the demands of development, which lasted a good portion of the first two decades of independence. Institutional growth and multiplicity that followed laid much of the basic infrastructure for Ghana's social and economic development but also led to waste, inefficiency, and near-collapse of public services in the 1970s and 1980s. This together with the political instability affecting Ghana in the late 1960s and early 1970s provided the needed rationale for pursuing structural adjustment policies characterizing the 1990s.

5:2.3 Structural Adjustment Phase

The structural adjustment programs begun in Ghana in 1983 aimed to transform the Ghanaian society in the most fundamental way by focusing on economic policies that create, rather than redistribute wealth. Based on the neo-economic principles of efficiency and competition, adjustment programs have steadily replaced public administration with public management, and much of the state's responsibility with the market. This is a profound political-philosophical shift in development that has moved Ghana away from the more autonomous, nationalistic, inward-oriented, import-substitution, and state interventionist model toward liberal or laissez-faire capitalism.

Among the key elements of the adjustment package are (1) liberalization, (2) de-bureaucratization, and (3) decentralization. Trade liberalization entails de-regulating the economy in a manner that opens Ghana to international and global competition. This requires micro financial, legal, and managerial reforms, as well as business promotion and private sector growth. De-bureaucratization involves labor reduction and redeployment, divestiture, and privatization of state-owned enterprises. Decentralization aims to rationalize the functions of all organizations forming the Ghanaian public administration system (Public Administration Restructuring, Decentralization, and Implementation Committee (PARDIC, 1990, p. 3). For well over a decade and a half, Ghana has pursued these programs with vigor that attracts commendation from the World Bank and International Monetary Fund. In fact, the International Monetary Fund (1997) highlights "implementation of stronger macroeconomic and structural policies and improvements in governance" as reasons for the economic upturn in Ghana, among other African countries.

5:3 Liberalization and its Impact

The last quarter of the twentieth century saw the intense process of liberalization in Ghana, encompassing almost all spheres of Ghanaian society: economic production and exchange, knowledge and information, political reform and institutional capacity-building, as well as social behavior and lifestyle. The major economic dimensions affecting the state and public administration include (1) fiscal and financial reforms, (2) trade and investment, and (3) export promotion (IMF, 1997). The effect of these reforms has produced far-reaching changes concerning the role of the state and public administration in Ghana.

Since the 1980s, there has been a gradual shift in fiscal and financial policies from tight exchange rate controls and restrictive taxes to a more open market-based system. Ghana pursued more liberal exchange rate and financial policies, allowing the market to determine the cost of foreign exchange and the demand for business finance. This necessitated the institution of parallel foreign exchange markets (*Forex Bureaus*) and other finance companies to deal in foreign currency, as well as provide non-banking services. By the end of 1998, a total of 264 *Forex Bureaus* operated throughout Ghana. As a result of such liberalization, other non-banking financial institutions emerged including finance companies, savings and loan companies, leasing companies, discount houses, building societies, a venture capital company, and a mortgage finance company (ISSER, 1999).

Easing exchange and financial restrictions helped the Ghanaian economy in many ways. First, it improved financial services by reducing currency trafficking and illegal operations to the barest minimum. Second, it expanded the financial market in terms of accessibility to foreign exchange and credit. Finally, it established realistic and dynamic rates for the Ghanaian currency, which had an artificially high value relative to convertible currencies in the pre-adjustment period. The non-banking financial institutions cater especially for the needs of numerous small-scale businesses, which tend to miss out in the formal banking sector. Thus in addition to opening the financial market to competition, liberalization created opportunities that the Ghanaian economy can take advantage of. This is consistent with one primary aim of SAPs, which requires altering trade, investment and financial laws to facilitate the flow of goods, services, and money into, and out of the country. But exchange and financial reforms also led to depreciation of the cedi (see Chapter 2, this volume), with implications for the Ghanaian economy. As a result of the devaluation, prices of imports including crude oil, medical supplies and equipment, educational supplies and equipment, as well as machinery continue to rise astronomically. To a large extent, such price increases are not surprising because one primary goal of SAPs is to "get prices right". This has eroded the real value of earnings thereby affecting the quality of life especially for rural dwellers, 3.8 million of who live in absolute poverty (World Bank, 1997; Konadu-Agyemang, 1998). Because SAPs changed the role of the state from active to passive leadership, the government is unable to act even under such dire circumstances.

5:4 Privatization and its Impact

The influence of SAPs on the role of the state and public administration is evident in major government policies undertaken in Ghana. Since the 1980s, the Ghanaian government has adopted the policy of privatization based on (1) enhancing efficiency, (2) reducing fiscal burden, and (3) enhancing economic growth (Gyimah-Boadi, 1991; Divestiture Implementation Committee (DIC, 1999). Supervised by the Divestiture Implementation Committee, privatization has affected government corporations or state-owned enterprises including hotels, industries, banking, manufacturing, transport, and communication companies. The mode of privatization takes different shapes and forms: outright sale, partnership, deregulation, and contracting out. For example in 1996 the government created the National Communications Authority (Act 524) to set standards for, and license privatized communications service providers. In 1997, Ghana Telecom (GT) was privatized by the sale of 30 percent government shares to Gcom, a consortium led by Telecom Malaysia. Likewise, the Ghana Agro-Food Company Limited (GAFCO) is a joint venture between Ghana (25 percent) and Industrie Bau Nord AG of Switzerland. By December 1998, the divestiture of 212 corporations had been completed (Table 5:4).

Table 5:4 indicates that the sale of public assets (112) is the commonest mode of privatization while leasing (6) represents the least. The data also indicate that privatization started off strongly in 1991 perhaps with small relatively unimportant enterprises. Since 1995 divestiture has leveled off, reaching the lowest point in five years in 1998 probably resulting from the slower processes involving larger and "strategic" enterprises. But the total of 212 divested enterprises in a decade shows that though divestiture may have started slowly it has become the norm in Ghanaian society.

Progress so far shows that privatization has relieved the government of much fiscal burden, an "economic strategy" aimed to bring prosperity to Ghana. Proceeds from the sale of public enterprises and savings that the government makes through divestiture can help to provide needed public services to society. Table 5:6 indicates that annual proceeds from divestiture have risen since 1990, the highest being in 1993 (273, 292 million cedis). In addition, divestiture resulted in increased tax revenues. The joint venture among GNTC Bottling, Coca-Cola International, and Africa Growth Fund (USA) contributed 12.7 billion cedis in tax revenue in 1999 up from 6.8 billion cedis in 1995. In terms of employment, the Tema Steel Company increased its work force from 130 during pre-divestiture to 584 in 1999 while the Tema Food Limited increased the number of its employees from 494 to 1,600 (DIC, 1999).

But privatization raises political concerns of transparency and consensus, which are important if the programs are to work well (World Bank, 1996). First, to many observers in Ghana, the process of privatization does not seem transparent enough and little information regarding the processes and outcomes of

Table 5:4 Divestiture Progress: 1991-1998

Mode	1991	1992	1993	1994	1995	1996	1997	1998	1999
Sale of Assets	16	4	3	30	19	18	15	7	112
Sale of Shares	11	5	2	2	6	1	2	2	31
Joint Venture	6	3	11	4	0	4	1	2	21
Lease	3	1	0	1	0	0	1	0	6
Liquidation	24	2	5	5	6	0	0	0	42
Total	60	15	11	42	31	23	19	11	212

Source: *http://www/webstar.com.gh/dic/results*

divestiture is available to many Ghanaian taxpayers. This is associated with frequent suspicions about, and allegations of "giving away national assets" under the pretext of fiscal expediency. Second, there seems to be little or no consensus on how and whether privatization programs in Ghana serve the public interest. This seems to reflect in general comments that public officials use privatization programs to sell public assets to themselves, friends, relatives, and foreign accomplices at give-away prices. As a result, privatization programs began a decade ago seems to lack public input and support. With a weak domestic private sector and a slow development of the capital market, privatization seems not to be working as effectively as it should.

Finally, due to such market-oriented measures the roles of the state and public administration are diminishing in major economic sectors including the national airline, shipping line, telecommunications, petroleum-chemicals, electricity, water supply, railway, banks, and mining. Previous public policies, programs, and projects based on nationalism, as well as development are being replaced by businesses based on economic efficiency and competition. The state and public administration are losing their roles to private businesses, transnational corporations, and international financial institutions guided by principles inherent in contemporary market ideology. For example, the government is losing its grip on Ashanti Goldfields Corporation, one of the richest mining companies in the world. The crisis facing the company seems to reflect a loss of governmental control and with it the loss of adequate guarantees to protect the public interest of Ghanaians.

The issue of public concern in Ghana seems to be that privatization is not only weakening the role of the state and public administration but also is promoting the diversion of such a role in favor of market forces, private interests, and business firms. For example the withdrawal of the state's role in the provision of basic utilities may disproportionately benefit local and foreign interests at the expense of the needs and interests of ordinary Ghanaians. In fact, it is becoming difficult for the state to legitimate its active role in society because it will be an unexpected move if the state flexes its muscle to strengthen its social and economic role. In sum, the privatization programs render active leadership role of the state obsolete in Ghana.

Table 5:6   Annual Proceeds from Divestiture (millions of cedis)

Year	Total Proceeds
1990	258
1991	4,752
1992	6,000
1993	273,292
1995	81,977
1996	180,232
1997	103,200
1998	99,600

Source: *Ghana Statistical Services*

5:5 De-bureaucratization

The de-bureaucratization that began in Ghana in the mid 1980s consists of a package of political and structural components aimed to strengthen institutional capacity, as well improve administrative performance (Civil Service Performance Improvement Program (CSPIP, 1997; PARDIC, 1990). Politically, reform of Ghana's public administration system emphasizes a decentralized model, the notion that the central government cedes part of its power to sub national jurisdictions in terms of resource mobilization, allocation, and utilization (Ghana 1992 Republican Constitution). Structural requirements involve civil service labor reduction and redeployment, as well as functional rationalization. Philosophically, such an approach replaces public administration with public management, emphasizing performance and minimizing fraud, waste, and abuse. This is consistent with privatization and the market model because it circumscribes the appropriate functions of national and sub-national jurisdictions.

Specifically, decentralization:

> seeks to ensure increased management competence in the implementation of public decisions closest to the locus of implementation in a decentralized organization. It discourages and even prevents Ministerial organizations, which hitherto have had ministerial responsibilities, which have entailed, or enticed them into day-to-day management of government departments and parastatal organizations from undertaking such activities, and restricts them to their original task of policy planning and coordination (PARDIC, 1990, p. 4).

Begun in 1988, the decentralization program created 10 Regional Coordinating Councils, a four-tier Metropolitan, and a three-tier Municipal and District Assemblies (The New Local Government System, 1996, p. 11). This resulted in 3 Metropolitan, 4 Municipal, and 103 District Assemblies. Assemblies represent the basic units of Ghana's public administration system, forming a monolithic structure and unifying departments and organizations into a "single hierarchy model" at the sub-national level. Assemblies have responsibility for planning, deliberating, legislating, and executing development policies at the local level. This includes:

> the overall development of the District and shall ensure the preparation and submission . . . for approval the development plan and budget for the district (The New Local Government System, 1994, p. 8).

Government ministries have four distinct functional areas: (1) Policy planning, programming, budgeting, monitoring, and evaluation, (2) Research, statistics, and manpower development, (3) Information management, computer support, and public relations, and (4) General and financial administration. Directors head functional areas while Chief Directors coordinate them. Chief Directors have the expertise for supervising all agencies coming under the ministry and sector. They are appointed on contract or terms different from civil service standards and their remuneration is comparable to CEOs in the private businesses.

Ghana's civil service reform program has two aspects: (1) Redeployment of excess personnel, and (2) Improvement of managerial capacity. Shedding excess labor involves reducing 5 percent of civil service employees over five years (Adofo, 1990). Redeployment aimed: (1) to reduce the size of the civil service to an appropriate limit, (2) to re-deploy excess labor in productive sectors of the economy, (3) to enable government pay employees improved remuneration, and (4) to increase efficiency in public management (Amekudzi, 1993). Labor reduction rid the civil service of redundant and "ghost" workers. Table 5:7 indicates steady reduction of civil service employees between 1987 and 1990, the highest being 12,849 achieved in 1989.

Ghana's new civil service Act (PNDC Law 327) provides the legal framework and ethical code for guiding civil service reform and improvement. The law strengthens the hands of the political heads of sub-national jurisdictions over civil service employees. The first aspect of civil service reform involved training personnel in new management techniques and reassigning responsibility

Table 5:7 Labour Reduction: 1987-1990

Year	Number Re-deployed
1987	12,100
1988	12,100
1989	12,849
1990	7,789
Total	44,838

Source: *Amekudzi*, 1993, p. 96

while the second emphasizes the provision of infrastructure: supply of copiers, computers, and electronic typewriters to improve information processing and decision making.

In terms of performance appraisal, civil service reform emphasizes a results-oriented appraisal system. This is based on (1) specific measurable goals within specified periods, (2) participation, where superiors and subordinates agree on the goals and objectives, and (3) feedback to enable subordinates know how they are performing. In addition, Chief Directors sign performance-based contracts on annually, forming the basis for their stay in office.

On remuneration, civil service reform provides for a "pay and grading" system, where (1) grade is related to pay, and (2) officers of the same grade receive comparable compensation across Ghana's public administration system (Gyampoh Commission, 1993). Pay and grading involves first simplifying salaries and wages structure to eliminate overlap, distortion, and inequity. Second, it consolidates salaries wages, and allowances into a single compensation package. Third and final, pay and grading reclassifies job categories into 12 occupational groupings with each grouping allotted six separate incremental scales. This decompresses salaries and wages ratio from 4:1 to 13:1 between the lowest and highest paid civil service employees. Decompression creates meaningful differences in remuneration in terms of levels of responsibility, linking reward to responsibility. Overall, pay and grading links productivity with salaries and wages and increases in remuneration are based on levels of productivity.

Ghana's administrative reform including restructuring, decentralizing, and improving public services seems to have made qualitative changes in public administration. Decentralization–devolving political authority to sub-national bodies–addresses issues related to over-centralization in the national government. The Mills-Odoi Commission (1967) lamented about over-centralization, which had resulted in the disproportionate development of urban centers at the expense of rural and semi-rural areas. Decentralization redresses such inequity and improves the allocation and distribution of national resources among segments of society.

More important, decentralization empowers Ghana's communities to participate in national decision-making related to Ghana's overall development. The Minister for Local Government:

> I consider democracy, development, and decentralization as inseparable triplets in Ghana; they can be linked to Siamese triplets sharing one stomach. What they eat individually will go into the same stomach (Ahwoi, 1992, p. 26).

Decentralization creates opportunities for Ghanaians, especially at the rural community levels to take responsibility, exercise initiative, and impact the political process to their advantage. The social, political, and other interest groups have seized such opportunities to integrate state and society in creative interactions to promote their development (Ayee, 1997).

Restructuring ministries and refocusing their functions on policy planning and analysis redresses concentration of authority, where ministries concern themselves with both policy planning and implementation. Concentration resulted in a situation, where ministries concentrate on the nuts and bolts almost to the exclusion of the larger picture (Adu, 1973). This tends to lead to "congestion" in the communication between ministries and lower bodies (Ayee, 1997, p. 73). Ministerial reorganization reduces concentration of authority, minimizes congestion, and improves Ghana's policy planning and analysis.

Ghana's civil service improvement program addresses institutional capacity building and utilization, emphasizing management as the basis of improvement and better performance. Performance is based on goals and objectives determined between supervisors and subordinates thereby improving responsibility and accountability on the part of public employees (CSPIP, 1997). Labor reduction, job redesign, and managerial training infuse the sense of efficiency and effectiveness in the Ghanaian civil service in fundamental ways. The governance perspective seeks to develop consensus building and collaboration in the implementation of policies and programs by involving ministers, chief directors, sector programs, and more. Reform limits governmental intervention in the economy, scales down institutional growth, and eliminates overstaffing, duplication, and waste. In sum, reform seeks to grow, develop, and strengthen the capacity of the public administration system (CSPIP, 1994).

But reform remains a long way from moving the Ghanaian public administration system to meet the expectations of Ghanaians: policy makers, academicians, practitioners, and citizens. In terms of decentralization, frustration seems common. Ghana's Minister for Local Government sums it:

> Decentralization has not taken place in Ghana. The reason largely is
> that the bureaucracy. . . particularly the top management personnel . . .
> is not in favor of decentralization. Every impediment has been placed
> in the way of implementing the decentralization program. Top civil
> servants want to know. Some have deliberately confused it with an
> exercise in de-concentration (Ahwoi, 1992, p. 23).

Ghana's bureaucracy cannot and should not take all the blame for the decentralization problem. For while the bureaucracy is "adjusting" too slowly to, or resisting the decentralizing process, other factors of equal or even more importance should not be discounted. For example, decentralization is often viewed as an instrument for legitimizing governmental power: where the government uses decentralization for largely "political purposes" rather than the interest of economic and social development. Ayee (1997, p. 51) argues:

> Decentralization under the PNDC suffered largely because it was essentially a means to an end of legitimacy. When party politics became inevitable, the regime itself lost interest in the DAs as agents of development and mobilized rural resources towards the 1992 elections.

Second, although much has been done to build the capacities of ministries, departments, and agencies, this seems to proceed at a slower pace than required. Managerial competency of personnel at both national and sub national levels remains low, resulting in poor performance (Crook, 1994; Ocquaye, 1995). In terms of communications, many sub national jurisdictions especially District Assemblies continue to be inaccessible. Both intra- and inter-regional communication remains poor, problems many people associate with the government's reluctance or unwillingness to address. Public administration continues to be centralized with regions and districts taking directives on practically every matter from Accra. Ayee (1996, p. 49-50) states:

> The history of decentralization in Ghana since independence has been characterized by the deliberate and conscious machination of the center to control the sub national units either to promote its patronage or use them as support bases. The result is that the country has had what one might refer to centralized decentralization.

Ghana's experience indicates tension arising from pursuing decentralization policies under a unitary system of government, the phenomenon Ayee (1997) describes as "centralized decentralization". Such tension tends to generate difficulties for and in Ghanaian public administration practice. At the core of this issue is the need to find a balance between national authority and decentralized administration. This lies not necessarily in finding a final solution, but perhaps in promoting dialogue involving especially the central authority on one hand and the sub-national authorities and civil society on the other.

Ghana's reforms over-emphasize omnibus, macro-structural change at the expense of the individual citizen. As a result, reforms focus on procedures, rules, and regulations as the means to ensure public accountability, efficiency, and effectiveness (Civil Service Handbook, 1994). While much can be said in favor of this managerial perspective, its applicability to public administration even in industrialized society has limitations (Martin, 1993; Pollitt, 1990; Steers, 1977). Limitations are associated with the diversity, multiplicity, and ambiguity of public organizational purposes, goals, and objectives. Limitations are inherent also in the nature of public problems, which Weick (1979, p. 12) describes as "wicked", defying "quick fixes". Such limitations are likely more complicated in Ghana, where our social and cultural systems seem to be at variance with modern managerial theory and practice (Hyden, 1983).

5:6 Conclusion

This chapter has highlighted how SAPs affect the roles of the state and public administration in Ghana. To provide the context for discussion, it examined the changing roles of the state and public administration under the colonial state, post-colonial state, and structural adjustment. The colonial state and public administration played limited roles, confined mainly to the provision of law and order. Colonialists never allowed market economies to develop. Neither did they allow democratic traditions to grow. Rather, the colonial state and public administration promoted monopolies and centralized administration to further exploitation and promote security of foreign economic concerns. In short, neither the colonial state nor its public administration provided active leadership roles required for progress and development in Ghana.

The post-colonial state emerged at independence, expanding its role and that of public administration in Ghanaian society. This led to the institution of a command economy and an expansive hierarchical public administration. While the post-colonial state provided dynamic leadership in society, public administration remained as a passive instrument for implementing public policies. But the institutions through which the state and public administration operated were not only over-centralized but also inconsistent with traditional Ghanaian cultural values. Thus institutional growth and multiplicity failed to nurture the ethic and culture of purposive leadership in Ghana.

SAPs changed the roles of the state and public administration in Ghana to a more market-based, neo-liberal governance based on the diminished autonomy of the state in relation to transnational and global capital. There is an eclipse of the developmental state, forging an ideology based on neo-liberal beliefs. The role of the state to mediate the interests of various competing segments of society is submerged under its role to ensure a business atmosphere for local and foreign private capital. Previous state-sponsored development initiatives are transferred to domestic and international market forces under liberalization, privatization, deregulation, and devaluation programs. These changes dismantle the state apparatus and challenge the role of the state and its intervention in Ghanaian society.

With these changes there is a "paradigmatic shift" from public administration to public management, a performance-based philosophy emphasizing efficiency and effectiveness. The major initiatives of the managerial changes include Public Administration Restructuring, Decentralization, and Implementation (PARDIC), Civil Service Reform Program (CSRP), Civil Service Improvement Program (CSPIP), Public Sector Financial Management Program (PUFMAP), to name a few. The goals of such reform initiatives include streamlining the size and scope, as well as reinforcing a market perspective of the Ghanaian public sector. The role of public administration is shifting toward enhancing micro-level management efficiency rather than developing overall societal progress and development.

The roles of the state and public administration are changing from the perspective of public interest concerns to "customers", especially the top 5 percent of the population: government functionaries and affluent business executives. For the most part, the state is viewed as a facilitator, public administration as public management, and public administrators as managers. The Civil Service Handbook employs the term "manager" as an essential criterion to shape the role of public administration, especially in ensuring a business-like atmosphere for public services. The customer-oriented role of the state may improve services for high-income families but worsen the plight of low-income citizens.

One implication of the changes in the roles of the state and public administration from expansive to narrow, from developmental to managerial, and from active to passive is that it might worsen income inequities and widen the urban-rural disparities in Ghanaian society. For, gainers of the pro-market policies are institutional investors, executives of private companies, contractors, and private consulting firms (Vickers and Wright, 1988; Martin, 1993). The Ghanaian experience with SAPs reinforces salient issues in the literature on liberalization and privatization: (1) adjustment hollows out the state (Milward, 1994), (2) favors political and business elite (Jaysankaran, 1995; Briones, 1995), and (3) losers are low-income citizens, which is the case in many African and Latin American countries (Bello *et al.*, 1994).

The second implication of the changing role of the state is the erosion of our national political integrity and economic sovereignty in relation to the international and global capitalist system. The autonomy of the state in Ghana is diminishing primarily because of the IMF and World Bank "conditionalities". Such a diminished role weakens the Ghanaian economy, expands foreign ownership of our national assets and resources, and reduces the state's capacity to adopt policies favorable to poor and working class people. The weakening of state institutions tends to reinforce a low image of governance and erode Ghanaians' trust in their government.

Finally, reduction of public confidence in our state institutions implies a fundamental challenge to the legitimacy of the state and public administration. Such legitimacy problems are exacerbated owing to diminishing capacity of the government arising from (1) auctioning of public resources and assets, (2) withdrawal of social and welfare subsidies, (3) reduction in public employment, and (4) declining commitment of public administrators who find their jobs insecure. The problems of declining public confidence and worsening legitimacy imply that the Ghanaian government needs to critically re-evaluate changes in the role of the state and the functions of its public administration.

The question remaining is where does Ghana go from here? This is a question Ghana is addressing by launching a Capacity Development and Utilization Initiative (hereafter CDUI) to sustain her economic recovery program and SAPs. The CDUI is all embracing, involving all stakeholders in government, donor-agencies, non-governmental agencies, and private businesses. It emphasizes participation, consensus building, local commitment, ownership, and sustainability (World Bank, 1996). The strength of the capacity building model lies in integrating state and civil society, which goes beyond the post-colonial or developmental state, as well as raw structural adjustment programs. It recognizes strengths and weaknesses entailed in both government and market-led development efforts.

But the CDUI does require fine-tuning. First, capacity building and utilization should be grounded in our indigenous understanding of authority, government, and purpose, the missing link in the construction of political and economic institutions since colonialism. This avoids grafting new institutions on old and worn out structures that bear no relationship to our political, social, and cultural experiences as a nation. Second, capacity building should carefully define what the state and public administration can and should legitimately do to preserve our national integrity and economic sovereignty. Finally, capacity building should focus not only on efficiency and competition but also equity in our policy development, implementation, and analysis efforts. Sustainable capacity building and utilization are difficult to achieve but as Hyden (1983) prophesied, there may indeed be "no shortcuts to progress" in Africa.

References

Adofo, K. O. (1990). *Size and Cost of the Civil Services in Africa: The Ghana Case Study.* Unpublished Manuscript. Accra, Ghana.

Adu, A. L. (1965). *The Civil Service in New African States.* London: G. Allen Unwin.

Adu, A. L. (1973). 'The Public Service and the Administration of Public Affairs in Ghana'. Paper presented at the Ghana Academy of Arts and Sciences. Accra, Ghana.

Ahwoi, K. (1992). 'The Constitution of the Fourth Republic and the Local Government System'. Address delivered at the 5th Annual Workshop on Decentralization in Ghana held September 1992. School of Administration, University of Ghana, Legon.

Amekudzi, T. (1993). *The Size of the Ghana Civil Service and its Impact on Performance.* Unpublished Thesis. Achimota, Ghana: GIMPA.

Anyinam, C. (1994). 'Spatial Implications of Structural Adjustment Programs in Ghana'. *TESG 85* (5) 446-450.

Ayee, J. R. (1994). *Anatomy of public policy implementation: The case of decentralization policies in Ghana.* Aldershot: Avebury.

Ayee, J. R. (1997). 'The Adjustment of Central Bodies to Decentralization: The Case of the Ghanaian Bureaucracy'. *African Studies Review,* vol. 40, 2, pp. 37-57.

Bello, W., Cunningham, S. and Rao, B. (1994). *Dark Victory: The United States, Structural Adjustment and Global Poverty.* London: Pluto Press.

Briones, L. M. (1995). 'The Impacts of privatization on distributional equity in the Philippines' in Ramadham, V. V. (ed.), *Privatization and Equity.* (pp. 83-98). London: Routledge.

Chazan, N. (1991). 'The Political Transformation of Ghana under the PNDC' in Rothchild, D. (ed.), *The Political Economy of Recovery.* Boulder, Colorado: Lynne Reinner Publishers.

Cook, P. and Kirkpatrick, C. (1995). 'Privatization policy and performance' in Cook, P. and Kirkpatrick, C. (eds.), *Privatization policy and performance: International Perspectives* (pp. 3-270). London: Prentice Hall/Harvester Wheatsheaf.

Crook, R. C. (1987). 'Legitimacy, Authority, and the Transfer of Power in Accra'. *Political Studies,* vol. XXXV, pp. 552-572.

Crook, R. C. (1994). 'Four years of the Ghana district assemblies in operation: Decentralization, democratization, and administration performance'. *Public Administration and Development,* vol. 14 (3), 339-364.

Glickman, H. (1988). *The Crisis and Challenge of African Development.* Westport, CT: Greenwood Press, Inc.

Gold Coast (1948). *Report of the Commission of Inquiry into the Disturbances in the Gold Coast (Watson Commission Report).* London: HMSO.

Gold Coast (1951a). *Report of the Commission on the Civil Service of the Gold Coast, 1950-1951, vol. 1 (Lidbury Commission Report).* London: C. F. Hodgson and Son.

Greenstreet, D. K. (1971). 'The Post-World War II Integration of Departments with Ministries in the Commonwealth States of Africa'. *Journal of Management Studies*, vol. 5, 1, pp. 15-22.

Gyimah-Boadi, E. (1990). 'Economic Recovery and Politics in PNDC's Ghana'. *Journal of Commonwealth and Comparative Studies*, vol. 28, 3, pp. 328-343.

Haque, M. S. (1997). 'Recent changes in Asian public service in the context of privatization' in IIAS and United Nations (eds.) *Public Administration and Development: Improving Accountability, Responsiveness and Legal Framework* (pp. 43-65). Amsterdam: ISO Press.

Haque, M. S. (1998b). 'Legitimation crisis: A Challenge for public service in the next century'. *International Review of Administrative Sciences*, 64 (1), 13-26.

Herbst, J. (1993). *The Politics of reform in Ghana, 1982-1991.* Berkeley: University of California Press.

Huq, M. M. (1989). *The Economy of Ghana.* London: The MacMillan Press Ltd.

Hyden, G. (1983). *No Shortcuts to Progress: African Development Management in Perspective.* Berkeley and Los Angeles, CA: University of California Press.

Institute of Social, Statistical, and Economic Research (1999). *The State of the Ghanaian Economy in 1998.* Legon, Ghana: ISSER.

Jayasankaran, S. (1995). 'Privatization Pioneer'. *Far Eastern Economic Review*, January 19, 44.

Jonah, K. (1989). 'Social Impacts of Ghana's Adjustment Program-1983-1986' in Onimode, B. (ed.), *The IMF, the World Bank, and the African Debt* (vol. 2). London: Zed Press.

Kimble, D. (1963). *A Political History of Ghana, 1850-1960.* Oxford: Clarendon.

King, S.C. (1999). *The Possibilities for Communities and Democracy.* Administrative Theory and Praxis 21(3).

Konadu-Agyemang, K. (1998). 'Structural Adjustment Programs and the Perpetuating of Poverty and Underdevelopment in Africa: Ghana's Experience'. *Scandinavian Journal of Development Alternatives and Area Studies* 17 (2), 127-143.

Mackenzie, F. (1993). *Exploring the Connections: Structural Adjustment, Gender, and the Environment.* Geoforum 24 (1), 71-87.

Mahmud, B. T. (1992). 'Administrative Reform in Malaysia Towards Enhancing Public Service' in Zhang, Z., de Guzman, R. P. and Reforma, M. M. (eds.) *Administrative Reform Ttowards Promoting Productivity in Bureaucratic Performance* (pp. 37-49). Manila: Eastern Regional Organization for Public Administration.

Marini, F. (ed.) (1971). *Toward a new public administration: The Minnowbrook Perspective.* Scranton, PA: Chandler.

Martin, B. (1993). *In the public interest: Privatization and public sector reform.* London: Zed Books Ltd.

Mendoza, M. S. (1996). 'Public Service Innovation in the Philippines' in Salleh, S. H. (ed.) *Public Sector Innovations—The ASEAN Way* (pp. 147-189). Kuala Lumpur: Asian and Pacific Development Center.

Milward, H. B. (1994). *Nonprofit Contracting and the Hollow State.* Public Administration Review 54 (1): 73-77.

Mohan, G. (1996). 'Neoliberalism and decentralized development planning in Ghana'. *TWPR,* 18 (4), pp. 433-455.

Ocquaye, M. (1995). 'Decentralization and Development: The Ghanaian Case under the PNDC'. *Journal of Commonwealth and Comparative Politics,* 33 (2): 209-39.

OECD (1993). 'Organization for Economic Cooperation and Development'. *Public management Development Survey 1993.* Paris: OECD.

Office of the Head of the Civil Service (1993). *The Civil Service Management Handbook* (vol. 1). Accra: Universal Printers and Publishers, Ltd.

Office of the Head of the Civil Service (1997). *The Civil Service Performance Improvement Program.* Accra, Ghana.

Onimode, B. (ed.) (1989). *The IMF, the World Bank, and the African Debt,* Volume I: The Economic Impact. London and New Jersey: Zed Press.

Pai, S. (1994). 'Transition from command to market economy: Privatization in Brazil and Argentina' in Jain, R. B. and Bongartz, H. (eds.). *Structural adjustment, public policy, and bureaucracy in developing countries* (pp. 158-182). New Delhi: Har-Anand Publications.

Pollitt, C. (1990). *Managerialism and the Public Services: The Anglo-American Experience.* Cambridge, MA: Blackwell.

Rasheed, S. and Luke, F. D. (1995). *Development Management in Africa: Toward Dynamism, Empowerment, and Entrepreneurship*. Boulder, CO: Westview Press.

Report of the Commission on the Structure and Remuneration of the Public Services (1967) (Unpublished).

Report of the Committee on Public Administration Restructuring, Decentralization, and Implementation Committee (1990) (Unpublished).

Republic of Ghana (1967). *Mills-Odoi Commission Report*. Accra, Ghana.

Republic of Ghana (1988). *Local Government Law, PNDC Law 207*. Tema: Ghana Publishing Corporation.

Republic of Ghana (1993). *Gyampoh Commission Report*. Accra, Ghana.

Republic of Ghana (1993a). *Civil Service Law, PNDC Law 327*. Tema: Ghana Publishing Corporation.

Republic of Ghana (1993b). *Local Government Act, Act 462*. Tema: Ghana Publishing Corporation.

Republic of Ghana (1999). *Divestiture Implementation Committee Report*. Accra, Ghana.

Republic of Ghana *National Institutional Renewal Program (NIRP)*. Tema: Ghana Publishing Corporation.

Rothchild, D. (1991). *Ghana: The Political Economy of Recovery*. London: Lynne Rienner Publishers, Inc.

Salleh, S. H. (1996). 'Global challenges and local public sector innovations' in Salleh, S. H. (ed.) *Public sector innovations–The ASEAN way* (pp. 1-62). Kuala Lumpur: Asian and Pacific Development Center.

Steers, R. M. (1977). *Organizational Effectiveness: A Behavioral View*. Santa Monica, CA: Goodyear Publishing Co., Inc.

Vickers, J. and Wright, V. (1988). 'The politics of privatization in Western Europe: An Overview'. *West European Politics*, 11 (4), 1-30.

Weber, M. (1958). *The Protestant ethic and the spirit of capitalism*. New York: Scribner.

Weick, K. (1979). *The Social Psychology of Organizing* 2nd ed. Reading, MA: Addison-Wesley.

Werlin, H. H. (1994). 'Ghana and South Korea: Explaining Development Disparities'. *Journal of African and Asian Studies*, 3 (4) pp. 205-225.

World Bank (1995a). *Bureaucrats in business: The economics and politics of government ownership*. New York: Oxford University Press.

World Bank (1995). *Country Briefs*. Washington, DC: World Bank.

World Bank (1996). *Reports on ongoing operational, economic and sector work*. Washington, DC: Knowledge Networks.

World Bank (1997). *World Development Report.* New York: Oxford University Press.

Ghana's Economic Slow Reform under Structural Adjustment Programs 139

World Bank (1992). *World Development Report.* New York: Oxford University Press.

6 Cocoa Production under Ghana's Structural Adjustment Programs: A Study of Rural Farmers

Kwaku Osei-Akom

6:1 Introduction

Like many developing countries, which experienced economic difficulties in the 1970s, Ghana adopted the IMF and World Bank inspired Structural Adjustment Program (SAP), also called Economic Recovery Program (ERP) in 1983 to reverse the decline and stimulate growth of the economy. The adoption of the SAP and its implications for the social and economic development of the country has been the subject of many articles and publications (Rothchild, 1991; Kapur *et al.*, 1991; Roe *et al.*, 1992; Kofi, 1993; Sowa, 1993; Beaudry and Sowa, 1994; Armstrong, 1996). While many of these studies focus on the SAP in the context of agricultural production, most of their analyses deal with issues of macro nature. For example, Bateman *et al.* (1990) examined one of the policy dilemmas facing post-independent Ghanaian governments; namely, the provision of adequate incentives to cocoa farmers to boost exports while meeting economic objectives of foreign exchange earnings and revenue. In a related study, Frimpong-Ansah (1991) argues that "public policy in Ghana, dominated over the years by a predatory coalition of revenue-hungry governments, profit-seeking businesses, price-conscious consumers and defected farmers" (p.144). As a result, these groups were unable to recognize the structural impediments to accelerated development inherent in unproductive and fragmented peasant agriculture and instead have excessively taxed the cocoa industry, in particular, to finance the interests of the coalition under the guise of development. Thus he blames the predatory relationship of the state apparatus for the poor performance of the cocoa sector.

Why the majority of the studies on the agricultural sector adjustment reform do not deal with the voices of the cocoa farmers is somewhat surprising given the fact that an overwhelming majority of Ghanaians, particularly those in the southern part of the country, depend one way or another on the cocoa industry for their livelihood. Among the studies that have examined some aspects of the cocoa sector, I find the following works to be particularly relevant to the current study; Ewusi (1989), Asenso-Okyere (1989, 1990), Bateman *et al.*, (1990), Frimpong-Ansah (1991), Pearce (1992), Kofi (1993), Boahene (1995) and Sarpong (1997). The major difference between these earlier cocoa studies and the one I report here is that in most cases they base their assessments of the cocoa sector on reports and figures derived primarily from official sources such as Government reports, records from the Ghana Cocoa Board (Cocobod), the World Bank and the IMF. None of them, perhaps with the exception of Ewusi (1989), Bateman *et al.* (1990), Okyere-Asenso (1991) and Boahene (1995) included the perspectives of the cocoa producers on the ongoing reforms. Thus, even though the report presented here is a continuation of the earlier studies, this study diverges from most of the earlier works in its focus: the rural cocoa farmers.

6:2 Background

Ghana is highly dependent on the export of primary products, particularly cocoa, timber, gold and diamonds. Of these cocoa occupies a central place in the Ghanaian economy and in the whole scheme of national development. Cocoa revenues have for a long time exerted great influence on the Government's ability to spend and import. Until quite recently, sales of cocoa beans and other cocoa products (powder, butter, liquor) in foreign markets were responsible for generating about 10 percent of Ghana's Gross National Product (GNP) and, on average, 60 percent of the country's revenue from exports (Leslie, 1994). Apart from its predominance in export earnings, the cocoa sector uses about 25 percent of the cultivated land area and employs about 20 percent of the country's labor force (Asenso-Okyere, 1989; Leslie, 1994). As the largest source of export revenue to the government and the main source of wealth to the people of the south, Ghana's ability to raise resources for development has always depended on cocoa exports.

Between 1970 and 1983, however, Ghana's cocoa production experienced a precipitous decline, due in large part to inadequate incentives for agricultural production, poor agricultural pricing policy and the over-exploitation of cocoa farmers. Other causes for the cocoa production decline included swollen shoot virus disease, shortage of farm laborers, drought, bush fires, aging of trees, and costly marketing services (Armstrong, 1996). The cumulative effect of the poor agricultural policy over the years was that by 1983, cocoa production had fallen to a very low level of 159,000 tons (Ghana Cocoa Board, 1999). Even in the wake of the reductions in output, cocoa still accounted for about 47 percent of Ghana GDP and 50 percent of its export earnings in 1988 (ISSER,[1] 1993).

Recognizing the role of cocoa and its multiplier effects in the economy, the government decided, at the inception of the ERP, that increased cocoa production was the only way Ghana could generate the foreign exchange needed to buy spare parts, equipment and other imports which would rehabilitate and improve the country's productive and export capacity. Thus, to boost cocoa production, the Government launched a new five-year, $128 million Cocoa Rehabilitation Project (CRP) within the framework of the SAPs in November 1988. The project was co-financed by grants from the World Bank and other donors. The main assumptions behind the CRP were that falling world market prices for cocoa, inefficiencies in the internal marketing under the Cocoa Marketing Board (Cocobod) and high levels of taxation of the crop had over the years given rise to low producer prices, disinvestments, declining cocoa yields and output in Ghana (ISSER, 1993). In the short term, the project sought to improve producer incentives, increase the quantity and reliability of cocoa input supply, and also to rehabilitate road and transport infrastructure. The program set a production target of 300,000 tons to be produced by 1991 (Ghana Cocoa Board, 1988; Jacobeit, 1991; ISSER, 1994).

To motivate farmers to grow more cocoa and also to reduce the smuggling of cocoa to the neighboring countries, the government was determined to increase the producer price of cocoa until it reached an indicative level of, at least, 55 percent of the f.o.b. price. Consequently, yearly increases in the producer price of cocoa became a key feature of the CRP. These increases were facilitated mainly by the massive devaluation of the Cedi, the local currency. However, given the dwindling world prices, this mode of making resources available to support producer prices could not go on forever. The indications were that improved incentives for local cocoa farmers would have to come from reduced government reliance on cocoa export levy and greater efficiency in the internal marketing of cocoa. To do these, the government decided to augment the producer price adjustments with institutional and managerial reforms in the state-run handling of the crop. In this regard, the program placed new emphasis on extension services, the research in and the provision of drought- and disease-resistant, high yielding cocoa varieties, and the training of cocoa farmers in more timely and more accurate application of inputs such as fertilizers and insecticides. In addition, with a view of putting cocoa marketing on a more commercial footing and thereby promoting increased marketing efficiency, the Cocobod was reorganized. A restructuring of the Cocoa Services Division, with some of its functions going to the Agricultural Extension Service, the introduction of competition into the internal marketing of cocoa as well as the divestiture of the Cocobod's state-owned large-scale cocoa plantations are all part of the measures that have been taken since 1984 to resuscitate the industry.

From a sociological perspective, these incentives were geared at improving the living conditions of the people in the rural areas. As Levi and Haviden (1982) put it, the basis of this approach is the belief that agricultural and rural development can be generated significantly only if new resources are brought into the rural sector. Despite the resources pumped into the cocoa sector, Ghana's cocoa output did not reach the projected 300,000 tons until the 1994/95-crop season. In addition, the government's policy of gradually raising the farm-gate price to, at least, 55 percent of the f.o.b. price was not achieved until the 1996/97-crop season, almost two decades after the program begun. The inability of the programs to achieve their targets raises a number of questions: What happened? Why have all the investments in the cocoa sector failed to yield the anticipated results? While the answers to these questions are somewhat difficult to address given the changing global economy, the views of the ordinary producers provide a unique insight into some of the issues surrounding the reported problems confronting the agricultural sector despite the government's infusion of "support" under the SAP.

6:3 The Current Study

To gauge the views of the ordinary cocoa farmer on the SAP, a field survey was conducted during the summer of 1995 in the Tano north district in the predominantly cocoa growing region of Brong Ahafo.[2] The Tano North cocoa district was selected for a number of reasons. First, the climatic and soil conditions of the area typify those of the other cocoa growing areas in Ghana. The area lies within the tropical equatorial climate, which offers a high and well-distributed rainfall, with some dry months to stimulate fruiting, and provides favorable conditions for maturing pods. In addition, the area lies within the very limited "Susan" soil belt, which supports excellent cocoa more or less indefinitely without any manure (Simmons, 1976). Consequently, Tano North district is predominantly a cocoa farming community and one of the major cocoa producing districts in the Brong Ahafo Region. Besides the above reasons, the socio-economic organization of the cocoa farmers is not different from that of the other cocoa growing areas in Ghana. For instance, not unlike the other cocoa growing areas in Ghana, the smallholder farmer dominates cocoa production in the district. Frequently, the cocoa farms are not very large. The farmer lives on the farm and intersperses the cocoa farms with subsistence crops such as plantains and cocoyams.

The universe of the study was the individual cocoa farmer in the towns and villages in the district who satisfy our definition of cocoa farmers. A two-stage stratified cluster sampling technique was used to select 150 farmers from a list obtained from the district cocoa purchasing clerks (Secretary Receivers) (see Tabel 6:1 for the characteristics of the farmers surveyed.)

For the sake of representativeness, both large towns and small towns were sampled. Since most farmers live in villages and hamlets, the study over-sampled this population. Subsequently, two-third of the farmers were selected from the villages and hamlets. Again, the study over-sampled the male farmers because studies show that in Ghana, while women are the major food producers, cocoa production is, dominated by men (Mikell, 1989; Osei-Tutu, 1992). In addition to the farmers, a sample of Cocoa officials from the District was also interviewed. In the case of the second group, the district officers of the Cocoa Services Division, only 19 of the 27 field officers in the district were included in the survey. In general, the respondents in both groups were interviewed individually. The questionnaires used in this study employed both closed and open-ended questions. In the open-ended questions, the farmers were encouraged to provide information, which was not covered by the questionnaire but were considered important for their production activities.

Regarding the cocoa farmers, the survey inquired about their views on six specific areas: changes that had been introduced in the last 10 years by the government to boost cocoa production, the producer price of cocoa and the general price level, the availability and prices of agricultural inputs, and implications of SAP measures on their farming activities, and also on their own personal lives. Other questions investigated the main impediments to increasing cocoa production, and what the farmers thought the government should do for them to grow more cocoa. For the respondents in the second group (the District officers) the survey solicited the following data: information on government policies toward cocoa production and how these policies were being implemented at the farm level. Data on the cocoa officials' general opinion on incentives the government should provide to induce the farmers to grow more cocoa.

Table 6.1 The Characteristics of the Farmers Surveyed, Tano District, 1995

Age of Farmer	%	Sample Size
15-19	.067	1
20-39	6.00	9
40-49	36.67	55
60 and over	56.67	85
Sex		
Male	78.00	117
Female	22.00	33
Marital Status		
Married	72.67	109
Single	5.33	8
Divorced	7.33	11
Separated	2.00	3
Widowed	12.67	19
Level of Education		
None	45.33	68
Primary	22.67	34
Middle	20.67	31
Secondary/Training/Tech	10.67	16
Post-Secondary (Advanced)	0.67	1
Religious Affiliation		
Traditionalist	20.00	30
Christian	70.00	105
Muslim	9.33	14
Other	0.67	1
N	100.00	150

Because most of the cocoa farmers (about 90 percent) interviewed had no formal education, the questionnaires were administered orally in the Twi language. This was made possible because the researchers and all the research assistants were native speakers of the Twi language. For the second group, the interviews were conducted in either English or Twi, depending on the individual's level of fluency in either language. The written part, however, was entirely in English. Although a set of questionnaires was used, the mode of inquiry depended on how the interviews progressed. Thus, whenever an issue of interest cropped up, the interviewer probed further rather than strictly follow the questionnaire. In order to save time and reduce the possibility of distractions, the interviews were recorded on audiotapes. On the whole, the interviews were conducted smoothly and effectively without any major difficulty.

6:4 Findings of the Study

The evidence gathered from the survey indicates that many of the economic opportunities faced by the various agents in the cocoa sector have changed under the adjustment program. In particular, the expenditure switching policies of the program had greatly altered the relative prices of the variables, such as producer prices and input costs in favor of the cocoa sector. It was observed, however, that for most of the time, the direction and extent of the changes in the variables differed from what had been anticipated. However, there is no evidence that the reform program has brought any change in the subsistence mode of cocoa production. The organization of inputs and resources used in cocoa production still shares many of the similarities with the past policies, which the program intended to remove. The main findings of the study are shown in Table 6.2 and in the following paragraphs.

i. Producer price and volume of cocoa production One of the goals of the cocoa sector adjustment program was to raise the producer price to induce farmers to grow more cocoa. The evidence from the farmers indicates that cocoa price increase has indeed occurred under the SAP. Data from official records seem to support the farmers' assertion. Since the inception of the CRP, the government has adopted a policy of passing back to farmers an increasing proportion of the export price of cocoa to encourage production. As of 1995, the farmers were receiving 40 percent to 50 percent of the world price (ISSER, 1995). This stands in sharp contrast to the meager 25 percent they were receiving in 1984. As a result, the incentive for the farmers has improved and the declining trend in cocoa output, which characterized the early 1980s, has been reversed. This indicates that cocoa farmers in Ghana do respond to price incen-

Table 6.2 Some Problems facing Farmers under SAP*

Problems Cited	%	N
Inputs Supply	18.0	27
Payment Delay	16.7	25
Blackpod/Capsid	14.00	21
Timber Contractors[a]	10.0	15
Bad Weather	7.3	11
Bad Roads	6.7	10
Land constraints	5.3	8
Poor Soil	4.7	7
Bush-Fires	4.0	6
Lack of Medical Care	4.0	6
Lack of Extension Services	3.3	5
Other	6.0	9
Total	100.0	150

*These exclude finance and labor
a: Refers to the destruction of their farms by these contractors.

Source: *Author's Survey, Tano District, B/A, 1995*

tives.

The evidence gathered in the field also indicates that cocoa output has taken a longer time to reach projected levels, because compared to the general level of prices in the country, the new cocoa price was not high enough to induce farmers to invest in purchasing inputs like insecticides and fungicides which are necessary to increase output. It was also observed that because of the poor price the farmers were not replanting their old and diseased cocoa farms; instead they are using their land for growing staple grains, particularly maize, whose prices have also gone up under the SAP but do not require extensive use of purchased inputs. Almost all the farmers surveyed said a substantial proportion of their farm income for the last three years came from crops, which traditionally were grown for subsistence consumption.

ii. Agricultural inputs With regard to the distribution and prices of imported inputs, the findings indicate that as result of the improvements in transport and road networks during the initial years of the program, there has been marked improvement in the distribution of inputs such as insecticides, fungicides and spraying machines. The removal of subsidies on inputs has, however, shot up prices beyond the reach of most farmers. This is borne out by the fact that the price of insecticides had increased from 600 Cedis per liter in the 1989 cocoa season to 2,585 Cedis the following season. During the same period, the price of fungicides increased from 200 to 2,900 Cedis (ISSER, 1995). As a result of the price increases, the quantity of insecticides and fungicides purchased by farmers has steadily decreased. This partly explains why cocoa output has not gone up as expected. Thus, while farm inputs have become available on the market, most of the farmers cannot afford them.

iii. Agricultural labor Evidence from the study also shows that the farmers' complaint that there is labor shortage is not completely unfounded (see Figure 6.1). Even though the public sector retrenchment exercise of the adjustment program led to the loss of about 235,000 formal sector jobs between 1985 and 1990 (ISSER, 1995), this did not lead to an increase in the supply of labor in the rural areas as had been expected. It was observed that a majority of those made redundant in the public sector employment cutback ended up in the informal service sector with only a few of them actually going to the agricultural sector. Of those who went to the rural areas, many started farms of their own, thereby marginally increasing the demand for labor rather than adding to its supply. As a result, the agricultural wage in the rural areas had been pushed up with only a few of the farmers able to afford the wage rates demanded by the available labor. It was discovered that the wages demanded by laborers in the survey area were substantially higher than the minimum wage. Thus, as long as the farmers are able to pay the going daily wage (by-day in local parlance), they could get all the labor they require.

iv. Social and economic infrastructure Again, evidence gathered from the survey shows that despite the heavy investment, which was initiated at the beginning of the adjustment program, the government has not been able to follow through with the program of improving and rehabilitating infrastructural facilities in the rural areas. In particular, it was observed that poor roads and lack of vehicles are still making access to market difficult for the farmers. What worsens the situation is the rigidity and inefficiency, which characterize the Cocobod input distribution system. As a result, inputs are more expensive to obtain and transport costs to market are greater, and thus, reducing the farmer's income from cocoa. Similarly, the budget constraints of the government under the adjustment program have greatly reduced the provision of services such as health, education, and water in the rural areas. This has resulted in the frequently appalling health and education status of the rural people. The rapidly growing numbers of unhealthy and uneducated rural people are less likely to be agricultural innovators that the ERP anticipated.

v. Land for growing cocoa Another important observation of the study was

Figure 6:1 Availability of Labor Since SAP

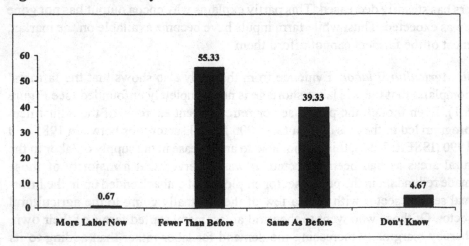

that land scarcity, which a few years back was considered a remote threat, was now a serious problem to increased cocoa production. As a result of increasing scarcity of land, farmers are compelled to remain on the same parcel of land, yet they continue to use their traditional production techniques of shifting cultivation. Another problem which makes it difficult for the young to take to cocoa farming is the system of land acquisition and the perennial nature of the crop. It was noticed that the majority of the people who have recently entered into farming are engaged in food crop farming with only a few of them going into cocoa. Almost all the new entrants into cocoa farming are using parcels of land, inherited from either relatives or land obtained as gifts, with little or no additional room for expansion.

6:5 Policy Implications of the Study

The findings of this study have both short-run and long-term implications. There are many short-run gains to be had from revitalizing the cocoa sector. Given the limited alternatives immediately available for increasing foreign exchange earnings, Ghana can still benefit from research that lowers the marginal cost of additional cocoa production. To this end, there should be a committed effort on the part of the government to find appropriate productivity-enhancing methods. On this score, Ghana can learn much from Malaysia, which has become a highly successful exporter of cocoa. The estate sector in Malaysia worked hard to develop new high-yielding varieties and to improve horticultural practices, and the government's research and extension services helped to diffuse the innovations to smallholders (World Bank, 1994a).

Additionally, there is a need to provide the necessary resources, which will restore the confidence of cocoa farmers. One remedy might be to reorganize cocoa marketing arrangement so as to raise the farmer's share of the f.o.b. price of cocoa progressively. Another short-term measure would be for the government to reduce the high taxes currently levied on cocoa to at least 60 percent, as obtains in other cocoa producing countries. These short-term measures will have to be complemented with improved distribution of inputs and services. Increases in cocoa prices will not only improve the financial well-being of the farmers but will also enable them to hire labor and purchase inputs to raise output.

In the long run, however, given the uncertainties that characterize cocoa exports and the increasing trade now going on between Ghanaian farmers and the neighboring countries in other commodities, there is a need to expand Ghana's commodity export base to include many of the crops hitherto cultivated for the domestic market. This calls for a comprehensive agricultural plan that sets forth the crops, which have high export potentials and the strategies that would be used to cultivate them. Such an agricultural plan would need to be drawn up by Ghanaian experts since they have a better understanding of local institutions, social, and economic variables as well as local political forces and constraints that influence the implementation and, therefore, the ultimate effectiveness of any reform. In addition, there is the need to create an enabling environment, harness new technologies, and build capacities.

Even though these ideas are not new, they have always not been given the attention, which would translate them into reality. Despite the numerous statements recognizing the importance of the agricultural sector in the development of the country, the successive Ghanaian administrations since independence have only paid lip-service to agricultural reform. One finds that controlled prices, restrictive marketing, disorganized research and weak extension services, poor rural infrastructure and non-effective farmers' associations are still the rule rather than the exception. It is only concerted action that will make it possible for Ghana to achieve and sustain the type of agricultural growth envisaged under the SAP.

6:6 Conclusions

The Cocoa Rehabilitation Project, which aimed at developing the agricultural sector to export commodities, earn foreign exchange, and invest it to achieve economic recovery, has not been able to generate the anticipated results. This was mainly due to the fact that the SAPs agricultural sector policy was based solely on macro-economic variables without taking into account the institutional context within which the policies were implemented. The ERP worked under the assumption that devaluation would automatically lead to increases in farm-gate prices of agricultural exports. It does appear that the authorities failed to realize that commodity taxes are not attractive to a fiscally strapped developing country such as Ghana, and that a larger part of the border price increase from devaluation is not always passed on to the farmers. Again, the program assumed that when there is cutback in the public sector employment, the rural-urban drift will be reversed and thus, create jobs in the rural areas in general and the cocoa farms in particular. Little was thought of the adverse effects of the recommended cutbacks on infrastructure and its effects on the rural communities in which the bulk of the cocoa farmers live. Furthermore, the program failed to take into consideration some of the peculiarities of the rural areas where it is supposed to operate. For instance, the planners failed to consider "acts of nature" like bush-fires, poor soils, and availability of land. It does appear that the general weakness of the cocoa sector adjustment program is that the most important people–the cocoa farmers–were never consulted at any stage prior to the application of the recommendations. It is not an overstatement for one to say that the program was elitist. This might well account for its limited success. The only conclusion that may be drawn from this is that the Bretton Woods institutions, which inspired the cocoa project, never really examined endogenous factors before prescribing it, or they were only concerned with immediate payment of Ghana's debts than to finding a lasting solution to the country's agricultural decline.

In order for the agricultural sector to play the role of poverty alleviation and principal foreign exchange earner, there is a need for the government to embark on a comprehensive reform program, which would boost not only the exports but also the entire agricultural sector. It is time a more holistic approach to planning is introduced. It is my hope that such recommendations would galvanize Ghanaian authorities into seeking better ways to sustain and increase the level of agricultural production by more liberalization of conditions, such as reallocating the bulk of the proceeds from agriculture to the farmers.

Notes

[1] Institute of Statistical, Social and Economic Research of Ghana.
[2] This survey was conducted as part of a study for the Ph.D. dissertation of the author.

References

Armstrong, Robert P. (1996). *Ghana Country Assistance Review: A Study in Development Effectiveness*. A World Bank Operations Evaluation Study. Washington, DC: World Bank.

Asenso-Okyere, W. (1989). *The Effects of Domestic Policies on Exportable Primary Commodities: The Case of Ghana Cocoa*. Research Report Number 1. Morrilton, Arkansas: Winrock International Institute for Agricultural Development.

Asenso-Okyere, W. (1990). 'The Response of Farmers to Ghana's Adjustment Policies' in *Long-Term Perspective Study of Sub-Saharan Africa* vol. 2, Economic and Sectoral Policy Issues. Washington, DC: World Bank.

Bateman, M. J., A. Meeraus, D. M. Newbury, W. A. Okyere and G. T. O'Mara (1990). *Ghana's Cocoa Policy*. Washington, DC: World Bank.

Boahene, Kwasi (1995). *Innovation Adoption as a Socio-Economic Process: The Case of the Ghanaian Cocoa Industry.* Amsterdam: Thesis Publishers.

Ewusi, Kodwo (1989). *The Impact of Structural Adjustment on the Agricultural Sector in Ghana.* Legon, Ghana: Institute of Statistical, Social and Economic Research.

Frimpong-Ansah, J. H. (1991). *The Vampire State in Africa: The Political Economy of Decline in Ghana.* London: James Currey Ltd.

Ghana Cocoa Board (1987). *Hand Book,* 7th edition. Accra.

Ghana Cocoa Board (1988). *23rd Annual Report and Accounts for the Period Ending 30th September.* Accra: Welmax Graphic Arts Ltd.

Ghana Cocoa Board (1999). 'Ghana Cocoa'. <http://www.tcol.co.uk/orgs/ghcoco/ghcobrd.htm> (November, 20).

Government of Ghana (1996). *Report of the Cocoa Task Force on the Study into the External Marketing of Ghana's Cocoa.*

Institute of Statistical, Social and Economic Research of Ghana (ISSER) *The State of the Ghanaian Economy.* Various issues, particularly 1992, 1994 and 1998 issues. Legon: University of Ghana.

Jacobeit, Cord (1991). 'Reviving Cocoa: Policies and Perspectives on Structural Adjustment in Ghana's Key Agricultural Sector' in Donald Rothschild (ed).

Kapur, Ishan, Michael T. Hadjimichael, Paul Hilbers, Jerald Schiff and Philippe Szymczak (1991). *Ghana: Adjustment and Growth, 1983-91.* IMF Occasional Paper No. 86, Washington, DC: International Monetary Fund.

Kofi, Tetteh A. (1993). *Structural Adjustment in Africa A Performance Review of the World Bank Policies under Uncertainty in Commodity Price Trends: The Case of Ghana.* Tokyo, Japan: UN World Institute for Development Economics Research (UNU/WIDER).

Leslie, Zurick (KPMG Peat Marwick Management Consultant) (1994). 'Commercialization of the Ghana Cocoa Board'. A paper delivered at Southampton University in October.

Levi, J. and M. Haviden (1982). *Economics of African Agriculture.* London: Longman.

Mikell, Gwendolyn (1989). *Cocoa and Chaos in Ghana.* New York: Paragon House.

Osei-Tutu, Baffour (1992). *The State and Small-farmer Production: A Study Based on Empirical Research in Pramso, a Small Village in Southern Ghana.* Unpublished Ph.D. Dissertation, Howard University, Washington, DC.

Pearce, Richard (1992). 'Ghana' in *Structural Adjustment and the African Farmer*. Alex Duncan and John Howell (eds.). London: James Currey/ Overseas Development Institute.

Roe, Alan and Hartmut Schneider (1992). *Adjustment and Equity in Ghana*. Paris: OECD Development Center Studies.

Sarpong, Daniel Bruce (1997). *Growth in Ghana: A Macroeconomic Model Simulation Integrating Agriculture*. Brookfield, VT: Ashgate Publishing Company.

Simmons, John (ed.) (1976). *Cocoa Production: Economic and Botanical Perspectives*. New York: Praeger.

Sowa, Nii Kwaku (1993). 'Ghana' in *The Impact of Structural Adjustment on the Population of Africa*. Aderanti Adepoju (ed.). London: James Currey.

World Bank (1993a). *Ghana 2000 and Beyond: Setting the Stage for Accelerated Growth and Poverty Reduction*. Africa Regional Office, West Africa Department. Washington, DC.

World Bank (1993b). *The East Asian Miracle: Economic Growth and Public Policy*. New York University Press.

World Bank (1994). *Adjustment in Africa: Reforms, Results and the Road Ahead*. World Bank Policy Research Report. Oxford: Oxford University Press.

World Bank (1994b). *Adjustment in Africa: Lessons from Case Studies*. Washington, DC: World Bank.

7 Structural Adjustment Programs and Ghana's Mineral Industry

ERIC ASA

7:1 Introduction

The introduction of Structural Adjustment Programs (SAPs) under the umbrella of the IMF has translated into considerable activity in Ghana's economy in general and in the mineral industry in particular. The change in the country's real GDP from 1994 to 1995 was 4.5 percent. Though the GNP per capita in 1995 stood at $360, the ratio of external public debt service to GDP for the same year was 95 percent. Ghana's debt level inflated from US $5,443.3 million in 1993 to US $6,550 million in 1997. The overall growth rate of Ghana's economy was 3.7 percent in 1990-1995. Under SAPs, the Government of Ghana established new institutional and legislative frameworks resulting in the promulgation of the Minerals Commission Law (1986), Minerals and Mining Law (1986), Additional Profits Tax Law (1986), Minerals Royalties Regulations (1986), Small Scale Mining Law (1989), Mercury Regulation (1989) and the Precious Minerals Marketing Corporation Law (1989). The Ministry of Mines and Energy has so far issued 224 exploration and 25 mining licenses to various national and international companies. Primary mineral (gold, diamonds, manganese and bauxite) exports increased from US $243 million in 1990 to US $679 million in 1995, a growth of 180 percent. In 1995, primary mineral products accounted for about 45 percent of Ghana's exports. Gold production amounted to 95 percent of the value of the minerals exported. The mineral industry now directly employs about 20,000 workers in the larger mines. An additional 30,000 are employed in small-scale mining.

The success of SAPs can be directly related to the success of the mineral industry, especially gold mining ventures. Ghana might have unknowingly fallen prey to the resource curse hypothesis. The dismal nature of the gold market and the resulting dwindling profits may stall new investments in gold and consider-

157

ably reduce the taxable income of existing operations. The speed at which the industry has been developed, the low education level of the locals, the lack of substantial human capital and of financial and economic incentives may prevent the country from reaping any significant benefits from increased activity in the mineral industry. Geologically, Ghana has a finite stock of gold-bearing formations and the discovery of new mines will taper off in future. The fundamental error in these strategies is that Ghana may have to have the size, population density and diverse geological endowment of Australia or Canada to succeed in a resource-based economic development strategy. The benefits from the present strategies may be short-lived and may not address the long-term needs of the country. Most of the mining and exploration operations are located in the sensitive rain forest areas, which support the agricultural industry (timber, cocoa, coffee, foodstuff, etc.). The major method of exploitation of the mineral deposits is surface mining, which can affect large tracts of land. Gold mining is therefore likely to crowd out agricultural production and timber harvesting/cultivation. As a natural consequence of increased mining in the absence of sound environmental regulation, pollutants (solid, liquid and gaseous) are discarded into the environment without rendering them innocuous. Over and above this, a new mining force, small-scale mining, which employs open-container mercury amalgamation as part of mineral processing, has generated a lot of activity. These small-scale mines are located in the valleys and near sources of water supply to the agrarian communities. Their indiscriminate mining and processing methods and wanton use of mercury pose a substantial threat to the environment as well as to the health and safety of locals.

The resulting externalities (gaseous effluents, liquid effluents, solid waste, blasting hazards, land degradation and deforestation and the dislocation/relocation of communities) are all too evident in and around the mining communities. The irreversible and accelerating loss of forest reserves, arable lands and water resources together with the degradation of human life may have to be addressed by an appropriate public policy framework. The ripple effect of some of the pollutants may even increase the operating costs; reduce stock levels of mineral resources and the net benefit to society.

In this chapter, the effects of SAPs on the mineral industry and the macroeconomic implications to the country as a whole will be analyzed. The gains, shortfalls and the social costs of SAPs as well as various public policy alternatives will be addressed. A workable solution and a strategy/framework for implementation will then be suggested.

Table 7.1 Mineral Legislation of Ghana

Legislation	Subject	Applicability
Income Tax Ordinance, 1943	Income tax	General Economy
Mineral Duty Ordinance, (1952 (No. 20)	Mineral Duty	Mineral Sector
NLC Decree 78, 1966	Income Tax	General Economy
Mining Regulations, LI 665, 1970	Mining	Mineral Sector
Mineral Duty Amendment Act Act 374, 1971	Mineral Duty	Mineral Sector
Diamonds Decree, NRC Decree 320, 1972	Diamond Mining	Mineral Sector
Mineral Duty Decree, NRC Decree 346, 1975	Mineral Duty	Mineral Sector
Income tax Decree, SMC Decree 5, 1975	Income Tax	General Economy
SMC Decree 48, 1976	Mineral Duty	Mineral Sector
Act 444, 1981	Mineral Duty	Mineral Sector
PNDCL 96, 1984	Income Tax	General Economy
Investment Code, PNDCL 116, 1985	General Economy	Mineral Sector classified as distinct
Additional Profits Tax Law, PNDCL 122, 1985	Windfall Profits Taxation	Mineral Sector
Minerals and Mining Law PNDCL 153, 1986	Mining regulation	Mineral Sector
Minerals Commission Law, PNDCL 154, 1986	Establishes Minerals commission	Mineral Sector
Minerals (Royalties) Regulation LI 1349, 1987	Royalties	Mineral Sector
Mercury Regulation, PNDCL 217, 1989	Legalization of Mercury	Mineral Sector
Small-Scale Mining Law, PNDCL 218, 1989	Legalization of Small-scale Mining	Mineral Sector
Diamonds (Amendment) Law, PNDCL 216, 1989	Diamond mining	Mineral Sector
Precious Minerals Marketing Corporation,k PNDCL 219, 1989	Establishes PMMC as market for small-scale miners.	Mineral Sector
Minerals Commission Act, Act 450, 1993	Replaces Minerals commission Law, PNDCL 154	Mineral Sector
Minerals and Mining Amendment Act, 1994	Reduced income tax from 45% to 35% economy-wide as in other industries	Mineral Sector

7:2 Overview of Ghana's Mineral Industry

Since the early 1900s, over 78 Acts, Ordinances, Decrees, Codes and Laws concerning Ghana's mineral industry have been passed. Prominent among these are the administration of Lands Act; Act 123 (1962); the Minerals Act; Act 126 (1962); Minerals Commission Law; PNDC Law 154 (1986); Minerals and Mining Law; PNDC Law 153 (1986); Mineral (Royalties) Regulations; LI 1349 (1987); and the Additional Profits Tax Law; PNDC Law 122 (1985).

Most of the legislation passed prior to 1943 was aimed primarily at controlling the technical and organizational aspects of mineral development. Ordinances, like the Mercury Ordinance, Explosive Ordinance (1928), and Mine Health Ordinance (1928) are examples of such laws. Legislation affecting the fiscal aspects of the mineral industry came into force after 1943. The Mineral Duty Ordinance of 1952 and its two amendments–Mineral Duty (Amendment) Act 167 (1963) and Act 374 (1971)–were largely regarded as the only legislation specific to the mining industry. Table 7:1 summarizes legislation that has directly or indirectly affected the mineral industry. As is evident from the table, various governments attempted to capture the perceived economic rents from mining to finance budget deficits. Some of the legislation was aimed at specific mineral operations or companies. Prior to 1985, the minerals sector had been grouped with other industries in the determination and application of fiscal instruments and economic incentives.

Table 7.2 shows the important fiscal instruments of the past–S.M.C.D.5 of 1975, Investment Code (Act 437 of 1981), P.N.D.C. Law 96 (1984), P.N.D.C. Law 116, (Investment Code, 1985)–as compared to present legislation. S.M.C.D.5, a general income tax law enacted in 1975, integrated most of the prior fiscal instruments affecting the mineral industry into a single law.

The financial/economic burden resulting from SMCD.5, coupled with the political instability of the country probably crippled an already struggling minerals industry. No new major investments were made in Ghana's mineral sector between 1938 and 1986.

Two new laws, the Investment Code of 1981 and P.N.D.C.L. 96 of 1984, were passed to recognize the uniqueness of the mineral industry, as well as eliminate the economic burden of past fiscal instruments on the industry. Thus the mineral sector was formally recognized as distinct with special characteristics and growth needs. As a strategic niche, the Minerals and Mining Law (P.N.D.C. Law 153. 1986), Additional Profits Tax Law (P.N.D.C Law 122. 1985), and the Mineral (Royalties) Regulations of 1987 (L.I. 1349) marked the debut of a new wave of favorable fiscal instruments, which have attracted considerable investments in the mineral industry.

Table 7.2 Fiscal Instruments Applicable to the Mineral Sector

ITEM	SMCD 5, 1975	ACT 437 I.C. 1981	PNDCL 153 M&M LAW, 1993	M&M Amendment Act, 1994
1. INCOME TAX	50-55%	45%	45%	35%
2. ALLOWANCES:				
Initial Capital Allowance	20%	20%	75%	
Subsequent Cap. Allowance	15%		50%	
Investment Allowance	5%		5%	
R&D Allowance		25%		
3. ROYALTY	6%	2-6%	3-12%	
4. MINIMUM TURNOVER TAX	2.5%	2.5%		
5. MINERAL DUTY	5-10%	5-10%		
6. IMPORT DUTY	5-35%	5-35%		
7. FOREX TAX	33-75%	33-75%		
8. IMPORT LEVY	10%	10%		
9. GOLD EXPORT LEVY	C3/oz oz . 100,000 oz	C3/oz >100,000		
N.A.- Not Applicable.				
I.C.- Investment Code.				
SMCD- Supreme Military Council				
M&M Law- Minerals and Mining Law				
C- Ghana's currency, the cedi				

7.2.1 Minerals and Mining Law, P.N.D.C Law 153, 1986

The eleven-part law enacted by the Provisional National Defense Council (P.N.D.C) on July 4, 1986, stands out as the single most comprehensive document directly related to Ghana's mineral industry. A summary of the main features of the law is as follows:

• All minerals are regarded as the property of the Republic and, therefore, vested in the President/ruling body for and on behalf of the people of Ghana. The Government has the right of pre-emption of all minerals obtained in Ghana and mineral products derived from refining or treatment of such minerals.

The Government also has the right to acquire a 10 percent interest in the rights and obligations of mineral operations that operate free of charge. Government can acquire a further 20 percent right in economic mineral operations (45 percent in the case of salt) at a fair market value.

• The four types of mineral rights/licenses, which can be granted by the Secretary of Mines and Energy, are reconnaissance license, prospecting license, mining lease and restricted mining lease. A summary of the characteristics of these mining rights is presented in Table 7.3.

• The holder of a mining lease is entitled to a number of capital allowances, benefits and financial responsibilities/taxes. These range from accelerated depreciation of capital investments, exemption from the payment of customs import duties on mine plant to capitalization of all exploration and pre-production expenditure upon the acquisition of a mining lease. Disputes may be subjected to international arbitration in accordance with the procedures of the U.N. Commission on International Trade Law.

Table 7.3 Types of Mineral Rights

Type of Mineral License	Objective	Areal Extent	Duration
Reconnaissance	Regional mapping and exploration incl. geophysical, geochemical but not drilling methods	Unrestricted	12 months with reduction in size
Prospecting	Drilling and other methods can be used to search for and determine the extent and value of mineralization	150 sq. km.	Renewable up to 2 years at a time with reduction in initial area by half
Restricted Mining Lease	Building materials for local consumption		
Mining Lease	Exploitation of econoic mineral deposits	Maximum of 50 sq. km per lease and a total of 150 sq. km per company	30 years

7:2.2 Additional Profits Tax (APT) Law, P.N.D.C. Law 122, 1985

An additional profits tax, which is a cash flow tax, shall be paid on the carry-forward cash balance only after a mineral project has achieved a specified/threshold rate of return. The threshold rate of return is set at 35 percent for new projects, while the level for existing operations is 17 percent. A number of accounting items are added to and/or deducted from taxable income in order to arrive at the carry-forward cash balance.

7:2.3 Royalty/Regulations, L.I. 1349, 1987

A variable royalty payment is levied on the profitability of mineral operations. A profitability index or operating ratio, which varies from a minimum of 30 percent to a maximum of 70 percent is used as the basis of the royalty calculation. The operating ratio is the ratio of the "operating margin" to the value of the quantity of minerals per annum. The value of the operating ratio establishes the applicable tax rate (see the royalty schedule shown in Table 7.4):

The operating ratio is defined as,

$$OR = \frac{VMW - OPC}{VWM} \times \frac{100\%}{1}$$

Where VWM is the value of mineral won and OPC is the operating costs.

In marginal projects, where the operating ratio could fall below 30 percent, the difference between the actual operating cost and the operating cost that would have made the operating ratio exactly equal to 30 percent is added to the operating cost of the following year for the purpose of calculating the operating ratio. In such a situation, the carried forward difference should not exceed the permissible capital allowance for the accounting year.

7:2.4 Small-Scale Mining Law

Table 7.4

Value of Operating Ration (OR)	Rate of Royalty
OR< 30%	3%
30%<OR>70%	3% + 0.225*(OR-30%)
OR > 70%	12%

Policy in the small-scale mining sector was aimed at market liberalization and the creation of a simple licencing system. Eight small-scale mining centers have been created in the major production areas. Mining engineers who are well equipped to monitor the activities of the miners manage these centers.

Under the small-scale mining law, Ghanaian small businesses and individuals are given licenses to mine gold from the various areas of the auriferous belt, which are not licensed to large-scale mining companies. Thus mining has been allowed in areas, which were hitherto considered illegal to win gold from in the past. The mining methods employed range from the digging of small pits to the washing of river gravel. Most of the gold accessible to small-scale mining is free milling and require gravity concentration methods to concentrate the gold prior to amalgamation.

The small-scale mining industry employed about 30,000 people in 1995. This sector produced about 3.7 kg. of gold and 443,244 carats of diamonds in 1996. Precious Minerals Marketing Corporation and Miramex buy the products of the small-scale miners.

7:2.5 Mercury Law

With the advent of small-scale mining, a new law called the Mercury law was promulgated to legalize the use of mercury. Gold winning by small-scale miners involves the amalgamation of the final concentrate with mercury. In this crude process, the gold-mercury amalgam is separated from the gangue material and the amalgam is squeezed of excess mercury. The amalgam is then heated in an open flame to vaporize the mercury. The entire process is done with very little protection, and the mercury vapour can be directly inhaled into the lungs. The part that escapes is lost to the environment and ends up in the water system or the food chain.

7:2.6 Institutional Framework

The various institutions directly related to the mineral industry are briefly described in the following paragraph.
• Ministry of Mines and Energy: the ministry has overall responsibility of the mineral and energy industries.
• Chamber of Mines: it is an independent organization representing the mineral industry. It consists of representatives from the various mining companies.
• Minerals Commission: the Commission acts as the mineral policy making and management arm of the government. On one hand, it advises the government on all mineral-related matters and serves as the initial contact and information source for prospective investors on the other.
• Mines Department: this department administers certification (trade licenses), occupational health and safety issues. It also maintains a record of mining activities and operations.

• Geological Survey Department: this department is involved in geological studies and mapping as well as production of the requisite maps. It is responsible for maintaining all geological records.

• Land Commission: it is responsible for the legal examination of new applications and keeping of records of leases, licenses and land titles.

• Environmental Protection Agency: it has overseeing capabilities of mineral-related environmental issues.

7:3 Structural Adjustment Programs and the Mining Industry

Under the auspices of the World Bank, structural adjustment programs have been implemented in Ghana, since 1983. The 3-phase program was divided into stabilization, rehabilitation, and liberalization and growth. The stabilization part (1983-1986) involved arresting and reversing the decline in production, controlling inflation, reforming prices and restoration of production incentives and confidence in Ghana, improvement in the foreign exchange availability and allocation mechanisms, and mobilization of resources to improve the living standards of Ghanaians. The rehabilitation phase (1987-1989) involved resolving balance of payment issues, improvement in public sector management, stimulation in investments and savings, and achievement of a 5 percent growth rate. The liberalization and growth phase (1989-1993) involved reduction in inflation rate, promotion of private investments, promotion of growth in the agriculture sector, and generation of balance-of-payment surplus. The planned growth rate was 5 percent (or about 2 percent per capita). Ghana is now believed to be in the enhanced mode of its structural adjustment or economic recovery program.

The new institutional and regulatory framework developed under the structural adjustment programs have translated into considerable economic activity in Ghana's mineral industry. The Ministry of Mines and Energy has so far issued 224 exploration (reconnaissance and prospecting) licenses and 25 mining leases have been granted to various international and national companies.

Thus, the new mining policies and legislative framework have translated into considerable interest in the mineral resources of Ghana, especially gold. Investments in the industry amounted to about US $460 million by 1990 (Barning, 1990). Present estimates indicate that the mineral industry now employs about 20,000 workers and has attracted about $1 billion in new investments in the form of exploration expenditure, new mining operations and refurbishing of existing operations.

7:3.1 Mineral Production

Under the liberalization and privatization programs, Ghana Government has reduced its share in most of the mining operations in the country.

Mineral production in Ghana has increased considerably since the inception of the structural adjustment program. In fact the structural adjustment program can be seen as the mineral sector development program. Both new and existing operations have accounted for growth in mineral production. Companies like Ashanti Goldfields Corporation, Ghana Bauxite Company, Ghana National Manganese Corporation, which were in operation prior to 1983 have substantially increased their production levels. Since the inception of the structural adjustment program, all the minerals produced in Ghana have at least doubled in output. Production levels increased from 283,593 oz. of gold in 1983 to 1,747,018 oz. in 1997; from 338,769 carats of diamonds in 1983 to 562,651 carats in 1997; from 70,235 metric tons bauxite in 1983 to 536,732 metric tons in 1997 and from 169,840 metric tons of manganese in 1983 to 355,232 metric tons in 1997.

As is apparent from Figure 7.1, the most dramatic increase is in gold production. Gold production has increased by about 500 percent under the structural adjustment program (1983 to 1997). In 1996, a total of 4 major operations accounted for a greater proportion of the gold produced in Ghana. Ashanti Goldfields Corporation–Obuasi (54.2 percent), Teberebie Goldfields Corporation (13 percent), Ghanaian Australian Goldfields (7.4 percent), and Gencor Limited–Bogosu (6.7 percent) accounted for about 81 percent of Ghana's production in 1996 (Coakley, 1996). Gold production will continue to grow as new operations commence production. Even if only 10 percent of the existing exploration companies are successful, gold production in Ghana could double again in the near future. This trend may continue into the early part of this century. However this may not yield much at the present metal prices.

7:3.2 Metal Prices

Though Ghana's mineral output has been increasing under the structural adjustment program, prices have been fallen concurrently. Figure 7.2 depicts the gloomy picture of falling metal prices. Manganese has experienced the most dramatic reduction in price. Manganese prices fell from $72/metric ton in 1991 to $32.5/mt in 1997. It is evident from Figure 7.2 that prices of bauxite and diamonds have also been falling, gently but steadily over the years.

Some of the factors contributing to low metal prices are the economic woes of the Asian tigers, better price and quality of substitutes, new industrial manufacturing technologies, supply of metals, advances in technology of production of metals, the fall of the iron curtain, trade liberalization and good investment

Figure 7.1 Mineral Production in Ghana (1990-1997)

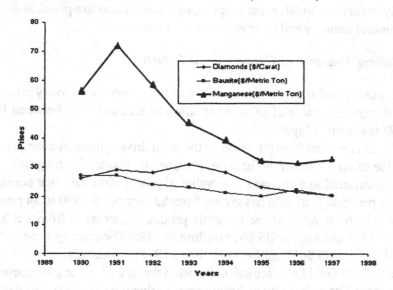

Figure 7.2 Prices of Bauxite, Manganese and Diamonds

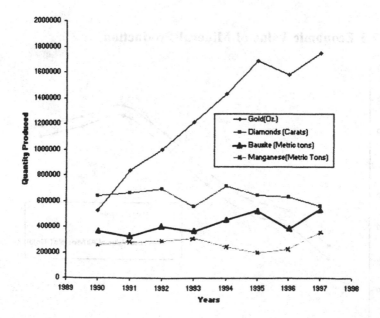

climate in most developing economies including the Soviet block. Metal prices are likely to continue to fall in the long run, exposing Ghana to a possible decline in the mineral industry and in its economy as a whole.

7:3.3 Salient Features of Ghana's Gold Industry

Ghana's mineral industry is at the forefront of the economic recovery program. The total economic value of all the minerals produced in Ghana between 1990 and 1997 is shown in Figure 7.3.

Gold, as is evident from the graph, is the main driving force. A change in the total value of all the minerals (bauxite, manganese, diamonds and gold) is directly proportional to a change in the value of gold within the same period of time. The percentage of gold has grown from 83 percent in 1990 to 94 percent in 1997. The total value of the minerals produced increased from US $243 million in 1990 to a high of US $679 million in 1995. The country exported US $642 million and US $613 million in 1996 and 1997 respectively.

The Ashanti Gold fields are still considered the largest single gold operation in the country. The probability of discovering another mine of the size of Ashanti is fading. Most of the new mining operations are medium scale mines. Ashanti Goldfields itself is beginning to feel the effects of falling prices in the commodi-

Figure 7.3 Economic Value of Mineral Production

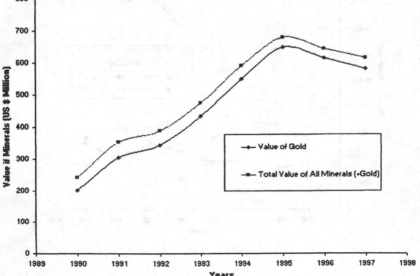

ties market. Depending on the price of gold, most of the new mines may run out of economic deposits in 10-15 years. This is an indication that the new economic development engine (mining) may only be able to support the country in the short to medium term.

The presence of good economic gold reserves is fundamental to the success of Ghana's gold industry. However, many of the new operations are small to medium-scale operations, possess limited reserves, and employ leaching technology to recover gold from of low-to-medium grade oxidized ore. The technologies used by the new companies are surface mining and leaching or dredging, for example, Teberebie Goldfields (heap-leaching), Goldfields of South Africa/Tarkwa Mine (heap-leaching), Obenemasi Gold Mines Ltd. (mobile carbon-in-pulp), Bonte Gold Mines (small alluvial operation), Abosso Goldfields Ltd. (carbon-in-leach) gold recovery systems. The average mine life is from 7 to 10 years. This involves low initial investments, shorter breakeven time and higher future costs. As the oxidized ore is depleted, new investments may be required vis-à-vis increasing cost of mining and processing complex ores, especially for the mines within the Birrimian system. Those operations in the Tarkwaian may still be able to get away with lower processing costs. This together with failing gold prices may decrease new investments in gold and slow down Ghana's economic engine.

The availability of large, efficient and low cost capital is necessary since mining ventures involve high risks. Ghana's mineral industry was developed ahead of the financial sector reforms making it impossible for Ghanaians to take full advantage of the present boom. As with most World Bank sponsored development strategies, the boom in Ghana's mineral industry was financed by money from the developed world at high rates of interests due to the risk factors associated with investments in Africa. The direct benefits to the country are therefore limited to 10 percent retained interest in the mining ventures, employment and the taxation collectable. As argued by Asa (1994), the additional profits tax may never be realized in any of the mining ventures. Ghana may therefore have to look elsewhere for supplementary income.

The swiftness with which the mineral sector development program was launched, together with the new technologies being applied, the requirement for skilled manpower was filled by the importation of skilled labor. Ghanaians would otherwise have filled these jobs. Operating costs at one of the youngest gold operations, Teberebie Goldfields Ltd., were affected by lack of skilled labour and by high equipment downtimes (Coakley, 1996).

7:4 Energy Crisis

The mineral industry is energy-intensive and consumes high levels of electrical energy and oil supplies. Mines have been known to be the largest customers of power plants since their processes involve the handling of high volumes of input material. From Figure 7.4, it is apparent that the quantity of oil imported has increased from US $199 million in 1990 to US $253 million in 1996. This increase is due to both growth in the quantity of oil imports, and fluctuations in the price of oil.

As depicted in Figure 7.5, the government of Ghana increased the price of a litre of premium gasoline by 1500 percent between April 1983 (Cedis 5.5/litre) and March 1990 (Cedis 88.1/litre). The price was further increased by 784 percent between March 1990 (Cedis 88.1/litre) and February 1997(Cedis 778.9/litre).

Electricity generation rose to a high of 6603 billion KWH in 1992 before dropping to 6134 billion KWH in 1995 due to fluctuations in rainfall patterns. Electricity generation increased again to 6885 billion KWH in 1997. Consumption of electricity has been growing, reaching 7342 billion KWH in 1997. The mines quota has more than doubled from 315 billion KWH in 1990 to 748 billion KWH in 1997. Most of the increases have been taken from VALCO and Exports. The exports quota was reduced to zero in 1996. A new generating capacity is under development in the Western Region.

Figure 7.4 Total Value of Oil Imports (1989 to 1996)

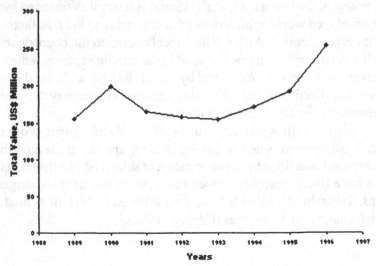

Figure 7.5 Price of 1 Litre of Premium Gasoline

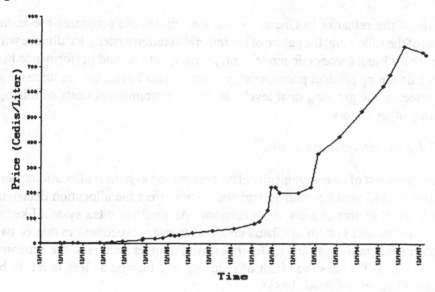

Figure 7.6 Electricity Generation and Consumption

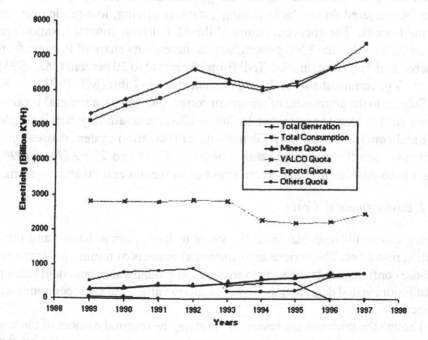

7:5 Setbacks to the Structural Adjustment Program

Some of the setbacks to Ghana's structural adjustment programs are in the form of the following: the nature of the mineral taxation system; its alliance with the World Bank (economic model, competency, actions and performance history); declining product prices; energy crisis; Ghana's size; the resource curse hypothesis; the growing debt levels; and the environmental costs of mining, among other factors.

7:5.1 Mineral Taxation System

The litmus test of an economically efficient taxation system is allocational neutrality. An efficient taxation system should not affect the allocation decisions made on all sectors of a country's economy. An inefficient tax system like the mineral taxation system of Ghana encourages over-investment in one or two sectors (mineral sector) and under-investment in other sectors of the economy (Asa, 1994). This has the effect of reducing real income as less is left to be taxed (Garnaut and Ross, 1983).

Like Papua New Guinea (PNG), Ghana's mineral taxation system favors the investor if the additional profits tax (APT) is never triggered. If the APT comes into play, the system favors the government. In reality the APT may never be triggered due to "gold plating", transfer pricing, low-grade ores and low gold prices. The apparent failure of the APT-driven mineral taxation system might have led the PNG government to increase its share of Porgera from 10 percent to 25 percent and Ok Tedi from 20 percent to 30 percent (MJ, 1993). The PNG government also took a 50 percent share in Lihir (MJ, 1993).

Subject to the attainment of minimum compliance levels, a mineral taxation system must be easy to administer. In fact an effective taxation system is simple, clear and concise (Asa, 1994). Ghana's mineral taxation system causes some problems when the operating ratio is between 27.99 and 29.99 (Asa, 1994). This is probably due to the complex structure of the mineral taxation system.

7:5.2 Environmental Costs

Mining can result in damages to the value of land, human health and other existing resources. The various environmental impacts of mining in Ghana are land-use conflicts, contamination to surface and groundwater, land degradation, air pollution, visual degradation, loss of property values and socioeconomic imbalances.

Though the minerals are vested in the state, the original owners of the land are supposed to retain the surface rights. These owners can lose the surface rights as in the case of Ashanti Goldfields, which has been holding its mineral rights for more than a century. In such a situation, the land rights are lost permanently and the effects are inter-generational. The issues of land ownership, control, and property damages led to the closure of Bougainville in PNG (AMM, 1994) and have resulted in considerable turmoil in Ghana (Asa, 1999a).

Most of the new ventures are surface mining and leaching operations. They can affect large tracts of land and water resources. There have been situations where entire villages were relocated to make room for mining. Dust from the mining operations and gaseous emissions from roasting of pyritic ores are the most prominent air pollutants.

Due to the absence of appropriate mineral beneficiation technology, most of the small-scale miners have resorted to mercury amalgamation of the gold concentrate. This process releases a large quantity of mercury into the environment. Mercury, with its detrimental effects to humans, can remain in the environment for decades.

Aside from the funding problems faced by the Environmental Protection Council (EPC), an effective environmental action plan for Ghana can only be drawn after a study to measure willingness-to-pay (WTP) for the cleaning of the environment or willingness-to-accept (WTA) some level of pollution as a consequence of development (Asa, 1999a). The implementation of the wrong policy can have the unintended effects.

7:5.3 Ghana's Size/Resource Base/Geological Endowment

The present growth in Ghana is quite deceptive. In a resource-based economic development, deteriorating primary product prices can erode a country's economic base or competitiveness. The prices of all the primary mineral products of Ghana, manganese, bauxite, and gold have been falling and so marginal benefits of new investments are deteriorating. The situation may not be any different for the agricultural produce. In such a situation the country may be better off directing new investments into other sectors of the economy.

Therefore, Ghana may not be able to pay its debts in future under the present economic development plan and product prices. This will be especially true if production gains continue to be eroded by failing product prices.

7:5.4 Ghana's Debt Levels

The injection of funds from the multilateral and bilateral sources totaled about $8 billion in the first 7 years (1983-1990) of the SAP. Ghana received about $1.75 billion from the World Bank within the same period. The gross domestic product grew at an average of about 3.88 percent within this period. Ghana's debt, however, increased from $5.4 billion in 1993 to $6.5 billion in 1997. In spite of the injection of billions of dollars into Ghana's economy under the auspices of the structural adjustment program, only moderate growth has been experienced. What makes one suspicious are the recent actions and performance history of the World Bank. While the World Bank seems to be promoting investments in Ghana, it is quietly withdrawing support for Ghana's economic development plan.

The long and medium term debts owed to bilateral and multilateral organizations increased by 37 percent from US $4.5 billion in 1993 to US $6.2 billion in 1997. Within the same time the IMF debts decreased by 52 percent, from US $738 million in 1993 to US $354 million in 1997. Thus the IMF debt is not only a fraction of the total debt, but it has been declining whereas the long and medium term debts owed to bilateral and multilateral organizations are rising. In spite of this, payments to the IMF are also rising vis-à-vis a drop in short-term debts by 80 percent within the same period. This indicates that most of the IMF debt may be medium to short term. The IMF is trying to get its pound of flesh before Ghana's economy grinds to a halt. Their actions may be tantamount to insider trading. The IMF's diagnosis, prognosis and economic panacea have failed in Mexico and East Asia. The bank was still showering praises on these countries just prior to the unprecedented financial collapse. The world may no longer be able to rely on the IMF for the economic emancipation of the developing world. The economic model employed by the IMF is alien to the developing world and dates back to Adam Smith's invisible hand of God. It may not be the right remedy for present day problems of economic policy in national development.

7:5.5 *Marginalization of Africa*

Africa may have been reduced to a source of cheap raw materials for the rest of the world. Ghana, like most of Africa, is experiencing dwindling natural resources, and higher social and environmental costs. The price disparity between primary commodities exported from Ghana and finished products imported from the developed world may constrain Ghana's development.

7:5.6 *Resource Curse Hypothesis*

Evidence seems to suggest that countries like Ghana with rich natural resources tend to rely on resource-based economic development plans, whereas countries with a poor resource base like Japan, rely on manufacturing and human capital development. Work by Mayer (1997) suggests that countries with a rich natural resource-base and poor human capital endowments cannot diversify their economy. However, countries having a rich human capital and a poor natural resource endowment, as well as an open trade policy have successfully diversified their exports (Mayer, 1997). Ghana may therefore have fallen prey to the resource curse hypothesis.

7:5.7 Horizontal Export Diversification

Horizontal diversification is the situation whereby other raw materials that can be produced under similar technological conditions are substituted for the production of the main product. The tricky aspect of the SAPs is that Ghana is also dependent on other primary products (bauxite, manganese, diamonds, pineapples, etc) for export diversification. These industries are likely to fall victim to the same declining prices on the world market.

7:5.8 Ghana's Size and Resource Potential

There is no gainsaying the fact that the success of a resource-based development is dependent on the size of the country, diversity of resources, access and business environment. Some researchers have suggested that the total land area of a country can be used as a proxy of its natural resource endowment. This becomes even more pronounced in the case of Ghana, where the timber, mineral, water and fertile soil resources are mostly in the western part of the country. An increase in the production of gold may crowd out the production of cocoa and the availability of good drinking water.

7:5.9 Lack of Accountability, Competitiveness and Reward System

Like most African countries, the main driving engine in Ghana's economy is the government. Business in Ghana is done on the basis of party, friends and clan. This may have to give way to competitiveness and a good reward system if Ghana is to develop.

7:5.10 Dutch Disease and De-industrialization

This phenomenon re-emphasizes the problems associated with the development of exportable natural resources (Lindert, 1991). The manufacturing of traded goods declined as the Netherlands developed its North Sea natural gas fields. Natural resource development bids land, labor and capital away from the manufacturing sector, eroding profits and manufacturing capacity in other traded goods. This may partly explain the spiraling growth in the gold industry as compared to the stunted growth in other sectors of Ghana's economy.

Ghana also has a problem with immiserizing growth. In this situation, growth in a country's productive capacity in a good already exported can lower world market prices and hurt the growth efforts (Lindert, 1991). Though this may not explain the present low gold prices, it can happen in future especially as various gold projects come on stream throughout the world. This increased production coupled with sustained gold trade by some of the central banks can further lower gold prices and trigger immerising growth in Ghana.

7:6 A New Economic Development Paradigm

It is obvious from the preceding analysis that Ghana cannot develop under the World Bank's structural adjustment program. The failure of the present program is due to the absence of a viable economic development strategy/program. The new economic development paradigm calls for a shift from the resources-based to knowledge-based economic development program.

Under the new economic development strategy, an Economic Development Institute (EDI) will be set up within the UNDP as a new organizational wing for developing, monitoring and evaluating the effectiveness of development programs in the developing world. This is in view of the economic and political neutrality of the UNDP as well as its experience in development programs worldwide. Under no circumstances should the World Bank be allowed to single-handedly drive development programs. The World Bank, bilateral and multilateral organizations and companies are to be encouraged to invest in developing economies with the use of various incentives. The EDI will also seek to introduce competitiveness and various market indices to value a country's suitability for investment capital. Some of the indices should be efficiency of its government, laws protecting human rights, level of abuses of state funds by rulers, fiscal regime, banking system, university systems, level of research, development of human capital, etc. In the final analysis, bonds, bills and other financial instruments from well-rated countries can be traded in the market. Individuals, companies, and banks can purchase these development financial products (DFPs) to support economic development in developing countries. This will ensure accountability and efficiency in the distribution of development capital.

7:6.1 *Strategic Economic Development Plan (SEDEP)*

The first step is for the Government of Ghana to set up a new specialized industrial category within the Ministry of Science and Technology called Knowledge-Based Development Commission (KBDEC). KBDEC is to be an advisory, regulatory and promoting organization and should advise the Minister of Science and Technology on all matters relating to a knowledge-based industry. Various laws should be passed to set it up and to support its activities. Other aspects of the KBDEC have been developed under a separate document (Asa, 1999b). Among other things KBDEC will develop the requisite economic incentives and packages to attract new investments into knowledge-based industries. KBDEC should comprise well trained and groomed learned visionaries who can help Ghana develop its technological base. The areas of concentration should include computer hardware, computer software, robotics and machine learning, satellite technology, consumer electronics, and medical technology, systems, hardware and software.

Under the strategic economic development plan (SEDEP), it is recommended that part of the loans from the IMF, bilateral and multilateral organizations will have to be channeled through knowledge-based companies like Intel, IBM, Microsoft and Sun Systems. These companies are expected to use the money to invest directly in the knowledge-based industry of Ghana.

7:6.2 *Knowledge-Based Development*

If Ghana is to really develop it should embark on an aggressive knowledge-based economic development plan. Some of the highlights of the plan are as follows:
• Development of human capital/resources (5-year, 10-year, 50-year plans).
• Loans and tax incentives to be given to high tech companies like IBM, Intel, and Microsoft to do business in Ghana.
• Establishment of knowledge-based villages with appropriate tax incentives. As a standard, these villages are to be equipped with all the paraphernalia required by a high-tech company in Silicon Valley. They are to be well protected and promoted.
• Establishment of knowledge-based research projects in universities with adequate funding by government and/or international organizations. These projects are initially limited to the departments of physics, biology, chemistry, mathematics, computer science, medical sciences and engineering in the universities. These are the seven core areas. As the economic pie gets larger it can be spread to other sections.
• Strengthening of the entire educational system with the promotion of "Science for Economic Development Now" (SCIDEN) programs.

7:7 Summary

Under the umbrella of the World Bank, Ghana commenced a structural adjustment program in 1983. This led to considerable economic activity in Ghana's economy in general and in the mineral industry in particular. Though the GDP was 4.5 percent and the GNP per capita in 1995 stood at $360, the ratio of external public debt service to GDP for the same year was 95 percent. Ghana's debt level increased from US $5,444 million in 1993 to US $6,550 million in 1997. The overall growth rate of Ghana was 3.7 percent in 1990-1995. The IMF is reducing its involvement in Ghana and this may be a warning sign of the possible run down of Ghana's economy.

The success or failure of SAPs can be directly related to the success of the mineral industry, especially gold mining ventures. New institutional and legislative frameworks were established in the mineral industry; resulting in the promulgation of the Minerals Commission Law (1986), Minerals and Mining Law (1986), Additional Profits Tax Law (1986), Minerals Royalties Regulations (1986), Small Scale Mining Law (1989), Mercury Regulation (1989) and the Precious Minerals Marketing Corporation Law (1989). The various organizations responsible for the mineral industry are Ministry of Mines and Energy, Chamber of Mines, Minerals Commission, Mines Department, Geological Survey Department, Land Commission and Environmental Protection Council. The Ministry of Mines and Energy has so far issued 224 exploration and 25 mining licenses to various national and international companies. The total value of mineral exports increased from US $243 million in 1990 to US $679 million in 1995, a growth of 180 percent. The small-scale mining sector produced 3.7kg. of gold and 443,244 carats of diamonds in 1996. In 1995, primary mineral products accounted for about 45 percent of Ghana's exports. Gold production amounted to 95 percent of the value of the minerals exported. The mineral industry now directly employs about 20,000 workers in the larger mines and 30,000 workers in small-scale mining. Like the other mineral products exported by Ghana, gold prices have been falling. The dismal nature of the gold market and the resulting dwindling profits to mining companies may stall new investments in gold and considerably reduce the taxable income of existing operations. Geologically, Ghana has a finite stock of gold-bearing formations and the discovery of new mines will taper off in future. Since the mineral industry is energy-intensive (and inefficient), Ghana's energy consumption is likely to skyrocket with increase in production of minerals.

The fundamental error in all of these is that Ghana will have to possess the size, population density and diverse geological endowment of Australia or Canada to succeed in a resource-based economic development strategy. The development of a single product like gold or cocoa may lead to the Dutch disease and/or immiserizing growth. The benefits from the present strategies may therefore be short-lived and may not address the long-term needs of the country.

Most of the mining and exploration operations are located in the sensitive rain forest areas, which support the agricultural industry (timber, cocoa, coffee, foodstuff, etc.). The major method of exploitation of the mineral deposits is surface mining and heap leaching, which can affect large tracts of land. Gold mining is likely to crowd out agricultural production and timber harvesting/cultivation. Externalities (gaseous effluents, liquid effluents, solid waste, blasting hazards, land degradation and deforestation and the dislocation/relocation of communities) are all too evident in and around the mining communities. Increase in mining activities has resulted in environmental pollution. Small-scale miners are located in the valleys and sources of water supply to the agrarian communities. Their indiscriminate mining and processing methods and wanton use of mercury offer a substantial threat to the environment as well as health and safety of locals.

The IMF may be considering itself as a bank and not a development organization and has not performed well in other countries. Ghana and the rest of the developing world may therefore have to rely on a new economic development paradigm. A proposed Economic Development Institute (EDI) under the auspices of the UNDP will foresee economic development issues in the developing world. Among other things, EDI will seek to introduce competitiveness and various market indices in order to value a country's suitability for investment capital. In the final analysis, bonds and other financial instruments from well-rated countries can be traded in the market. Individuals, companies, and banks can purchase these development financial products (DFPs) to support economic development in developing countries. This will ensure accountability and efficiency in the distribution of development capital.

Under the proposed strategic economic development plan (SEDEP), resource-based economic development strategy will have to give way a knowledge-based strategy. Instead of all the money been given directly to the governments of developing countries, part of the money has to be channeled through knowledge-based companies like IBM, SUN, Microsoft, HP, etc for direct investment in Africa. Ghana may have to set up a Knowledge-Based Development Commission (KBDEC) in the Ministry of Science and Technology to advise, regulate and promote her knowledge-based industry.

A knowledge-based development plan involving the introduction of science for economic development now (SCIDEN) program, creation of standard Silicon Valley villages and various incentives will help propel Ghana's economic development into the 21st century.

References

Asa, Eric (1992) 'Gold Smuggling in Ghana'. Unpublished Report, Colorado School of Mines, Golden, Colorado.

Asa, Eric, (1994) 'Comparing Alternative Mineral Taxation System for Gold Mining in Ghana'. Thesis, Colorado School of Mines, Golden, Colorado.

Asa, Eric (1999a) 'Effects of Mining on Ghana's Natural Environment'. International Week Seminar. University of Alberta, Edmonton.

Asa, Eric (1999b) 'A New Economic Development Paradigm for Africa: the Case of Ghana'. Working Paper. University of Alberta, Edmonton, Alberta.

Barning, Kwasi (1990) 'A Review of Mineral Exploration Activities in Ghana, 1984-1990. Symposium on Gold Exploration in Tropical Rainforest Belts of Southern Ghana'. Mineral Commission, Accra, Ghana.

Coakley, G.J. (1996) 'The Mineral Industry of Ghana. U.S. Geological Survey – Minerals Information – 1996'.

Garnaut, R. and Ross, A. C. (1983) *Taxation of Mineral Rent*. Oxford: Clarendon Press.

Gold Institute (1999). Internet Site.

Green, T. (1982) *The New World of Gold*. New York: Walker and Company pp. 167-177.

Greer, Jed (1993) 'The Price of Gold: Environmental Costs of the New Gold Rush'. *The Ecologist*, vol. 23 no.3 May/June pp. 91-96.

Kaufmann, T. D. and Hammond, D. R. (1991) 'Selling Gold Forward Increases Profits for Gold Producers'. Mining Engineering, November 1991, Society of Mining Engineers, Littleton, Colorado, pp. 1317-1319.

Lindert, P. H. (1991) *International Economics*. 9th Edition. Richard D. Irwin Inc. Homewood Illinois U.S.A.

Mayer, J. (1997) 'Is Having a Rich Natural Resource Endowment Detrimental to Export Diversification?' Geneva: UNCTAD Report No. 124, 46 pages.

Mining Journal(MJ) (August 13, 1993) 'P.N.G. Looks to Control Lihir' vol. 321, no. 8237, Mining Journal Limited, London.

Mining Journal(MJ) (October 15, 1993) 'P.N.G.: OK Tedi Restructuring' vol. 321, no. 8246, Mining Journal Limited, London.

8 Impact of Structural Adjustment Policies on Forests and Natural Resource Management

NOBLE T. DONKOR

8:1 Introduction

The past two decades have seen economic stagnation and negative growth rates in many poor developing countries of the world. Among the symptoms attending this crisis have been unsustainable deficits in the balance of payment accounts, accumulation of external debt, large deficits in government budgets and inflationary pressures (Subramanian *et al.*,1994). Many of these countries are using similar structural adjustment policies (Structural Adjustment Programs) prescribed by international agencies and donors, to improve their trade balances, stimulate economic growth and reduce inflation. These policies include export promotion, currency devaluation, reduction in government spending, tax increases, privatization of public enterprises and land reform (Kaimowitz *et al.*, 1998).

How these programs affect the physical environment under various institutional and developmental conditions and a range of management regimes is uncertain (Kaimowitz *et al.*, 1998). Adjustment programs have implied superposition of two effects that have sometimes been cumulative but often contradictory. On the one hand, proponents of Structural Adjustment Programs claim adjustments help generate resources needed to protect the environment and eliminate economic distortions that cause environmental degradation, and improve living standards. They deny that adjustment increases poverty or diminishes governments' ability to implement environmental policies (Warford *et al.*, 1997). On the other hand, critics argue that economic growth and higher primary commodity exports increase pressure on natural resources, rising poverty leads poor families to depend more on overexploiting natural resources to sur-

181

vive, and cuts in public spending limit the governments' capacity to protect the environment (Reed 1996).

Ghana began one such program in 1983 and has since experienced a dramatic socio-economic transformation that includes a GDP growth of 4 to 6 percent over most of the past 15 years, increase in export revenue and the revamping of her industrial capacity. As a result of these achievements the World Bank/IMF has hailed Ghana as the most successful case of Structural Adjustment Programs in Africa. To answer the question whether or not structural adjustment has worked, most policy papers attempt to find out to what extent it has succeeded in reducing current account balances and government deficits, controlled inflation, restored economic growth and prevented the worsening of poverty and income distribution. This chapter however, focuses exclusively on the issue of Structural Adjustment Programs' impact on forests. It is worthy to note that Ghana's forest cover has dwindled from 8.3 million hectares at the turn of the twentieth century to 1.6 million hectares now, and the rate of exploitation of her forests in the last two decades has been extremely fast. This chapter attempts to provide a preliminary insight into how Structural Adjustment Programs might have stimulated or discouraged sustainable management, deforestation and degradation of forests in Ghana. However, I agree with Kaimowitz *et al.,* (1998) that it is difficult to fully distinguish the impact of Structural Adjustment Programs on forests from the many non-Structrual Adjustment Program factors that affected forests during the same periods. I also agree that it is better to make decisions based on partial information than to ignore key issues. I reserve the right to change my mind on any of the conclusions as research progresses and more data become available. I do not consider this inconsistent nor indecisive, but rather flexible and adaptive. I believe the sustainability of our forests depends on flexible, local, fine-tuned management.

8:2 Generalized Model of Structural Adjustment Programs and Forests

Policies of structural adjustment involving measures of trade liberalization, fiscal reforms, financial sector reforms and a reduction in the pervasiveness of government intervention (Williamson 1990) have particular relevance for agriculture and forestry which are dependent on the supply of public goods (Subramanian *et al.,* 1994). A number of variables associated with Structural Adjustment Programs (e.g., changes in prices, cost, incomes and government services) might either stimulate or discourage forest clearing and degradation (Kaimowitz *et al.,* 1998) (Table 8.1). In some cases this is because it is not certain how Structural Adjustment Programs will impact on the economic variable in question; in others it is because of doubts regarding the effect of differ-

Table 8.1 Expected Effects of Selected Structural Adjustment Programs on Deforestation and Forest Degradation, based on Previous Econometric Models

Type of Policy	Effects	Comments
Devaluation	Increases	Raises agricultural and timber prices
Removing price controls on food	Increases	Raises agricultural prices
Removing agricultural price supports	Decreases	Lowers agricultural prices
Removing agricultural import restrictions	Decreases	Lowers agricultural prices
Increased agricultural export taxes	Decreases	Lowers agricultural pricees
Increased road investments	Increases	Lowers transport costs
Removing fuel subsidies	Decreases	Raises transport costs
Lower spending on settlements	Decreases	-
Lower credit subsidies for crops	Indeterminate	has conflicting effects
Reduced input subsidies	Indeterminate	Has conflicting effects
Reduced spending on forestry	Indeterminate	Depends how money is spent
Reduced public employment	Indeterminate	Has conflicting effects
Restricted monetary supply	Indeterminate	Has conflicting effects
Industrial trade liberalization	Indeterminate	Has conflicting effects

Adapted from Kaimowitz et al. 1998.

ent economic change on the forest, and hence suggest further modeling work with additional research (Kaimowitz *et al.*, 1998).

Economic models tend to show that higher aggregate agricultural and forest product prices encourage deforestation and logging in unmanaged forests (Barbier and Burgess 1996). For example, when farming and logging become more profitable, landowners and migrant landless workers fleeing from unemployment in urban areas expand these activities to additional land.

Adjustment policies, such as currency devaluation, removing price controls on food, and fiscal incentives for agricultural and forest exports that improve the terms of trade for agriculture and timber increase forest clearing and degradation (Cruz and Repetto, 1992; Weibelt, 1994). Those that reduce average agricultural prices by eliminating price supports, opening domestic food markets to imports or taxing exports to obtain revenue should have the opposite effect. However, Structural Adjustment Programs are more likely to generate additional forest clearing when the main source of these problems is production for export (Kaimowitz *et al.*, 1998).

Investments in roads in forest areas in developing countries put pressure on them by making it cheaper to reach those areas and transport goods there to markets (Chomitz and Gray, 1996; Mertens and Lambin, 1997). Policies of structural adjustments may either decrease such investments to curtail public spending or raise them to promote exports.

Policies of structural adjustments generally reduce spending on subsidies for fuel, livestock and agricultural credit and inputs. Less spending on fuel subsidies and livestock should discourage forest clearing since it then costs more for families to move the agricultural frontier, transport their products and engage in extensive cultivation and cattle raising (Kaimowitz *et al.*, 1998). Higher agricultural inputs and credit costs have ambiguous effects. They discourage forest clearing by making agriculture less profitable, but encourage it by favoring more extensive production systems that utilize more land (Holden, 1996).

In the first few years after implementation, Structural Adjustment Programs tend to depress economic growth, employment and personal consumption because they restrict government spending and money supply, cut public sector employment and force inefficient local industries to compete with cheaper and higher-quality imports from abroad. The net effect of this on forests is uncertain. On the one hand, fewer job opportunities may induce underemployed families to move to the agricultural frontier and clear additional forest to produce cash crops such as cocoa for export, or engage in illegal chain saw operation to fell more timber. On the other hand, falling urban incomes depress food prices and make land clearing for agriculture less profitable, especially if crops under cultivation are arable ones meant for local consumption (Jones and O'Neill, 1995).

Over the longer term, Structural Adjustment Programs may lead to economic growth, more job opportunities and higher consumption of agricultural and forest products. This could reduce forest clearing by providing alternative employment outside agricultural frontier areas. However, it could also promote clearing by increasing demand for products produced in those areas (Kaimowitz *et al.*, 1998).

Across-the-board reductions in public spending associated with Structural

Adjustment Programs could diminish governments' ability to manage forests or make it more difficult to establish such capacity. However even before the Structural Adjustment Program, most developing countries had little capacity in this regard. In addition, greater foreign assistance, if partially channeled into forestry projects, may compensate for any decline that might occur (Kaimowitz *et al.*, 1998).

8:3 Ghana and Structural Adjustment Programs

Ghana's natural forests, which cover 35 percent of the total land area (Table 8.2), fall within two ecological zones: tropical high forest covers one-third of the country and supports two-thirds of the population and the majority of its economic activities. The savannah zone covers the remaining two-thirds of the country, and is a source of building poles, and fuel wood. Annual deforestation rate during the past two decades have been around 1.3 percent, which is equivalent to about 1,172 square kilometers per year (World Development Indicators 1997). Ghana's population is about 19 million, with a population density of 75 persons per square kilometer. The urban population is about 37 percent (see Table 8.2).

Ghana had a per capita GDP of US $1,960 in 1994. Ghana's per capita

Table 8.2 Background Information on Ghana

Variable	Year	Quantity
Land are (1000ha)		23,834
Forest & woodland (1000 ha)	1993	8226
Forest & woodland (%)	1993	35
Unproductive forest (1000 ha)	1993	6173
Productive forest (1000 ha)	1993	2053
Plantations (1000 ha)	1993	78
Population (millions)	1998	19
Population density (per km²)	1997	75
Population growth rate (%)	1990-97	2.7
Urban Population (%)	1999	37
GDP per caput (US$)	1994	1,960
GDP growth (average annual)	1980-90	3.0

Source: *World Development report (1998-99)*

income has been increasing since 1980: 1980-1990 (3.0 percent), 1990-1995 (4.3 percent). During the 1970s and early 1980s Ghana entered into economic crisis and depended heavily on cocoa, timber and gold exports. Political instability, excessive borrowing, macroeconomic mismanagement, falling international cocoa prices, depreciation of the cedi (Ghanaian national currency) and severe drought in 1983 caused GDP to fall annually from the early 1970s to the early 1980s, and inflation to rise dramatically. The average annual rate of inflation was 28.6 percent from 1984-1994 even when the economy had started picking up gives an indication of the deplorable nature of the economic crises in the early 1970s and 1980s.

Adjustment policies used to respond to these economic difficulties in developing countries include: currency devaluations, reduced public spending, increased taxation, trade liberalization, financial market liberalization and promotion of foreign investment (Kaimowitz *et al.*, 1998). Ghana sluggishly implemented most of these policies in the mid-1980s (Table 8.3), a couple of years after its crisis began.

8:4 Impact of Structural Adjustment Programs on Forests in Ghana

Forest clearing and degradation in Ghana has increased since the introduction of Structural Adjustment Programs. However, like the situation in many other developing countries such as Cameroon, Bolivia, and Indonesia (Kaimowitz *et al.*, 1998), this trend cannot be entirely attributed to the Structural Adjustment Programs because of past mismanagement, the Structural Adjustment Programs seem to have been an important contributing factor creating the environment for deforestation. Nevertheless, the effects of most of these Structural Adjustment Program measures on forests in Ghana have varied from minor, uncertain and mixed to negative (Table 8.4). Logging for export has increased considerably following the Structural Adjustment Programs. This is against the background that the forest cover has dwindled to a mere 1.6 million hectares at present. Figures for timber production for the pre-Structural Adjustment Program period compared to post-Structural Adjustment Program period show dramatic increasing trend of forest utilization in the latter (Figure 8.1a). Sawn wood, veneer and plywood exports have also increased in the last decade (Table 8.4 and Figure 8.2c). These figures do not take into account chip and particleboards, fiberboards, wood based panel, and, of course, the massive felling of trees by illegal chain saw operators. It is always difficult to quantify illegal operations. Ghana produced 37 percent more logs in 1994, than in 1991. The government banned all log exports in November 1995 as a result of the chaotic log export

Table 8.3 Structural Adjustment Policy Measures and their Effects on Forests in Ghana

Measure	Specific adoption	Effect on forests
Exchange Rates		
	Official exchange rate devalued	Negative
	Eliminated multiple exchange rates	Mixed
Government Spending		
	Public employment reduced	Mixed
	Government salaries frozen	Mixed
	Investment projects suspended	n/a.
	Agricultural credit subsidies reduced	Minor
	Agricultural input subsidies reduced	Minor
	Elimation of agricultural price supports	Minor
	Food subsidies reduced	Minor
	Spending on settlements reduced	n/a
	Closure/sale of parastatal enterprises	Mixed
	Reduced spending on forestry	Uncertain
	Public road construction maintained	Minor
Government Revenue		
	Value added taxes created/raised	Mixed
	Agricultural export taxes raised	Mixed
	Fuel prices raised	Mixed
	Fiscal incentives for exporters	Negative
Trade		
	Domestic agricultural markets liberalized	Mixed
	Agricultural export markets liberalized	Negative
	Import tariffs reduced	Minor
Other		
	Financial markets liberalized	Negative
	Foreign investment promoted	Mixed

n/a= not applicable

Table 8.4 Production and Trade of Tropical Timber (100m³) by Ghana

	1991	1992	1993	1994	1995	1996	1997	1998
Logs	1229	1218	1682	1682	1194	1166	1000	1100
Sawn	420	538	546	550	582	520	575	590
Veneer	30	28	61	57	58	95	75	80
Plywood	15	20	26	20	35	40	65	71

Source: *ITTO* (1998)

situation that existed during 1993-1994 (Figure 8.1b). Despite the ban on log exports in 1995, the value of the country's wood products exports grew between 1996 and 1997 by 37 percent to a total of US $170.5 million.

Currency devaluation has often been the most important factor increasing forest clearing and degradation under the Structural Adjustment Programs, at least in the short term (Kaimowitz *et al.*, 1998). Devaluation, which usually goes with Structural Adjustment Programs, greatly encouraged forest products and cocoa exports in Ghana (Figure 8.2a). Government subsidies for agricultural credit and fuel have continued to decline under the adjustment. Though these worsen the economic base of most of the people they do not slow agricultural expansion.

In Ghana, farmers have cleared increasing amounts of forests for cocoa, oil palm and other food crop production. There are no reliable national statistics for areas of forests cleared for agriculture. The lack of urban employment opportunities resulting from economic crisis, cuts in public spending, closure/sale of parastatal enterprises led some urban dwellers to return to farming and may have added to forest clearing related to rural population growth. The quest for fertile land was exacerbated in 1983 by bush fires that destroyed many farms across the country. The expansion of cocoa cultivation has been at the expense of the tropical forest. This implies that there are no longer vast areas of uncultivated forests awaiting development by enterprising cocoa farmers except in forest reserves. In fact, illegal settlement and farming has been reported in some forest reserves. Typical examples include Desiri forest reserve in the Brong-Ahafo and Ashanti Regions, and Tano-ehuro, Bodi, Bia Tawya, Manzan, Sukusuku and Tano Suraw forest reserves in the Western region (Forestry Department of Ghana and International Institute for Environment and Develop-

Figure 8.1 (a) trends in round wood production
(b) round wood exports
(c) sawn wood exports for pre- and post-SAP periods

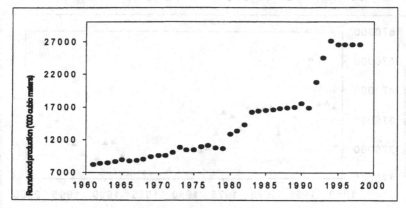

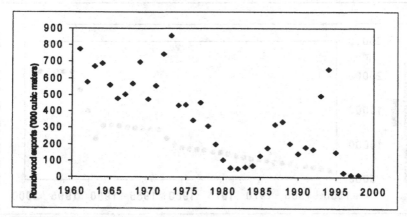

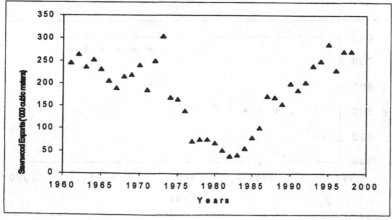

Figure 8.2 (a) trends in cocoa production
(b) wood fuel production
(c) wood charcoal production for pre- and post-SAP periods

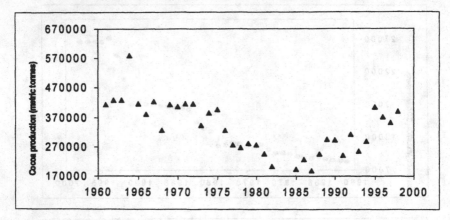

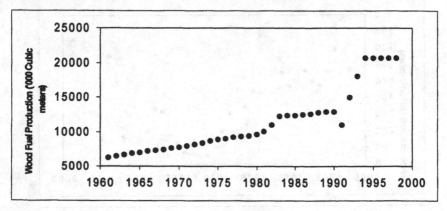

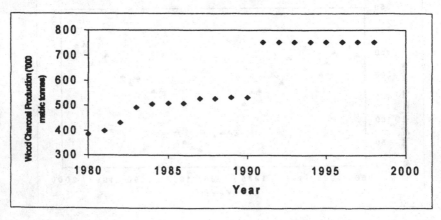

ment 1993). Other activities that put pressure on the forests include the harvesting of wood fuel and charcoal production. The household energy from fuel wood and charcoal in Ghana is between 75-86 percent (UNDP 1997; FAO 1999). Since the early 1980s, the production of wood fuel and charcoal has been on the increase (Figure 8.2). Though this trend might be due to the high population growth rate, the worsened economic base (even with Structural Adjustment Program, annual rate of inflation between 1980-1994 was 28.6 percent: UNDP 1997) of most forest areas dwellers might have pushed them to exploit anything to survive.

The proposition that adjustment programs make it more difficult for governments to control inappropriate forest clearing and unsustainable logging practices may receive mixed reactions in Ghana. On the one hand, the government has shown courage and political will to put in place mechanisms to ensure sustainable management of the remaining forest. On the other hand, some of the governments efforts may have come a little bit too late or may have been ineffective in controlling logging activities because of the economic crises and dependence of people on clearing forests for farms, timber exports and products for survival. In other words, reduced government spending might have resulted in reduced support for forest protection/management and inefficient monitoring, hence there has been an increase in illegal activities in forest reserves.

There has been a steady evolution of measures to maintain sustainable levels of log harvesting and to promote in-country downstream processing the last decade. Initially, the government banned the export of 14 commercial timber species, and later increased the number to 18. In March 1996 new export levies on logs of other 25 species were introduced, ranging between 10 to 30 percent to facilitate a shift from the processing of primary to lesser-used species (LUS). The Ghana forestry Department has termed these LUS as the "pink star species". These species, commercially exploited at very low rates, constitute less than 50 percent of the annual allowable cut. In the same year export levies of some air dried lumber species were suspended as a result of agitation from a section of the timber industry, in particular lumber producers, who called for a review of the export levies. The ban on export of all logs in 1995 remains operational. The ban on log exports was for three reasons: (1) over-capacity in the mills; (2) the mills employ thousands of people and there was a risk of there being insufficient raw material for them to operate; and (3) time was needed to undertake an inventory of the nation's forests. The annual allowable cut was reduced from 1.2 million cubic meters to 1.0 million cubic meters. There has been an emphasis on the relevance of high value, low volume production in exports, as opposed to the high volumes and low value in the past.

It remains to be seen whether this will work or not.

The current growth in the economy (measured by GDP) has sparked off a series of development projects, most of which require the heavy use of wood (ITTO 1998). It has also induced private sector participation in large-scale development and investments in forestry sector, including illegal chain saw operators. The Forestry Department in conjunction with the Ministry of Lands and Forestry recently launched a nationwide educational program to eliminate chain-sawn lumber from the markets in the country.

In the face of dwindling forest resources, the government is vigorously pursuing a plantation project with the aim of reducing pressure on the natural forest. The project is intended to increase the annually planted area to 8000 ha by the end of the establishment program.

The Forestry Department has assumed responsibility for the management of the nation's timber resources outside the forest reserves and has begun an inventory of the timber in these areas. This measure is intended to bring to sustainable levels the harvesting of timber outside the reserves. On a serious note, one wonders how much timber is left on lands outside the forest reserves that warrants such colorful declaration.

Since 1994 the government of Ghana has introduced four major legislative reforms as an exhibition of political will in order to achieve sustainable management of the meager remaining forest (Prebble, 1998). These are the Timber Resources Management Act No. 547, Legislative Instrument for stumpage fees and royalties, Logging manual to guide operators in the management of the forest, and Draft bill on the transformation of the forestry department into a forest service.

Prior to the passing of Act 547 in December 1997, the 1962 Concessions Act (Act 124) governed forestry operation. Under the Act 124, the timber concessionaires had no corresponding obligation to the forest. With the new law, however, operators enter into a contract, a Timber Utilization Contract (TUC), with the government through which they are allowed to utilize the forest on the understanding that they also manage the forest's sustainability. Only registered companies are eligible to apply for a TUC and before qualifying for this, they must submit to the Timber Rights Evaluation Committee (TREC) a reforestation plan. For TUC covering ten square kilometers, contractors are obliged to reforest a minimum of ten hectares. As an incentive to encourage companies to establish these forest plantations, the timber harvested from these forests will be the property of the concessionaire or TUC-holder and not the government, and contractors may export it as they like, including in log form. Secondly, loggers and millers have to provide an inventory of efficient equipment that ensure their effectiveness in forest exploitation in order to reduce wastage. Thirdly, they must provide a list of qualified and competent staff to run the mills and logging companies, i.e. they must employ professional foresters. The question is, in Ghana who is a professional forester? Is it one with a Bachelors or Masters Degree in forestry or is it by virtue of some professional accreditation? Fourthly, plans to address the likely environmental impacts due to exploitation must also be submitted. Fifthly, contractors must enter into an agreement with the landowners and undertake some social responsibilities e.g. providing amenities, such as a school or clinic, etc. Such measures are intended to involve landowners or the community in the management of the forests to reduce, if not eliminate, illegal encroachment and chainsaw activities.

The second policy is the legislative instrument that is intended to introduce stumpage fees that constitute an increase in royalty and forest management fees and to provide the procedure for the implementation of Act No. 547. Ghanaian timber tends to be very cheap on the international market and that is why there is a lot of waste–the recovery rate is below 40 percent. If the raw material becomes more expensive, exploitation and management of the forest will be efficient and recovery will be improved.

The new logging manual addresses issues of sustainable forest management and takes into account of all these new measures, as mentioned previously. It is hoped that implementation of these laws will improve sustainable management of the remaining forest landscape and to achieve the ITTO's Year 2000 Objective.

To date, the Forestry Department has been a part of the civil service and has been under-resourced. With its change to a Forest Service, it will now become semi-autonomous, more commercial, self-financing and have a better-motivated staff. The staff of the former department will be reduced from about 4,000 to 2,700 to become more efficient. It is possible that the Forest Service will be able to embark on a reforestation program in partnership with the private sector and other organizations, and provide technical services to forest plantation developers on a commercial basis. This author has some reservations about these "fine" objectives of the forest service to come. One, if the current work force is not resourceful, what is the guarantee that mere reduction of number of workers will ensure resourcefulness? Two, being more commercial and self-financing implies that the little forest left is going to be exploited more, unless someone tells me that being more commercial means the forest service is going to deal with other business activities outside forests. Three, I am afraid that the 1300 workers that will be made redundant could easily end up as forest encroachers and chain saw operators or collude with selfish timber contractors since they know more about the forests than the average Ghanaians.

8:5 Conclusions

Structural adjustment policies have important impacts on deforestation and forest degradation. Though the effects may not be direct at all times, the Structural Adjustment Programs seem to have been an important contributing factor to observed trends in forestry problems. In particular, when activities that negatively affect forests are mostly export-driven the struggle for survival and currency devaluation generally worsen these problems.

The Ghanaian case study lends little support to the idea that Structural Adjustment Programs lower deforestation and forest degradation by eliminating distortions associated with unsustainable management practices and previous government interventions. Ghana has had everything between distortions being maintained under the Structural Adjustment Programs to where some of the distortions have been eliminated with little effect on the remaining forests.

The indirect effects of Structural Adjustment Programs resulting from changes in overall economic growth, inflation rates, employment and consumption rates are difficult to measure (Kaimowwitz *et al.*, 1998). The situation in Ghana is that Structural Adjustment Programs have succeeded in creating more forestry-related employment opportunities that might increase deforestation.

In the face of government's spending restrictions associated with Structural Adjustment Programs, the government has also shown enough political will to attempt to promote sustainable forest management or control deforestation. However, one wonders whether these measures have not come too late or will be able to stem the tide of deforestation when most of the people are already dependent on the meager forests.

Acknowledgments

I thank D. Nanang and T. K. Nunifu for insightful discussions and comments on the manuscript.

References

Barbier, E.J. and Burgess, J.C. (1996). 'Economic analysis of deforestation in Mexico'. *Environmental Development Economics* 1 (2): 203-39.
Chomitz, K.M. and Gray, D. A. (1996). 'Roads, lands, markets and deforesta-

tion. A spatial model of land use in Belize'. Policy Research Working Paper No. 1444. Washington, DC: World Bank.

Cruz, W. and Repetto, R. (1992). *The Environmental Effects of Stabilization and Structural Adjustment Programs: the Philippine Case.* Washington, DC: WRI.

FAO (1999). ForestryProfile: Ghana.

Forestry Department of Ghana and International Institute for Environment and Development. (1993). 'Study of incentives for the sustainable management of the tropical high forest of Ghana'. Accra: Forestry Department.

Holden, S.T. (1996). 'Effects of Economic Policies on Farmers' Resource Use in Northern Zambia: Overview and Some Policy Conclusions in: Structural Adjustment Policies and Environmental Degradation in Tanzania, Zambia and Ethiopia'. Preliminary seminar report. Agricultural University of Norway.

ITTO (1998). Country Profile: Ghana.

Jones, D.W. and O'Neill, R.V. (1995). 'Development Policies, Urban Unemployment and Deforestation: the Role of Infrastructure and Tax Policy in a Two-sector Model'. *Journal of Regional Science* 35 (1): 135-153.

Kaimowitz, D., Erwidodo, Ndoye, O., Pacheco, P. and Sunderlin, W. (1998). 'Considering the impact of Structural Adjustment Policies in Bolivia, Cameroon and Indonesia'. *Unasylva* 49: 57-64.

Mertens, B. and Lambin, E.F. (1997). 'Spatial Modeling of Deforestation in Southern Cameroon. *Applied Geography* 17 (2): 1-20.

Prebble, C. (1998). 'Interview with HE Mr. Cletus Avoka, Minister of Lands and Forestry'. *Tropical Forest Update* 8(4): 21-23: Ghana.

Reed, D. ed. (1996). *Structural Adjustment, the Environment and Sustainable Development.* London: Earthscan.

Subramanian, S., Sadoulet, E. and de Janvry, A. (1994). *Structural Adjustment and Agriculture: African and Asian Experiences.* Food and Agriculture Organization: Rome, Italy.

UNDP (1997). 'Human Development Report'. New York: Oxford University Press.

Warford, J.J., Munasinghe, M. and Cruz, W. (1997). *The Greening of Economic Policy Reform.* Washington, DC: World Bank.

Weibelt, M. (1994). 'Stopping Deforestation in the Amazon: Trade-off Between Ecological and Economic Targets?. *Weltwirtschaftliches* Archive Review of World Economics, 131.

Williamson, J. (1990). 'The progress of policy reform in Latin America' in *Policy Analysis in International Economics*, Report No. 28. Washington, DC:

Institute for International Economics.

World Bank (1997). *World Development Indicators*. New York: Oxford University Press.

World Bank (1998-1999). *World Development Report 1998/99*. New York: Oxford University Press.

9 Structural Adjustment Programs and the Mortgaging of Africa's Ecosystems: The Case of Mineral Development in Ghana

CHARLES ANYINAM

9:1 Introduction

Since 1980 the economies of about 75 countries in Eastern Europe, Asia, Latin America, the Caribbeans, and Africa have been managed by the International Monetary Fund (IMF) and the World Bank by means of a series of processes summed up as structural adjustment. These programs have come under intensive scrutiny. The literature is now full of studies that have tried to answer such questions as: To what extent have structural adjustment programmes (SAPs) helped to alleviate the socio-economic hardships of the people of these countries? To what extent have these programmes affected such services as education, transportation, and health? What has been the social costs of SAPs? In what ways have SAPs affected food production or agricultural development, manufacturing industries, and the lumbering industry? (see Kraus, 1991; Anyinam, 1994; Konadu-Agyemang, 1998; Owusu, 1998).

In spite of the large compendium of literature on adjustment programmes, few studies have explicitly investigated ways in which SAPs have contributed to the disruption, deterioration, and destruction of the eco-social systems of these countries. Despite the intentions of IMF and World Bank towards the sustainability of the environment of these countries, it is generally recognized that structural adjustment programmes have been carried out primarily without

197

any respect for the ecological systems of the countries. Increased destruction of forests and watersheds, high rate of soil erosion and desertification, loss of indigenous genetic species, and increased industrial and agro-chemical pollution are evident in all the countries which have come under SAPs. Apart from environmental and human health costs, the economic costs of the environmental degradation are also mind-boggling. For example, a World Bank Report estimated the cost of neglecting environmental problems in Nigeria at about US$5 billion annually, the great bulk of this figure being accounted for by soil degradation (US$300,000 m), water contamination (US$1,000 m), and deforestation (US$750,000 m). On the basis of these figures, environmental neglect of development could cost as high as US$20 billion per year or more in the African region (Iwugo, 1992).

One of the economic sectors that have received much attention under SAPs is the the mining sector. In Africa, mining accounts for half of all exports and one third of total tax receipts. Commercial-size mining operations take place in 19 African countries while small-scale mining provides a living for nearly one million miners and their families in more than 30 African countries. Many of these countries have, in the last decade or so, recorded substantial increases in minerals production. Such increases have, however, had an adverse impact on the environment.

The deterioration of the ecosystems is not peculiar to African countries. In countries like Bolivia, Brazil, Columbia, Ecuador, Guyana, Indonesia, the Philippines, and Venezuela, there is enough evidence of environmental degradation as a result of increased mining activities (Greer, 1993; Ndi, 1993; Hobson, 1993; Shakesby and Whitlow, 1991). Mining, of course, has a long history of environmental destruction in these countries. But recent upsurge of mining activities has tended to intensify environmental pollution problems without significant appropriate measures being taken to control environmental pollution.

To contribute to the literature on the role of structural adjustment programmes to the deterioration of environmental impacts of mining activities, this chapter focuses on mining development in Ghana since the introduction of Structural Adjustment Programs in the country. The objective here is to provide a preliminary examination of the environmental impacts of recent investments in the mining industry, with special focus on gold mining. The purpose is to demonstrate ways in which the mining industry under Structural Adjustment Programs is contributing to the disruption, degradation and pollution of Ghana's eco-systems – its aquatic, territorial, and social systems. It will be argued that there is ample evidence to show that mineral developments under Structural Adjustment Programs have intensified the rate of pollution, land degradation, and destruction of Ghana's eco-social systems.

The next section provides a brief introduction of the economic situation in Ghana before the introduction of structural adjustment programs. This is followed by a summary of the main components of the economic recovery strategies undertaken under the aegis of the World Bank and International Monetary Fund. Before focusing on the mining industry under structural adjustment, a brief historical account of mineral development in the country before 1980 will be provided. The extent to which the aquatic and terrestrial ecosystems and the lives of the local people have been affected by the new "gold rush" and upsurge of investments in other minerals will then be examined. In particular, the impact of liquid effluent discharges, acid mine drainage, air emissions of sulphur dioxide, metals, and particulate matter, the employment of mercury and its consequent poisoning, and disfigurement of the local landscape will receive attention in the discussion. Efforts being made by the government and the mining companies to minimize environmental pollution in gold mining areas will also be assessed.

9:2 Overview of Structural Adjustment Programs in Ghana

During most of the 1970s and early 1980s, Ghana suffered an economic malaise marked by shrinking output, high and accelerating inflation, and growing balance of payment problems. By 1983, foreign exchange reserves were nearly depleted and the country had incurred large external payment arrears. Cocoa production, the mainstay of the country's economy, fell from 413,000 tons in 1970 to 159,00 tons in 1983 and as a principal export crop, accounting for over 70 percent of total exports in the mid-1970s, the cocoa decline dominated the deterioration in export earnings. In the mineral sector, the second major source of foreign earnings for the country, overall production fell by 55 percent between 1975 and 1983. Over the same period, the output of timber and timber products, which accounted for over 9 percent of exports in the 1970s, also fell by 57 percent. As cocoa and other exports earnings declined, the balance of payments and government finances suffered tremendously. By 1982, Ghana had suffered a fall of 41 percent in her terms of trade relative to 1971, caused largely by oil price increases and decline in world cocoa price. As well, by the end of 1983, the capacity utilization of the once flourishing manufacturing sector had fallen to 20-25 percent and value added in real terms accounted for only 8 percent of GDP compared with 14 percent in 1970. The real volume of imports also fell steadily in the 1970s and by 1983 was under one-fifth its 1971 level (Huq, 1989; Anyinam, 1994).

The food self-sufficiency ratio for Ghana fell from 83 in the mid-1960s to only 62 in 1982. By the end of 1983, per capita production of cereals, cocoyam and cassava was 26 percent, 45 percent, and 52 percent respectively of their 1970 levels. Inflation rates increased in the 1970s and became very high in the early 1980s, well above 100 percent per annum. With a population growth of 3 percent a year, per capita income declined by almost half during the 1970-80 period. The average wage earner experienced a decline of 30 percent in real earnings between 1980 and 1983 bringing the overall decline since 1975 to 86 percent. Domestic savings rate and rate of investment collapsed, the latter from 12 percent in 1975 to 4 percent in 1983. Emigration of skilled labor eroded rapidly productive base of Ghana's economy.

Such a gloomy picture of Ghana's economy was due to the cumulative effect of years of unsuccessful attempts to design policies appropriate to the economic problems of the country. The Provisional National Defence Council (PNDC), under the chairmanship of Flight Lieutenant Jerry Rawlings, therefore, took over the administration of Ghana on 31st of December 1981 at the time when Ghana had stumbled into its worst economic crisis. It was the time when the systems of health care, education, transportation, and communication were all in disarray.

It is against this background that the Rawlings' regime, in an attempt to rescue a virtually bankrupt country, designed an economic recovery programme in close collaboration with the IMF and the World Bank. Initially launched in January 1983, Ghana's adjustment programmes have been supported by successive arrangements with the IMF, including a three-year arrangement since 1988, under the enhanced structural adjustment facility (ESAF). Overall, the total amount of IMF financial resources committed to Ghana during 1983-91 amounted to SDR1,208 million. The World Bank and other creditors and donors have also provided substantial technical and concessional financial assistance.

The key elements of the adjustment strategy have been (1) a realignment of relative prices to encourage productive activities and exports and a strengthening of economic incentives; (2) a progressive shift away from direct controls and intervention by the government toward greater reliance on market forces; (3) the restoration of fiscal and monetary discipline; (4) the rehabilitation of the economic and social infrastructure; and (5) the undertaking of structural and institutional reforms to enhance the efficiency of the economy and encourage the expansion of private savings and investment (Kapur *et al.*, 1991). These key elements continue to influence current economic policies of the government.

Asked about the achievements of the PNDC government, the Secretary for Finance and Economic Development summed them up as follows:

It is in the area of economic policy improvement as well as macro economic performance that I think the PNDC has registered its most important achievements. Take GDP growth, GDP per capita. What is happenning in the external sector (import/export), the budget itself, revenues, grants, expenditure, monetary policy, national savings, investments, there is no doubt at all in these areas there has been a very important turn around in national development against a background of negative growth spanning over a decade before the programme began (West Africa, 1992).

Undoubtedly, the implementation of these policies resulted in a major turn-around in Ghana's overall economic and financial performance as indicated by macro-economic indicators. Between 1983 and 1991, growth in real GDP recovered, allowing gains in real per capita incomes, inflation declined, and the overall balance of payments position switched from deficit to surpluses, facilitating the elimination of external payments and a buildup of gross official reserves. In particular, real GDP expanded at an annual average of 5.4 percent. National savings rate recovered from almost zero in 1983 to more than 5 percent. Gross national investment rose in relation to GDP from a historically low level of 3.7 percent in 1983 to an estimated 16.0 percent in 1990. Both the overall budget and domestic budget deficits improved considerably even though the latter was still in the red by 1988. Inflation rate declined significantly, from an average annual rate of 73 percent in the early 1980s to about 25 percent in the first five months of 1991. The infusion of an appreciable amount of foreign exchange into the cocoa, mining, and timber industries significantly reversed the downward trend in export earnings which had characterized these industries the late 1970s.

As well, there was a surge in Ghana's non-traditional exports. Non-traditional exports (such as pineapples, cocoa waste, yam, tuna, kolanuts, fruit, and vegetables and processed and semi-processed goods like aluminium sheets, coils, household utensils, furniture,) increased from $5.4 million in 1983 to $24 million in 1986 and to $42m in 1988. Ghana's apparent success in increasing its earnings from this sector stemmed from an export promotion strategy with a myriad of incentives and export support services introduced by the government. Manufacturing experienced some recovery with capacity utilization estimated to be in the 40-50 percent range. Wages/salaries of all levels of the civil service rose significantly after 1984. Average monthly earnings per employee increased sharply between 1983 and 1988, though the levels reached in 1988 were still less than 50 percent of those in the mid-1970s.

The overall performance of the economy in the 1990s also was encouraging. Growth stabilised and the budgetary and monetary performance recorded substantial improvements which contributed to a marked decline in the domestic inflation rate. GDP grew by 5.1 percent in 1997 as against 5.2 percent recorded in 1996 while the overall real GDP rose by 4.6 percent in 1998. The GDP in 1998 was accounted for by 5.3 percent in agriculture and 2.5 percent and 6 percent growth in industry and service sectors, respectively. The downward trend in inflation which characterized 1997 continued in 1998. From a rate of 20 percent in 1997, inflation fell to 15.7 percent in 1998. Total exports in 1998 increased by 17 percent and total imports also increased by 4 percent (ISSER, 1998;1999).

A detailed analysis of the macro- and micro-economic impacts of Structural Adjustment Programs is not the focus of the chapter. Thus, the discussion will now shift to mining activities under Structural Adjustment Programs. What follows is a brief account of mineral development in Ghana before the 1980s and changes that have occurred in the mining industry since the introduction of Structural Adjustment Programs.

9:3 The Mining Industry

Minerals are generally found in the south-western part of Ghana, particularly in the Birimian-Tarkwaian system of intensely folded metamorphics and argillaceous sediments. Gold has been mined in Ghana since the 15th century. Long before European contact, gold was actively mined by the local people by using the "panning method" as well as the shaft method. From 1493 until 1600, the Gold Coast (now Ghana) was responsible for about 35 percent of the world's gold production. The first European mining entrepreneurs negotiated concessions for mines in the early years of 1870s and by the 1890s, gold mining in Ghana was an exclusively expatriate enterprise. While there were only eight mines listed for 1885 and three for 1895, the number increased to 153 in 1905. By 1915, there were 67 mines but this number declined to 43 in 1925.

Gold alone held the mineral field until 1914 when manganese was discovered. Even though manganese is found at a number of places in the south-west of the country, only the large and rich deposit at Nsuta is mined. In 1919, the first diamonds were also discovered. There are two areas in the country that contain diamondiferous ores in substantial quantities, the Birim basin and the Bonsa basin, but only the Birim deposits are being mined. In 1921, large deposits of bauxite were discovered on the crest of Mt. Ejuanema, near Nkawkaw. There exists large reserves of bauxite in the Eastern, Western, and Ashanti regions of the country and the present rate of exploitation is small in relation to total known deposits (Anyinam, 1982).

Throughout the colonial period, mining played an important role alongside cocoa, in the economy of the country. However, in the decades before independence in 1957, the mining industry steadily declined. For example, the 50 gold mining companies of the 1930s shrunk to 11 in 1948. In pursuance of the government's socialist development strategy, the Nkrumah government nationalised five gold mines in 1961 marking the beginning of the participation of the state in the mining industry. In 1972, the military government under Acheampong took over a fifty-five percent share of the Obuasi gold mine managed by the Lonrho and Ashanti Goldfields Ltd. The 20 percent of the government's share has since been sold to the highest bidder on the stock exchange.

For a variety of reasons, mineral production declined between 1960 and 1980s. Gold production fell from 915,000 ounces in 1960 to 246,000 in 1987; diamond production fell from 2.6 million carats in 1966 to 0.4 million by 1987 (Jackson, 1992); bauxite fell from 200,000 tonnes in 1960 to just over 50,000 tonnes in 1983; managanese from about 600,000 in 1960 to about 240,000 tonnes in 1983 (Ansah, 1990). Overall, the value of mining fell by more than 60 percent between 1971 and 1983. Similarly, employment in the mining sector fell from 48,000 in 1960 to 26,500 in 1986 (Jackson, 1992).

Kesse identified ten different reasons for this decline in mineral production in Ghana between the time of independence and 1983. These included: inadequate ore reserves at most of the operating mines of the State Gold Mining Corporation; severe financial constraints; deterioration of physical plant and machinery due to age, poor maintenance and lack of special spare parts; operational difficulties due to shortages of essential consumable mining stores; drop in productivity and production because of sudden introduction of five-day working week in the mining industry without introduction of mechanization; decreased labour productivity; deterioration of management ability and know-how, poor supervision, compounded by high turn-over of senior staff; exodus of skilled junior staff and workers from the mining sector; poor industrial relations; inadequate security system resulting in theft; and political instability and the general detrioration of the country's economy (Kesse 1985).

9:4 Mining under Structural Adjustment

Under structural adjustment programs, the mining sector has received much boost. The government took a number of actions that have resulted in the increase of gold, diamonds, manganese, and bauxite. On the recommendations of

the World Bank, the government took a number of actions including:
• promulgating a new mining legislation;
• incorporating generous investment incentives for potential investors;
• strengthening the surpervising and regulating institutions responsible for development in the mineral sector.

In 1984, a Minerals Commision was established to advise the government and administer minerals-related legislation and to find ways to resuscitate the mining sector of the economy. A new Minerals and Mining Law of 1986 (PNDC Law 153) was promulgated. The law has a general objective of creating a positive enabling climate for both local and foreign investment in the mining industry. The Law provided several incentives and benefits for mining companies:
• companies were to pay reduced royalties on gold production from 3 to 12 percent depedning on the rate of returns;
• a mining lease attracted an income tax of 45 percent but where the rate of returns exceeded certain agreed levels, the company paid an additional profit tax;
• a holder of a mining lease qualified for a capital allowance of 75 percent of the capital expenditure incurred in subsequent years;
• companies are also granted allowances on capitalization expenditure for reconnaisance and prospecting;
• exemption from payment of custom import duties in respect of plant and machinery imported for mining operations;
• gold mining companies were permitted free transfer of dividends or net profits and were also allowed to retain 45 to 69 percent of their foreign earnings;
• gold export is done direct by the producers, giving investors more control over the marketing of their output.

The consequence of the PNDC Law 153 has been (1) the rapid expansion in existing mines, (2) re-activation of abandoned mines, and (3) escalation of new exploration in the country.

A three-stage license system was introduced for reconnaisance, prospecting, and mining lease. The most dramatic effect of the new government mining policy was on investments in exploration leading to development of new mines and in expanding existng operations. While between 1940 and the latter part of the 1980s, no new mine was opened in the country, more than 6 new gold mines had been established since 1988. By the end of 1994, over 150 companies were operating with 25 of them in active gold exploration and production.

Gold continues to be the main focus of activity for mining companies in the country. As at the end of 1986, 126 local and 77 foreign companies had been granted gold reconnaissance and prospecting licenses. The most important company is the Ashanti Goldfields Corporation (AGC) which is now transfered into a multinational group following a series of acquisitions. AGC has an exploration portofolio consisting of 35 properties covering major greenstone belts in 12 African countries (Mining Journal 1998).

The government also passed a new mining law in 1989 intended to encourage small-scale miners and to bring under control the many miners who have been operating clandestinely. On April 19 1989, the PNDC promulgated the Mercury Law (PNDCCL 217) which repealed the Mercury Ordinance instituted by the colonial adminstration. Under the provisions of the Mercury Ordinance, mere possession of the chemical mercury (vital in the extraction of alluvial gold) was made a criminal offence. Such licences for the importation of mercury and other chemicals required in the mining industry were issued to the expatriate mining companies only. Under the Provisions of the Rivers Ordinance the diversion of river courses for the purpose of extracting gold, diamonds and other minerals was also forbidden. The only gold dredging operation carried out in Ghana by the mid-1980s was at Dunkwa (on the Offin) by the State Gold Mining Corporation.

While PNDCL 217 maintained the requirement of a licence for importation of mercury, it made provision for licenced small-scale gold miners to purchase from licensed mercury dealers such reasonable quantities of mercury as may be shown to be necessary for the purpose of their mining operations. The legalization of small-scale mining has led to a resurgence of alluvial gold digging (locally known as galamasy) and diamond mining in Ghana which were widely carried out illegally and clandestinely. Currently, in addition to the major new gold operations, there are more 650 licensed operations under the small-scale mining programme and more than 150,000 small-scale miners are expected to tap the enormous mineral potential of the country. Small-scale gold mining production has been increasing since 1992 because of the favorable market existing in the country whereby the miners are paid 98 percent of the world market price.

According to the Mines and Energy Minister, over US$3.0 billion had been invested in the mining sector by the end of 1997 (Mining Journal 1999). The government is commited to the development of other mineral potential in the country and in pursuit of this objective, Ghana's World Bank-funded Mining Sector Development and Environment Project has continued to pursue the identification of areas in the country likely to host mineral deposits. The Geological Survey of Finland completed an airborne geophysical survey in November 1997, covering a total of 48,000 sq. km in the Brong Ahafo, northern, north-western, north-eastern, the Volta, and eastern parts of the country.

Under structural adjustment programs, the mining sector has, thus, become the most dynamic industrial sector of Ghana's economy. In terms of gross revenue, the mining industry is currently the single biggest foreign exchange earner for the Ghanaian economy, accounting for 30 percent of export earnings. Gold has now overtaken cocoa as the country's principal foreign exchange earner. Ghana has become the second largest producer of gold in Africa as a result. Production of gold has risen three-fold over the past ten years. As can be seen from the Table 9.1, there has been a general increase in mineral production in the past few years. The tremendous increases in minerals production have, however, come at a high environmental price.

To discuss the environmental impacts of mining in the country, it is pertinent to understand the physical environment of the areas in Ghana where minerals are mined to appreciate the nature and extent of the seriousness of environmental damage being caused by the various processes of mining. As indicated in section 3, almost all the minerals produced in Ghana come from the Southwestern part of the country. Gold is generally found in the south-western part of Ghana, particularly in the Birimian-Tarkwaian system of intensely folded metamorphics and argillaceous sediments.

Even though manganese is found at a number of places in the south-west of the country, only the large and rich deposit at Nsuta is mined. With regard to diamonds, there are two areas that contain diamondiferous ores in substantial quantities, the Birim Basin and the Bonsa Basin. Large deposits of bauxite occur on the crest of Mt. Ejuanema, near Nkawkaw. There exists large reserves of bauxite also in the Eastern, Western, and Ashanti regions of the country. Existing geological reports suggest that some other parts of Ghana above latitude 7 degrees where mining was carried out in the 1930s, may still have promising prospects. Thus, mining takes place in the rain-forest and semi-deciduous forest zones of the country, which are also areas of very high humidity and heavy rainfall. The mining areas are drained by some of the larger rivers of the country—the Tano, Ankobra, and Pra/Ofin river systems. The areas where mining takes place are also noted for the cultivation of most of the food crops and cash crops the country produces. Some mining activities do occur in the forest reserves.

9:5 Impacts of Mining Activities on the Environment

Until quite recently, environmental impact of mining in Ghana has not received attention; hence, the lack of government guidelines and regulation for control of environmental pollution in the mining industry for many years. The older mines were established at a time when there was generally less awareness and con-

Table 9.1 Ghana's Mineral Production, 1970, 1980-1993

YEAR	Gold (oz 000)	Diamonds (carats 000)	Bauxite (mt 000)	Manganese (mt 000)
1970	714.4	2,550.0	337.0	392.0
1980	353.0	1,149.0	225.0	250.0
1981	341.0	837.0	181.0	223.0
1982	331.0	684.0	64.0	160.0
1983	276.5	279.2	70.2	172.6
1984	228.0	341.9	227.4	267.9
1985	299.6	631.8	180.2	357.2
1986	287.1	557.1	226.4	338.9
1987	323.7	411.7	229.4	295.0
1988	372.8	339.4	299.9	301.0
1989	430.1	339.4	371.5	364.8
1990	543.0	636.5	368.6	246.8
1991	846.0	687.0	324.3	319.7
1992	1,004.0	694.0	399.1	279.0
1993	1,200.0	699.0	364.0	309.0
1996	1,7356.0	714.7	383.4	266.4
1997	1,9164.0829.5	536.7	332.4	n/a
1998	2.6158.0805.7	341.1	384.2	n/a

Source: *Euromoney (1992, 1995); Engineering and Mining Journal Annual Review (1994, 1999)*

cern about the impact of mining on the environment. Consequently, mining practices and treatment plants were not required to meet the environmental standards currently desired. The result is that a century of mining has left an enduring and profoundly destructive legacy to the environment and communities located close to mines in the country. Not surprisingly, a few studies exist to provide a glimpse of the extent to which years of mining have damaged the ecosystems of mining areas. The rest of the chapter examines specific impacts of mining on the ecosystems of the areas where minerals are exploited.

9:5.1 Mining and Aquatic Ecosystems

One of the most striking consequences of mining is the adverse change that it creates in aquatic systems. Metal contamination from mining activities occur in many rivers flowing in southwestern Ghana. Mine effluents which contain such metals as arsenic, cadmium, copper, iron, lead, nickel, and zinc have been reported. A study published in 1976 by Water Resources Research Unit illustrates the extent to which the river systems in the mining areas have been polluted (Mensah, 1976). This report indicated that the Ankobra, Ofin, Owen, and Birim river systems have been polluted by effluents of cyanide, arsenic, iron, copper, zinc, gold, silt, and suspended solids. The volume of these effluents ranged from 2.75 million litres per month into the Birim river by the Cayco alluvial diamond mining (Kade) to 283.67 million litres per month into the Ankobra river by the Ghana National Manganese Corporation (Table 9.2).

An examination of the environment of the Prestea Goldfields shows some serious adverse effects on the ecology within the project area as well as on the stretch of River Pra into which effluents are directed (Osafo,1983; Ofosu, 1987). The ecology of the river is greatly affected and this also makes it too costly to use the river downstream of the mines for portable water supply. Residents of the Adansi West District complained that three streams which serve nine villages in the district have been heavily polluted with cyanide used in gold production in the area. In another example, the Goldenrea Mining Company operating in the Kwabeng area in the eastern region of Ghana has partially diverted the course of the Awaso stream to ensure that the stream is not polluted. But this has affected the flow of water in the town, causing undue hardships to the population in the area.

A more recent study of five old mines observed the effects of serious water pollution in two of the mines which have no pollution control devices. In the affected area the deterioration of water quality had led to severe, transitional, and minor effects on resident aquatic organisms, fisheries, and other water users at progressively greater distance from the discharge (Mate, 1992). In the last six years or so, the water supply of Kwabeng in the eastern region of Ghana has been destroyed by the mining of gold in the basin of the stream from which the town draws its water.

The role of small-scale mining operators is also as damaging as corporations. For example, mercury pollution by small-scale gold mining are responsible for perhaps the most pernicious aspect of gold mining in recent years. Residents of Tontokrom in the Amansie district of the Ashanti region maintained that a strange disease, identified as *Buruli ulcer*, that swept through the district was due to chemical pollution produced by small-scale gold miners in the area. The district health officials who were aware of the incidence of the disease, however, did not believe any link between the disease and alluvial mining. The disease which starts as a small boil on any part of the body, can cripple its victims within days if proper treatment is not provided. The disease which can result, in some cases, in disfigured limbs, affected 33 of the 135 villages in the district (Sarpong, 1993) (see Chapter 16 in this volume).

Local rivers are the main sources of fish for the people and many people have been or will be affected by mercury contamination, often through eating contaminated fish. The potential risk of mercury contamination to the enivronment and local population is very large. It is generally known that symptons of acute mercury poisoning include breathing difficulties, nauses, diarrhoea, and skin irritation. Where chronic poisoning occurs, it can lead to insomnia, memory loss, vision problems, loss of smell, difficulties in hearing and ability to speak properly, severe tremors, brain damage, birth defects, and death.

The impact of small-scale mining can be summed up by the following statement:

> Ten years ago, the Woara Woara river was the main source of drinking water for the people of Mpohor in the Western Region. In the rainy season, it watered farmland along its course and maintained soil fertility, Today, the river banks are bare, the water muddy and only an occasional fish darts by. The river has been poisoned by mercury and other chemicals used in mining (Sarpong, 1993).

Compared to the older mines, the tailings dumps of new mines create greater amount of waste rock per unit of gold won. A recent survey indicates that mine imposed sediment loading of natural drainage is the most significant and pervasive environmental impact resulting from mining in Ghana. Depending on

Table 9.2 Mining and Effluents Discharged into Inland Waters

Industry	Product Effluent	Volume 106 1/month	Composition	Rivers Affected
Ghana Consolidated Diamond Akwatia	diamond very turbid	202.93	Suspended solids mostly silt approx. 17,790 mg.	Birim
Cayco, Kada	alluvian diamond	2.75	suspended solids, silt, very turbid	Birim
GH Consolidated Diamond Ltd, Edubiase-Oda	diamond	35.35	suspended solids, silt, very turbid	Birim
State Gold Mining Corp. Konongo	gold bullion	81.83	Cyanide, arsenic silt, iron, gold	Owen
Prestea Gold Mines	gold bullion	95.50	Cyanide, arsenic, silt, iron, copper, zinc, gold	Ankobra
Tarkwa Gold Mines	gold bullion	25.00	cyanide, aresenic, silt iron, copper, zinc, gold	Ankobra
Ghana National Manganese Corp.	manganese ore	283.67	fine clay and laterite traces of oil	Ankobra
Ashanti Goldfields Corp.	gold silver bullion	100.00	cyanide, aresenic, silt, copper, gold	Offin

Source: *Adapted from Mensah, G. (1976) Water Quality and Pollution Survey of Inland and Coastal Waters of Ghana. Accra. Water Resources Research Research Unit, Council of Scientific and Industrial Research*

their nature and concentration, these solids do interfere with the self-purification of water and reduce light penetration and, thus, obstruct the process of photosynthesis. Fish can be injured by sharply-edged and irregularly-shaped pieces of crushed rock taken in through the gills and become vulnerable to fungi and other infectious agents. Where there is excessive growth of aquatic weeds, fishing and other activities are affected. Many inland waters have been rendered unusable as a result of years of discharge of mine wastes. Surface mining has exposed large areas of unconsolidated material to the intense tropical rainfall, characteristic feature of the western region of Ghana. Rainfall-based erosion and scour are considered to be primarily responsible for mobilizing fine sediments and dumping them into receiving waters. Bed sedimentation and high suspended sediment loads downstream of many mine sites had rendered rivers turbid and unsuitable for drinking and other domestic uses by local people.

9:5.2 Mining and its Impact on Terrestrial Ecosystems

One of the earliest environmental impact studies of gold mining in Ghana revealed gross contamination of the atmosphere, soil, water, and vegetation by arsenic discharged into the air during extraction processes at the gold mine at Prestea (Derban, 1968). The contamination with arsenic was found in the surrounding area up to 40 miles away. The mill at Prestea goldmine was spewing out 4 tonnes of arsenic and 32 tonnes of sulphur dioxide into the atmosphere every day while an estimated 6 tonnes of sulphur dioxide were emitted daily from a single gold ore processing plant and in another 1,419 tonnes of arsenic trioxide and three times as much sulphur dioxide were emitted. Broad plantain leaves in the area contained up to 650 ppm and the dry vegetation around a great deal more. Hair samples from affected people showed up to 500 ppm and urine samples 2 ppp, indicating a high intake of arsenic (Derban, 1975). This is a serious pollution problem as values exceeding 2 to 3 PPM indicate arsenic poisoning (Farmilo *et al.*, 1968). Poisoning was found to be due to skin contact with arsenic dust, drinking of water containing arsenic, or inhalation of arsenic dust in the atmosphere.

For gold and other non-ferrous metals, surface mining can generate between two and five times as much waste as it does in the case of ore, while up to 90 percent of this ore ends up as tailings. The waste and especially the tailings can contain contaminants such as residual cyanide, acids, toxic organic substances, nitrogen compounds, oils and metals or metalloids including iron, copper, zinc, lead, arsenic, nickel, mercury and cadmium (UNEP, 1991). Tailings ponds allow solids to settle so that process water can be decanted off. As the surface of the tailings dries, it is subjected to wind erosion, particularly during the dry season. Such dusts are not only a nuisance but also may adversely affect environmental and human health, depending on the composition of the tailings and mitigation procedure in place.

A survey found that dust, sulphur dioxide and arsenic trioxide are the most significant mining-related effects on air quality (Mate, 1992). The dust problem relates particularly to the surface mines and have been causing a nuisance to local residents at two of the mines. The Adansi West District Health Management Team (DHMT) expressed its concern about the health hazards confronting the people of Sansu, one of the many villages and towns in the vicinity of Obuasi, the site of the Ashanti Goldfields Corporation (AGC). Surface mining operations have caused the whole town of Sansu to become extremely dusty while violent vibrations caused by blasting of rocks in the area has cracked many buildings in the surrounding areas of the mine. The recent call to the local assembly by the Adansi West DHMT to engage the services of an environmentalist to determine the extent of pollution in the district from gold mining activities demonstrates the extent of the seriousness of the environmental problems facing communities in mining areas in Ghana.

A study of five old established gold mines showed that the harmful effects of sulphur dioxide and arsenic trioxide were limited to two of the gold mines where sulphide ores were being produced. The best known effect of sulphur dioxide on the evironment is its major contribution to acid rain which adversely affect aquatic life, forests and other types of vegetation. Tremendous damage has already been done to the surroundings of Obuasi area, where atmospheric pollution from the mines has virtually killed the forest and its components within the middle section of River Birim around Takroase, Saabe and Wenchi (Osafo, 1983; Ofosu, 1987). The land is no more useful for local food crop production because the land is virtually dead. Residents in the Obuasi area have complained about plume of white powder that spews from the gold refinery and contain arsenic trioxide as well as sulphur dioxide and suplhur trioxide. These compounds cause coughing because they irritate the respiratory tracts and damage lung tissue of people living nearby. The chemicals have also killed all vegetation around the refinery, and cotton and alfalfa on surrounding farms seem particularly vulnerable.

9:5.3 Mining and Land Surface Destruction, Disfiguration and Degradation

A recent study estimates that the total disturbance to land caused directly by the existing ten large mines was about 0.025 percent of the country's surface area. Considerable deforestation and woodlands in mining areas to create support facilities and infrastructure (eg. pits props, railway sleepers, and fuel for roasters and for domestic purposes) has occurred. The direct and indirect effects of mining have caused not only a loss of both primary and secondary forest, but also farmland and crops.

The traditional mining concession area in Ghana has been large, usually in excess of 100 sq. km to allow ongoing exploration. As long as mining was underground, this meant that pre-existing surface land uses were relatively free to continue operating. To place a similar area in a strip-mining situation poses rather greater problems of insecurity for other land users. All the new mines are surface operations and this means that there would be higher ratios of land use loss than was previoulsy the case. Most of the new mines involved open pit of low-grade surface gold deposits. The total land requirements are generally significantly greater than the older underground mines and it is, therefore, expected that their impact on the other users of land, especially smallholder farmers would be larger than previously.

Recently, a whole village near Tarkwa in the western region was forced to resettle in a newly built township as a result of the gold mining activities of Teberebie Goldfields Ltd (TGL). The old village, according to the company, is too close to its mining area and given the hazards associated with mining, the new township was built to ensure the safety of the people.

9:5.4 New Environmental Policies and Meausres

In the past decade or so, the government has taken certain measures in attempt to reduce environmental damage that mining and other economic activities have caused the nation. The newly-established Ministry of the Environment is charged with the responsibility of developing an effective system for the management of environmental resources in the country.

The body overseeing environmental management issues in the country is the Ghana Environmental Protection Council (EPC) which was established in 1973, as a statutory body for coordinating environmental protection activities. The Council which is now called Environental Protection Agency (EPA) advises the government generally on all environmental matters relating to the social and economic life of Ghana. It is responsiblie for conducting and promoting investigations, studies, surveys, and research relating to the improvement of Ghana's environment and the maintenance of sound ecological systems. The Agency is also supposed to embark upon general environmental educational programs for the purpose of creating enlightened public opinion regarding the environment and an awareness of the public's individual and collective role in its protection and improvement.

Currently, Environmental Impact Assessment (EIA) enjoys at least statutory status in Ghana. In July 1989, the Enivronmental Protection Agency (EPA) issued draft guidelines for Environmental Impact Assessment (EIA). The Minerals Commission of Ghana now requires that before a new mine lease is granted, environmental impact assessment should be undertaken. Notwithstanding this commitment, the Environmental Impact Assessment procedures have turned out to be largely superficial and perfunctory in nature, excluding as much as possible local community participation (Ndi, 1993). There is a legitimate concern that such exercises will become mere formalities which are geared towards giving clearance required for the inception of a new mining project. It is important to ensure environmental impact study is conducted at a level which is not too highly technical to either discourage or exclude local community participation and should be geared more towards identifying the potential hazards posed to the environment as well as to the social milieu of the locality (Ndi,

1993).

To assist the development of a National Environmental Action Plan, the government appointed six committees which included one on "Mining, Industry, and Hazardous Chemicals" in 1990. As yet, however, a formal framework and guidelines regulating the conduct of mining companies in environmental matters do not exist in Ghana. For the first time in the history of the mining industry, the Ghana government commissioned a study on the effect of mining on the environment in November-December 1990 under the auspices of the Minerals Commision. It involved visits to the operating major mines in the country – four old established gold operations, five new gold operations, the existing diamond mine and the manganese and bauxite mines. As well, for the first time, in 1996, Ghana's Mining Sector Development and Environment Project which is funded by the World Bank earmarked three areas that have been degraded by small-scale mining for reclamation. These areas were Akwatia in the eastern region for diamonds, Nueng Forest reserve in the Western region, and Achiaman in Greater Accra region. It is reported that if these physical reclamation projects succeed, the scheme will be replicated in other areas degraded by small-scale miners (Mining Journal 1999).

At the industry level, after years of polluting the environment with sulphide dioxide, Ghana's richest goldmine at Obuasi in the Ashanti region has introduced a new environmental-friendly method which uses bacteria in the production of gold, instead of the old methods of pressure leaching and roasting which have caused acid rain in the Obuasi area. The bacteria eat the raw ore and liberate the gold. This bio-oxidation plant in the Sansa Sulphide Treatment plant is expected to end pollution in the Obuasi area. The Ashanti Goldfields Corporation has also acquired a vibration-monitoring equipment to monitor the effects of blasting rocks on towns and villages within the Corporation's concesional areas. The Corporation has also acquired an analyser which will be used to monitor the level of sulphur dioxide in the atmosphere to make sure it is within acceptable environmental levels. One of the new mines located in an area of sulphide ores has installed a baghouse to extract an estimated 99 percent of the arsenic trioxide. The same company has made a commitment to retrofit its roaster with a sulphur dioxide scrubber if ground level concentrations of sulphur dioxide should exceed levels permitted by the World Health Organization.

The measures so far taken by the government and the mining sector are steps in the right direction. Having these environmental measures in place is not enough. Concrete steps should be taken to enforce them. All the mining companies, not just a few ones, must be required to institute the appropriate pollution control devices at the mines.

9:6 Conclusion

There is every indication that the mining industry in Ghana will continue to expand in the foreseable future. The current rate of mineral exploration in the country testifies that more mines will be established and the number of small-scale miners is also projected to run into several thousands. The critical question is: what kind of environment will the increased mining activities leave for the future generation? All indications are that environmental degradation will grow from bad to worse. This is because the number of mines continue to increase without significant precautions being taken to reduce environmental pollution and degradation. Surface mining and open-pits operations are now more favoured than deep mining. The mushrooming of gold-dredging by small-scale miners, the irresponsible use of mercury and cyanide, and the discharge of acidic or toxic waste directly into rivers are more likely to increase environmental pollution. The introduction of cyanide heap leaching methods to work low-grade mineral ores will also produce more pollution than in the past.

The current privatization process under Structural Adjustment Programs, has undoubtedly, increased production of minerals but it does not address the problem of inefficient natural resources management. Many foreign companies are eager to invest in the mining sector of the economy partly because of the increasing high cost of production and stringent environmental policies in their home countries.

It is easy for the extremist viewpoint to suggest that new mines should not be established at all but that is not the economic reality. The mining sector is critical to Ghana's economic development and therefore, minerals development cannot cease. A pertinent question that needs to be addressed then is: how can Ghana promote the mining industry without negatively affecting the eco-social environment? Several measures can be taken.

It is important for the government to ensure that the mining industry internalizes environmental costs in the planning process. Mineral-bearing lands should be opened to mining only if the mining companies can fully pay all social and environmental costs involved. Funds needed for environmental improvement in mining areas must be generated from within the mining industry. For example, Miner's Reclamation Fund can be created to be used to restore all mineral lands to conditions prior to use or to another acceptable and sustainable form and use. Miner's Forest Protection Fund is also required to be established because damages caused by mining roads and the resultant loss of aesthetic value go beyond land reclamation and restoration.

The government should negotiate terms of investment and mining technology to be employed at the same time. High levels of investment alone are not sufficient to ensure a clean industry. Successful negotiation depends on how the country is prepared to enforce environmental policies, in addition to how amenable corporations are to comply with national environmental standards. An environmental code of practice for mining operators which will give guidelines on the duties and obligations of gold and other mining companies is needed to safeguard the people who inhabit areas where exploitation of minerals takes place. A publication of the concessions under which the mining companies operate can help in the monitoring of their activities in respect of the environment (Kwarteng, 1993).

Many countries, like Ghana, do not have comprehensive environmental laws. They lack stringent air and water quality standards. Even though sustainable development and environmental protection are central to current development strategies in Ghana as recent measures show, few steps have actually been taken to incorporate these ideas into the mining industry. The environmental guidelines and laws for environmental protection which were recently established ignore the current level of contamination caused by the mining industry.

Lacking in Ghana also is the institutional support necessary to prevent environmental damage caused by mining operations. There is a need for a collaboration of mining firms, workers, the scientific community, NGOs and public institutions like Ministries of the Environment and Health. The ability to minimize environmental damage clearly calls for a macro-approach. Over the past decade or so, many communities in developing countries have been mobilizing themselves to defend ecological balance and fight against the destruction of resources upon which their survival depends. Grassroot movements, with all their limitations, have emerged as cultural and political antidotes to the dangers of technological nihilism and unaccountable bureaucracy. Classic examples are the Chipko Movement in India, the rubber tappers movement in Brazil, and Ngoni people in Nigeria. Communities in mining areas in Ghana can mobilize themselves for their voices to be heard and, thus, put more pressure on the government and the mining companies to respond to their demand for a clean environment.

More research must be conducted to monitor the environmental impacts of mining activities in the country. Acting singly or interactively, air, soil, and water contaminants that result from mining cause direct and indirect disturbances and stresses in the ecosystem health of mining areas. Gradual and subtle changes occur over time in mineral-rich environments, especially changes in the metabolism, growth, and composition of forests, cummulative effects of contamination in fish and other organisms, poisoning of soils and humans. The cumulative effects of these stresses that slowly but steadily erode the integrity of ecosystems are often very difficult to evaluate. The impacts overlap numerous types of ecosystems and habitats. It is very imperative that Ghana designs and implements long-term strategies to identify and quantify ecosystem pathology in order to find more lasting solutions to environmental degradation and pollution.

Unless the mining industry effectively addresses the issues of forest protection, mining pollution, land restoration, and compensation for loss of aesthetic value, and social impacts of their activities, there is every likelihood that the future generations will face not only mineral resource depletion, but also landscape damage and serious environmental contamination and pollution.

References

Anin, T.E. (1987). *Gold in Ghana.* Accra: Selwyn Publishers Ltd.

Ansa, K. (1990). 'Rehabilitation of the Ghanaian Mining Sector'. *Natural Resources Forum* vol. 14(3), August, p. 240-243.

Antwi, H. and D.W. Gentry (1992). 'Ghana makes changes to revive its gold mining industry'. *Mining Engeineering* 44(10) 12421244.

Anyinam, C. (1982). *The Role of the Mining Industry in Ghana's Economy.* A Research Paper submitted to the Faculty of Graduate Studies and Research in partial fulfilment of the requirements for the degree of Master of Arts in International Affairs. The Norman Paterson School of International Affairs, Carleton University, May.

Anyinam, C. (1994). 'Spatial Implications of Structural Adjustment Programs: The Ghanaian Experience'. *Tidjdschrift voor Economische en Sociale Geografie (Journal of Economic and Social Geography).* 85(5). p. 446-460.

Derban, I.K.A. (1968). *An Outbreak of Arsenic Poisoning in a Gold Mine* (mimeographed). Accra: Ministry of Health.

Derban, I.K.A. (1975). 'Some environmental health problems assoicated with Industrial development in Ghana'. *Health and Industrial Growth.* Ciba Foundation Symposium 32, Associated Scientifica Publishers, Armsterdam.

Ephson, B. (1991). 'Pause for Fresh Air'. *West Africa,* January vol. 52, p. 21-27.

Farmilo, C.A. *et al.* (1968). 'Arsenic Poisoning of Prestea People' 5th International Meeting of Forensic Science. Toronto.

Greer, J. (1993). 'The Price of Gold: Environmental Costs of the New Gold Rush'. *The Ecologist,* 23 (1), May-June, 91-96.

Hobson, S. (1993). 'The 1990s: The Environmental Decade'. *Engineering &*

Mining Journal 194(1) 30-31.

Iwugo, K.O. (1992). 'Sustainable Development in Africa: Some Institutional Requirements for Action and Implementation'. *EcoDecision* Sept 24-27.

Jackson, R. (1985). 'New Mines for Gold: Ghana's Changing Mining Industry'. *Geography* 77(335). Part 2. April.

Kapur, *et al*. (1991). *Ghana: Adjustment and Growth 1983-91*. International Monetary Fund: Washington DC.

Kesse, G.O.K. (1985). *The Mineral and Rock Resources of Ghana*. A.A. Balkema/Rotterdam: Boston.

Kwarteng, A. (1993). 'Privatising Ashanti Gold Mine'. *West Africa* 11-17. October. 1806-1808.

Maim, O. *et al*. (1990). 'Mercury Pollution Due to Gold Mining in the Madeira River Basin, Brazil'. *Ambio* vol. 19(1), Feb. 11.

Mate, K. (1992). 'Environmental Impact of Mining in Ghana: Issues and Answers'. *Natural Resources Forum* (1) Feb. 49-53.

Mensah, G. (1976). *Water Quality and Pollution Survey of Inland and Coastal Waters of Ghana*. Water Resources Research Research Unit, Council of Scientific and Industrial Research. Accra.

Mining Journal (1993). 'New Growth and Investment in Ghana'. July 2 321(8231). 9-11. London.

Mining Journal (1998). 330(8487). July 3.

Mining Journal - Annual Review (1999). Jan 22, 68-69.

Ndi, G.K. (1993). 'Cooperation Towards Sustainable Development in the African ACPs'. *Environmental Policy and Law*. 23(1)18-33.

Ofori, Cudjoe S. (1991). 'Environmental Impact Assessment in Ghana: Current Administration and Procedures - Towards an Appropriate Methodlogy'. *The Environmentalist* 11(1) 45-54.

Ofosu, G.K. and A. Iddrisu (1987). 'Environmental Impact Assessment as a Planning Tool - an Overview' in *Planning for Development in the Third World: Problems and Solutions*. S.B. Amissah (ed.). LARC/UST, Kumasi. 147-151.

Osafo, S. (1983). 'The New Investment Code'. *Environment and Policy and Law* 10, 15-16.

Sarpong, S. (1993). 'Ghana: All that Gliters is not Gold'. *The New African* October, 20.

Shakesby, R.A. and J.R. Whitlow (1991). 'Failure of a Mine Waste Dump in Zimbabawe: Causes and Consequences'. *Environmental Geology and Waste Science* 18(2) 143-153.

West Africa (1991). 27 May - 2 June, 875.

West Africa (1993). 4 -10 October, 1978.

10 Structural Adjustment Programs, Human Resources and Organizational Challenges Facing Labor and Policy Makers in Ghana

KWAMINA PANFORD

10:1 Introduction

This chapter addresses three concerns with respect to the implementation of Structural Adjustment Program (Structural Adjustment Program) in Ghana:

- The organizational challenges the Structural Adjustment Program poses to workers, mainly those who are unionized under the umbrella of the Ghana Trades Union Congress (TUC).
- How Ghana's Structural Adjustment Program is impacting the status of women employees, the education, training and employment opportunities available to girls and women, and
- Some of the general social, and economic outcomes of the Structural Adjustment Program as they relate to human resource development with emphasis on education, training and employment opportunities.

In assessing the impact of Structural Adjustment Program on workers, some of the issues treated are the shrinkage in the number of workers employed, the size of the unionized sector of the economy and why levels of employment, especially formal sector jobs are dwindling fast, is summarily described. In addition, how various labor organizations led by the TUC are responding to these new labor market changes are examined.

The profuse literature on Structural Adjustment Programs in Africa, and more so on Ghana, indicates that there are several criteria for ascertaining their success. In the donor community, Ghana for example, is hailed as one of the

219

successfully adjusted countries. Proponents of Ghana's Structural Adjustment Program allude to the fairly impressive macro-economic gains, especially 1984 to 1992.[1] However, this study focuses on the impact of Ghana's Structural Adjustment Program on workers who belong to trade unions and their women counterparts in the formal sector of the economy. This study also examines the extent to which the Structural Adjustment Program has constrained opportunities for vertical social mobility. It appears that the Structural Adjustment Program is limiting social mobility due to limited educational opportunities (from primary through tertiary education) and limited employment avenues. The disparate impact of limited education and formal sector employment for girls, women and youth in general is also evaluated.

Looking at the Structural Adjustment Program from the point of view of "the average Ghanaian", "the poor" or "ordinary workers", this study applies Professor Kwesi Botchway's (Ghana's long-serving Minister of Finance) definition of the goals of Ghana's Structural Adjustment Program to assess the impact of the Structural Adjustment Program on the bulk of the Ghanaian populace:

> Improving the living standards of the rural populace and the working classes... is in the final analysis, the very essence... and the ultimate standard by which its success will need to be judged (Jonah, 1989).

10:2 The Shrinking Formal Sector and Declining Union Membership

Ghana is experiencing an unemployment crisis. This crisis stems partly from the fact that the Structural Adjustment Program has not engendered jobs to replace the ones its implementation has eliminated. In Vision 2020 (Government of Ghana, 1995:6) (the government's blueprint for turning Ghana into a "middle-income country") the government states that over 65,000 public sector positions were abolished between 1986 and 1995 as part of the Structural Adjustment Program. The 65,000 jobs cut represent more than 25 percent of jobs in the public sector. According to Dr. Robert Dodoo, Head of Ghana's Civil Service, the public sector had approximately 250,000 employees in September 1994 (Dodoo, 1994: 1).

Both a reduction in the size of the government and severe restrictions on public infrastructure expenditure has led to declining jobs as the following illustrate.[2] Between 1987 and 1994, the Ghana Education Service (GES) laid off 48,000 instructors and non-teachers. The state-owned cocoa company, Cocobod retrenched at least 100,000 workers from 1983 to 1990 (Kraus, 1991: 32). Each of these organizations lost a minimum of half of all their employees as part of schemes related to the Structural Adjustment Program, which is designed to reduce the financial role of the state. It is not only the public sector that is shedding off jobs. The Government of Ghana (1995:6) reports that both public and private sector jobs declined from 464,000 jobs in 1985 to 230,000 jobs in 1990, that is, five years after the Structural Adjustment Program was adopted.

Based on the government's own data and other sources (Tutu, 1994; Kraus, 1991; EIU 1989; 1991; 1993; 1994; 1995) the redeployment plan implemented under the Structural Adjustment Program and other conditions partly induced by the Structural Adjustment Program had by the early 1990s culminated in a loss of between 40 percent-50 percent of all jobs in Ghana (see Table 10.1). The January 1995 government estimate for unemployment was 19 percent (Government of Ghana, 1995:10-11). The Ghana Trades Union Congress, the nation's major labor federation contends that unemployment rates might be higher. The federation's figure is closer to 40 percent.

A comparison of Ghana's levels of formal sector employment in the last decade may reveal the dire nature of the current unemployment caused in part by the Structural Adjustment Program. As shown in Table 10.1, combined public and private sector jobs peaked in 1985 at 464,300 with public employment of 397,100. By 1990, however, total formal sector jobs declined to 229,600. The impact of declines in public sector jobs on Ghana's overall employment is severe due largely to the state being the largest employer.

The data and tables alluded to depict a labor scene with high levels of unemployment unprecedented in Ghana. There are also some new dimensions to the phenomenon. Less than 30 percent of graduates of Ghanaian university graduates find jobs in the first 12 months after completing national service. For the first time Ghana is experiencing a new phenomenon: diminishing job prospects for university degree holders. This condition may have been ameliorated by the national service absorbing temporarily new graduates into the work force and hence avoiding a socially unacceptable situation of massive unemployment of university graduates. This new type of unemployment entails a waste of human resources, developed at high public cost and poses new and severe public policy challenges.

The redeployment of labor connected to Ghana's Structural Adjustment Program (also called "rationalization") has devastating effects on unions. As it is the case in most African nations, the public sector is not only the largest employer, but it is also the traditional trade union stronghold. According to the TUC, the rapid "down" or "rightsizing" of the public sector has resulted in

Table 10.1 Formal Sector Employment in Ghana, 1960-1990

Year	All	Private	Public
1960	332,898	148,547	184,351
1961	344,505	132,699	211,806
1962	350,890	125,881	225,009
1963	368,724	123,008	245,716
1964	379,667	131,904	247,763
1965	385,181	110,634	274,547
1966	356,054	94,128	261,926
1967	356,111	66,695	256,416
1968	385,291	103,150	282,141
1969	395,172	115,265	279,907
1970	396,000	106,600	289,400
1971	396,600	109,600	287,000
1972	424,500	108,800	315,700
1973	409,100	109,100	300,000
1974	457,400	127,700	329,700
1975	461,100	137,400	323,700
1976	483,400	140,600	342,800
1977	474,600	143,900	330,700
1978	481,800	131,100	350,700
1979	482,100	122,800	359,300
1980	337,200	46,300	290,900
1981	207,900	32,200	175,700
1982	279,600	43,600	236,000
1983	312,000	57,700	254,300
1984	450,800	64,500	386,300
1985	464,300	67,200	397,100
1986	413,700	66,300	347,400
1987	394,300	79,000	315,300
1988	306.900	55,300	251,600
1989	215,100	38,200	176,900
1990	229,600	40,200	189,400

Source: Tutu (1994:10)

drastic reductions in the number of trade union members. By the TUC's own estimates (TUC, 1996:54) in had 635,000 members in 1985. By 1996, membership had slid to 520,936, a loss of 18 percent of membership. It is estimated that nearly one out of every three unionized worker has left his or her union due to redeployment directly caused by the Structural Adjustment Program, or indirectly due to repercussions of the Structural Adjustment Program in the last decade and a half.

The steep decline in unionization has meant declining union revenues from members' dues. That has placed the TUC and its 17 affiliates in dire financial straits. The trade unions are becoming cash strapped. It is not only the unions that are experiencing the financial pangs of the Structural Adjustment Program. Ghanaian workers in general are being hit hard by rapidly eroding wages. Although prior to the Structural Adjustment Program (from 1975-1983) workers were stunned by ever-rising inflation with rapidly falling wages, wages have declined faster during the Structural Adjustment Program era. As it is the case in most of Sub-Saharan African countries, Ghanaian workers have lost as much as 10 percent per annum of their wages. Some estimates put the annual wage depreciation as high as 25 percent (EIU; 1989; 1991; 1994; 1995; Panford, 1999). It must, however be noted that the rate of decline in wages has been slow in the last few years although the objective of boosting public sector wages and improving other terms of employment and productivity through downsizing or "rationalization" is yet to be achieved (Dodoo, 1994).

A paradoxical situation that was not anticipated by the Structural Adjustment Program's advocates is that even though it was claimed that the programs would spur economic growth in Ghana, both endogenous and exogenous conditions have reacted to severely limit the availability of good paying jobs. In the Ghanaian context, and as well as elsewhere, these are jobs that yield high wages, and provide income security and benefits such as health and further education/training for employees. Such jobs allow workers to afford adequate housing, food, health and education for themselves and their dependents. On the contrary, the imposition of one of the most severe austere cost-recovery regimes has combined with declining job prospects and falling wages to exacerbate living and working conditions for the majority of Ghanaians, especially low-income workers and the unemployed. Why the Structural Adjustment Program has apparently failed is described in the next section.

10:3 Why Ghana's Structural Adjustment Program Has Not Led to a Boom in Jobs

Despite the designation "Structural Adjustment" given to Ghana's economic program supervised by the Bretton Woods institutions, the program does not address the fundamental weaknesses of the Ghanaian economy. These problems plague most Third World or African economies. After several years of promoting the Structural Adjustment Program, the President of the World Bank, James D. Wolfensohn, acknowledges that "...the World Bank has ignored the basic institutional infrastructure, without which a market economy...cannot function" (Schneider, 1999:326). Both the World Bank and the IMF seem to have been oblivious to some of the basic weaknesses of the Ghanaian and other African economies (Schneider 1999:327):

> An oligopolistic banking sector...a large agricultural sector dominated by peasant farming...inadequate, deteriorating infrastructure designed during the colonial era to extract resources from Africa, rather than to facilitate internal trade; an export sector dominated by primary products, many of which are facing declining terms of trade; poorly funded education and health care systems.

In addition to the fundamental problems identified by Schneider (1999), the Structural Adjustment Program has not addressed the problems inherent in a climate-dependent economy such as that of Ghana. As it happened during U.S. President William Clinton's visit to Ghana in March 1998, because of inadequate rainfall, the country faced a severe energy crisis. That exposed Ghana's vulnerability to dependence on expensive oil imports. Domestic and commercial users experienced power outages, which crippled several industries. Inflationary pressures that built up in the economy also adversely affected the gross domestic product.

Besides, the over a decade and a half of Structural Adjustment Program has not led to a fundamental diversification of the Ghanaian economy and exports away from precarious reliance on cheap export commodities. Structural Adjustment Program has not led to any appreciable levels of mechanization, irrigation and the application of both natural and synthetic fertilizers to Ghana's agriculture. Hence when there were adequate rains, food output was sustained and inflation was held in check. Conversely, when the rains failed to come, food crops failed also and the country experienced rising cost of living. In Ghana, during droughts, food alone often accounts for 25 percent to over 30 percent of the Consumer Price Index (Economist Intelligence Unit (EIU), Country Reports/Profiles, Ghana, 1996, 1997, 1998, and 1999). In addition to drought, poor storage, transportation and marketing pose serious challenges. In 1996, the Ghana government admitted that about 25 percent of food produced in the country was wasted due to inadequate transport, storage and preservation (Ghana Government, National Budget, 1996). The high cost of food has contributed substantially to the economy's low capacity to generate jobs and boost investments through domestic savings. It has also fueled the often antagonistic relations between organized labor and the government. Trade unions often advocate increased minimum wages to offset the high cost of living, especially food.

The *Financial Times* (November 3, 1999:4) also observes that the decline in world market prices for precious metals, especially gold "has...dealt a blow to the tortuous gains the country has made during...years of World Bank-sponsored structural adjustment". It quotes Ghana's Deputy Minister for Finance, Victor Selormey who said that: "Cocoa revenue have gone down, gold revenues have gone down, donor inflows haven't come and oil costs have gone up" (*Financial Times*, November 3, 1999:4).

In the last few years, Ghanaians have experienced the real life implications of what the *Financial Times* (November 3, 1999) refers to as the "volatile price movements" for Ghana's export. With gold providing approximately 20 percent of Ghana's export earnings, record low gold prices have combined with slumping cocoa prices to make it increasingly harder to create jobs to replace those lost due to Ghana's redeployment of labor, started in 1986. Due to a record, 20-year low in gold prices on the world market, in 1999, Ghana's leading gold mining company, Ashanti Gold Fields Corporation laid off 2,000 workers. These workers have joined their public sector colleagues who had been retrenched earlier due to the "rationalization" or redeployment in the public/civil service.

The adoption of Ghana's Structural Adjustment Program was motivated largely by the expectation that it would turn Ghana into an investor's haven. Through foreign direct investments, Ghana was expected to attract billions of dollars. As the Structural Adjustment Participatory Review Initiative (SAPRIN) (1999) has observed, the Structural Adjustment Program is designed to "create hospitable environments for foreign investors and for the movement of goods, services, and financial resources across national boundaries". As a trade off for the harsh measures related to the Structural Adjustment Program, it was expected that economic growth would be accelerated.

The World Bank, urged the Ghana Government (Panford, 1996 and 1999) to reduce labor costs, especially wage gains through collective bargaining, to make Ghana attractive to overseas investors. After over a decade and a half of vigorous efforts to constrain wages, Ghana, like most African countries has not benefited from massive infusions of capital from private and overseas investors. The Ghana Labor Department (March, 1998:10) reports that after three years of intense efforts to develop "a market-based economy", 593 projects, which employed 33,055 Ghanaians, were registered with a total investment of U.S. $1.10 billion from local and foreign businesses. The investment of U.S. $1.10 billion is hardly enough to create jobs for the 210,000 new jobs seekers who enter the labor market every year (Ghana Labor Department, March, 1998: 2).

A major thrust of the Structural Adjustment Program in Ghana that has obstructed job growth is what both the ILO and UNDP (1997) call "economic growth through contraction". The Ghana Labor Department (March, 1998:6) captures this policy's actual impact:

> The most significant labor market development has been the persistent decline of the population employed in the public and other state enterprises which have declined appreciably from 9.3 percent in 1992 to 6.4 percent in 1995 following the Government's commitment to withdraw from the productive sectors of the economy through redeployment, privatization and divestiture of state enterprises.

This adverse labor market condition is engendered by the intent of the advocates of the Structural Adjustment Program to seek economic growth by fiscally restraining the state and balancing the nation's budget, often at the expense of workers and other poor sections of society. One outcome with severe ripples in the entire economy and society is that job growth is stifled and living and working conditions deteriorate for the majority of Ghanaians, both urban and rural. The workers who are jobless are victims of the anti-job growth dimensions of the Structural Adjustment Program, especially what has become known as redeployment, retrenchment and more recently rationalization (Dodoo, 1994) (see Table 10.1). In terms of its role, the state under Structural Adjustment Program is discouraged from being an employer. This role differs radically from the early post-independence state that emphasized job creation and improved working conditions as essential to nation-building and economic and social development.

10:4 Why the Private Sector Has Not Been an Engine of Job Growth

Based on the assumption that the Ghanaian state (like its counterparts all over

Africa) was overextended (Rothchild, 1991) the IFIs and other western donors have pushed hard for privatizing most state-owned companies. According to the ILO (1995), Ghana by the mid-1990s was among a handful of African nations that were responsible for over 60 percent of the divestiture of state companies in all of Sub-Saharan Africa. Typically, when private investors assume ownership of publicly owned companies, they reduce the size of the work force to ensure higher rates of profits. The closure and sale of state-owned companies explain the high levels of joblessness in contemporary Ghana.

Trade liberalization under the new Structural Adjustment Program regime has also made it difficult for private enterprises to generate jobs to adequately compensate for those lost due to the closure and divestment of state companies. The Ghana TUC (Response to Questionnaire, International Relations Department, Accra, August-September, 1995) describes the role of trade liberalization in diminishing job prospects for Ghanaian workers and new entrants into the labor market.

The wholesale trade liberalization policy, which is being pursued under the ERP/SAP, has led to increase in the importation of goods from other countries. Presently, the Ghanaian market is flooded with both consumer and capital goods, which have crippled the development of domestic production. At the moment, many local manufacturing firms have closed down. Since our local industries are in their infant stages, they have not yet expanded over time to enjoy economies of scale. Consequently, they are out competed from the market by the big foreign multinational corporations, which produce at cheaper cost. Therefore, the Ghana Trades Union loses many of its membership as the affected organizations laid off their workers due to lack of market for their products.

Under pressure from donors to create an atmosphere attractive to multinational companies, the government of Ghana has been less stringent in the enforcement of labor market regulations that protect workers from layoffs (Tutu, 1994). Between 1986 and 1992 when the divestiture of state enterprises was accelerated, the government of Ghana repeatedly clashed with workers and unions over the dismissal of workers, the payment of severance packages and the use of the wage as a fiscal policy tool (EUI, 1998; 1999). Several of the disputes were litigated in the Ghanaian courts and the TUC filed complaints with the International Labor Organization in 1994 and 1995 (TUC, 1996).

While due to several internal and exogenous factors, the Ghanaian economy lost jobs that paid well and provided benefits, the only sectors of the economy that grew were the informal sector and the service sector. Both sectors do not provide secure and good paying jobs. In the view of the TUC, they often offer only "lousy" jobs (TUC, 1996). Besides, the great hope pinned on the informal sector as the new engine of economic prosperity has not materialized. Instead, it is flooded with excessive labor that earns meager incomes. In most cases, wages earned are not what trade unionists refer to as "livable wages".

An industry experiencing an appreciable boom in growth is tourism. But it is not yielding jobs on a sustainable basis because of reliance on imported inputs. In addition to boosting tourism, the government launched an export processing zone scheme to attract foreign capital. By 1997, approximately 2,500 jobs had been generated in these zones in the Accra-Tema region (Ghana Labor Department, March, 1998:11). Thus current labor market trends in Ghana indicate a situation in which the loss of jobs far outpaces the rate at which jobs are generated.

Demographic phenomena have also contributed to make the challenges of job creation more problematic in Ghana. While job opportunities dwindle, both the population and work force continue to grow at 2.5 percent and almost 3 percent per annum (EIU, 1998; 1999). In effect, Ghana for example, has to create at least 200,000 new jobs every year to meet the needs of the youth who join the labor market for the first time.

Thus although in using macro-economic indicators alone, we may observe that up to the mid 1990s, Ghana chalked impressive economic changes, in the area of jobs and incomes security for workers, the nation's record needs improvement. While some aspects of the economy improved, the rate of growth may not have been adequate to diminish the misery that has accompanied the massive job cuts of the last decade. Examples will support this point. From 1983 to 1992, the economy grew on the average at 5 percent per annum while the population grew at just under 3 percent per annum. While the work force was 8 million in 1996 it is expected to balloon to 12 million by 2006 at a 50 percent growth rate. Even with the most optimistic anticipated GDP growth of 5 percent a year, a 50 percent growth in the work force poses a severe human resource development and public policy challenge. Besides, the relatively impressive achievements of the last decade and a half, such as increasing the use of industrial capacity (which fell substantially from 1975 to 1985) the country has not gone beyond recovery to create an industrial sector capable of absorbing rapidly growing numbers of new workers and those redeployed or retrenched from the shrinking public sector.

10:5 Structural Adjustment Programs, Education and Human Capital in Ghana

In the last decade there have been tremendous shifts in emphasis and the public financing of education, especially at the secondary and tertiary levels. Major assumptions driving these policy shifts include donors' belief that primary/elementary schooling is inexpensive and the most cost effective investment for the country. The flip side of this assumption is that secondary and tertiary education are too costly, and yield dwindling social benefits. Thus relatively more funds are going into primary schools in Ghana. In 1993, 44 percent of the Ministry of Education's budget went into primary education compared to 38 percent in 1989 (EIU, 1998; 1999).

Since 1986, under a policy of cost recovery (what the government calls cost sharing) parents have had to pay more fees toward secondary and tertiary education. Parents now bear most of the cost for room and board, tuition and books in secondary schools. In the late 1990s, new fees were instituted in universities as a result of the state's fiscal crisis.

Several domestic factors have combined with IMF and World Bank policy preferences to generate several dilemmas for the policymaker in the area of public education and human capital development. Severely limited public funds and attempts to comply with spending limits demanded by western donors make it difficult to overcome complex issues of high cost; access to good quality and affordable education, the relevance of the curriculum and the utility of the educational system in terms of the job prospects of graduates and even their roles in society.

Prior to the ongoing crisis in the educational system, Ghana emerged as one of the leading African/Third World nations with a fairly developed educational system. In spite of its relatively small population, Ghana produced more than its fair share of university graduates. For example, as indicated in Table 10.3, barely two years after independence, enrollment in universities almost tripled. Ghana became known for the premium it placed on education and foreign governments, universities, and international organizations, including the United Nations, employed Ghanaians. However, with the economy's collapse in the 1970s (Panford, 1994, 1996, 1999 and Rothchild, 1991), the quality of education eroded rapidly as the percentage of the GDP allotted to education fell. With hard living and working conditions, elementary, secondary and university teachers and other educators fled in large droves to Nigeria, Liberia, Gabon, and the southern African nations, and to Europe and North America.

The PNDC initiated reforms aimed at revitalizing the economy. These reforms have affected the organization and financing of education. Financial changes included an increase of funding from 1.4 percent of the nation's budget in the 1980s to 3.8 percent in 1993 (EIU, 1998; 1999). There has also been the first major restructuring of education since the Nkrumah government's 1961

Table 10.2 School Enrollments in Ghana (1951-1959)

	1951	1957	1959
Totals	226,218	589,153	624,575
Primary	154,360	455,749	465,290
Middle	66,175	115,831	139,984
Secondary	2,937	9,860	11,111
Technical	622	3,057	2,782
Teacher Training	1,916	3,873	4,274
University	208	783	1,134

Source: *George, Betty S. (1976)* Education in Ghana. *Washington, U.S. Government Printing Office*

Educational Act. In 1986, the largely British-style educational system in Ghana was replaced by a Junior Secondary School (JSS) and Senior Secondary School (SSS) system which was placed on top of Ghana's six years of primary schooling. The JSS and SSS last three years each followed by four years of tertiary education. Expected additional changes include making the curriculum functionally relevant to the Ghanaian society. To achieve the latter, greater emphasis is placed on the practical application of knowledge and the development of graduates' technical competence. Thus there is a thrust away from more liberal arts type of education and emphasis on vocational skills. This is anticipated to enhance the job prospects of graduates. Financial repercussions of the post Structural Adjustment Program educational system have included cuts in government expenditure, in terms of actual cash outlays, plus the imposition of new and steep user fees to allow the state to recoup its expenditure.

The PNDC's reforms notwithstanding, Ghana's educational outlook has not improved substantially. For example after the launch of a mass literacy campaign to make 6 million people literate by 2000, it is estimated that close to 50 percent of the population is illiterate (EIU, 1998; 1999). Even the World Bank in a 1996 study indicated that the rate of recovery in school enrollment from 1989 to 1994 has not been spectacular or met expectations. The Bank's own Social Indicators of Development of 1996 suggest that secondary school enrollments are down (EIU, 1998; 1999). The rise in poverty associated with the Structural Adjustment Program and the high cost of education may account for the declining enrollment and the high drop out rates.

While real incomes and job security have not improved under the Structural Adjustment Program (for example, severe job cuts have not led to substantial pay raises for civil servants) most parents, especially the working poor, have to contend with rising tuition costs, buying expensive books and other fees. Due to poverty and parents' declining incomes and poor job prospects, Ghana's dropout rate is unprecedented.

The new regime of cost recovery imposed in the 1990s, and emphasis on vocational training could have severe consequences for the production and re-production of the elites in the Ghanaian society. The introduction of user fees of 186,500 cedis to 416,500 cedis ($200-$400 U.S. dollars) in the 1999-2000 academic year, could severely limit access to university education for the poor. In a country with a per capita income of less than $500, the true meaning of these fees will be that more Ghanaians would be blocked from receiving university education. Mostly children of the rich, the elite or those with access to foreign exchange through overseas remittances could afford to attend universities. Thus in a nation that in the past ensured fair access to higher education that led to reasonable levels of vertical social mobility and cohesion, education is steadily being priced out of the reach of the poor.

The user fees and other aspects of the post Structural Adjustment Program reforms in education could be eroding the significant gains made through the Accelerated Education Plan of 1952 adopted by the Transitional Government prior to full independence, and the Education Act of 1961. The latter created virtually free education and increased access to schools. There are additional potentially adverse social implications of the Structural Adjustment Program in the sphere of education, especially for girls and young women.

10:6 Women/Gender and the Adverse Potential Effects of Structural Adjustment Programs

The issues examined in terms of the impact of Structural Adjustment Program are levels of employment, literacy and school enrollment and working conditions for girls and women respectively. Assessing the impact of Structural Adjustment Program on women and girls is important because women constitute slightly more than half (approximately 51 percent) of Ghana's population.

We might begin the discussion of the effects of the Structural Adjustment Program on women and related gender issues by asserting that nowhere in the world was social progress (including reducing social and gender inequities) achieved solely through voluntary private sector initiatives. No society has relied on private individuals to overcome social disadvantages that accumulated over time and led to women and girls lagging in education, training and employment compared to men and boys. It is through appropriate public policy intervention in the form of legislation and social programs that women and girls are able to attain vertical social mobility. As it is in most societies, in Ghana in the late 1950s to 1960s, an expanded public sector and rapidly growing educational opportunities deliberately provided by the first Ghanaian government proved critical in allowing women and girls to improve their socio-economic status.

We cannot discount the social fact that even today women and girls are at a social disadvantage compared to men. However, we could discern trends that indicated that prior to the Structural Adjustment Program, women and girls were achieving substantial social gains. An example is increased school enrollment:

> During the last decade a rapid expansion has occurred in the educational system, not only in the number of new schools constructed but also in the number of pupils who have received full-time public education. Noteworthy elements in this have been the increasing number of girls receiving education and their ratio to boys in the total number of school places taken up. In 1958 there were 299,346 boys, or 65.8 percent, attending primary school and 155,707 girls, or 34.2 percent. A decade later the position of girls has improved, for there were 594,917 boys, or 55.5 percent, and 477,606 girls, or 44.5 percent (Greenstreet, 1979:26).

We might only add that the gains by girls were the results of the government's plan to attain social development by enforcing equal access to education.

10:6.1 Levels of Employment

Women used the expanded educational opportunities (heavily subsidized or provided cost free by the state) to acquire qualifications for employment in the civil service and in the public sector that grew exponentially. Initially most of the women took secretarial positions and teaching jobs. Later some became administrators and in the case of several state-owned companies, factory workers. The 1961 Education Act ensured access to education for girls through scholarships and low-cost boarding school facilities. Girls who enrolled in schools such as Wesley Girls (at Cape Coast); Mfanstiman (at Saltpond); St. Louis (in Kumasi); Holy Child (at Cape Coast); Aburi Girls (at Aburi) and Achimota (in

Accra) secured important professional positions. Some advanced in even some of the technical sciences and fields after receiving university education at home and abroad:

> ... women have risen to positions of professional importance in Ghana. Early 1990s data showed that 19 percent of the instructional staff at the nation's three universities in 1990 was female. Of the teaching staff in specialized and diploma-granting institutions, 20 percent was female; elsewhere, corresponding figures were 21 percent at the secondary school level, 23 percent at the middle level, and as high as 42 percent at the primary school level. When women were employed in the same line of work as men, they were paid equal wages.... (Berry, 1995:101-102).

Ghana's Structural Adjustment Program has begun or has the potential to erode some of the important gains made by women. The World Bank admits that in its own study of 1996 to 1998 (EIU, Country Report, Ghana, 1998), both school enrollment and level of literacy were falling for women in Ghana. Although the full effects of the Structural Adjustment Program are still unfolding, available evidence suggests that the World Bank's observations may be valid. Steep user fees, declining wages and poor job prospects for the graduates of schools have culminated in some parents taking the painful decision to withdraw or not enroll their dependents in school. Facing these conditions some parents may be reverting to a tradition common in most societies. That is, preference for educating males when there is some financial cost involved as Greenstreet (1979:26) notes:

> Generally, parents who have insufficient financial means give preference to boys over girls in respect of education, and particularly higher education. This reflects the idea prevalent in most societies that a boy, being a future breadwinner, has the first claim on education in order to better his chances of employment.

The large numbers of street children including school-age girls and young women who roam the streets of Accra and other urban centers hawking items, and using markets such as Makola (in Accra) as their permanent home may be indicative of the beginning of the erosion of the gains made by women in the first three decades of independence. Faced with the high cost of education coupled with poor job prospects, most parents and girls choose petty trading.

The pricing of formal education out of the reach of most Ghanaians, especially girls and young women might result in fewer women participating in the formal labor force. This in turn might slow women's vertical social mobility.

10:7 Structural Adjustment Programs and Women's Working Conditions

A point less known about Ghana's labor laws is that Labor Decree, 1967 (NLCD 157) was a pro-women piece of legislation. It aimed at ensuring, for example, that even when a woman was pregnant or breast-feeding a baby, her job and career were not jeopardized. NLCD 157 led to a situation, which became common in the formal sector of the Ghanaian economy as Berry (1995:102) describes, "When women were employed in the same line of work as men, they were paid equal wages, and they were granted maternity leave with pay". Greenstreet (1979) and Obeng Fosu (1991) made a similar observation concerning the equality of the sexes in formal job settings.

One of the provisions of NLCD 157 was the generous maternity and postpartum leave given to women employees. Before and after birth, women were given a total of 12 weeks leave with full pay for each child born. In case of pregnancies with complications, a medical officer's note entitled pregnant women to six weeks of rest from work with half pay. On top of that, women workers were allowed two-half hour breaks each day to breast feed their infants. According to Obeng Fosu (1991) who was a Chief Labor Officer, in practice most employers, especially those in the formal sector paid pregnant mothers for three months while on leave after childbirth.

Other aspects of Ghanaian labor laws proved pro-women. For example, Government Circular No. 1310/71 of September 13, 1971 made it unlawful to dismiss female employees who were ill due to pregnancy. This circular and similar legislative provisions and policy measures worked in tandem with expanded and inexpensive educational opportunities to facilitate the entry of large numbers of women into the formal sector in the first two and a half decades after independence. Thus as Greenstreet (1979:18) has aptly noted concerning equality of the sexes in the workplace in the 1970s: "complete legal equality in rights and privileges exists....". She adds, "Where there are special laws relating to women exclusively, these are aimed rather at improving the position of women in relation to men rather than at discriminating against them".

With emphasis on downsizing the public sector and reducing labor costs, Ghana's Structural Adjustment Program has the potential to limit severely both the educational and career prospects for girls and women, in particular those that come from rural areas and other socially disadvantaged backgrounds. In contrast to the first years after independence when women were able to fulfill dual roles as employees in the formal sector and as mothers with children (Greenstreet, 1979:18), women may now find such roles increasingly difficult. As private employers seek to maximize profits from the state-owned companies they acquire, they may not be inclined to grant women and pregnant workers the liberal benefits they enjoyed in the past. Also as Mhone (1995:51) has observed, there may be a disparate impact on women of the layoffs associated with Structural Adjustment Programs. There may be some correlation between retrenchment and declining women participation in the wage economy in Africa. He emphasizes that although more men have lost jobs under Structural Adjustment Programs, "that proportion of women...fortunate...to be employed... has diminished". This may be due to women finding it more difficult to seek jobs in economies that are experiencing contraction in the public sector while the private sector is not able to create enough jobs.

10:8 Trade Unions' Responses to the Structural Adjustment Programs

This section provides a summary description of some of organized labor's responses to the adverse living and working conditions resulting from the Structural Adjustment Program. One measure instituted by the unions in Ghana is the introduction of competition into union organizing. Sections of the labor movement have spearheaded a drive toward union plurality to permit workers more choices in union affiliation. The founding of the Textile Garment Leather Employees' Union (TGLEU) in 1991 marked the existence of the first union not affiliated with the main trade union – The Ghana Trades Union Congress (TUC). Also in 1999, Ghana's second labor federation was created when the Ghana Federation of Labor (GFL) was launched officially. According to the workers who lead these new labor organizations, they were motivated by the desire to inject "new blood" and competition into union organizing to meet the challenges of the Structural Adjustment Program (GFL, *Brochure for Founding Congress*, April, 1999).

The TUC through its own initiatives is becoming a trendsetter in union organization in Africa (Parry, 1995). In responding to the job losses resulting from the Structural Adjustment Program, TUC affiliates have mounted campaigns to unionize senior staff at Ghana's financial institutions including Standard Bank (TUC, 1996). Other unions are also aggressively recruiting members from Ghana's expanding informal sector to recoup the heavy financial and membership losses they have sustained under Structural Adjustment Program.

Four unions of the TUC have designated special divisions devoted to recruiting and retaining members from the informal sector (Parry, 1995). By 1995, an additional four unions were also negotiating with associations of workers employed in the informal sector. For example, the TUC's Public Service Workers' Union organized small-scale photographers into the Ghana Union of Professional Photographers. This union in addition is seeking official affiliation with the Locksmiths and Key Cutters Association. The Industrial and Commercial Workers' Union (ICU) of the TUC led hairdressers and barbers to form a formal professional association under the banner of the ICU. Thus one outcome of the Structural Adjustment Program is that unions in Ghana are expanding their activities into the informal sector to compensate for the losses they have incurred (Parry, 1995).

As a bold response to job cuts, in 1997 the TUC launched the Workers' Enterprise Trust. With a goal of raising 25 billion cedis as capital, it is intended to generate both jobs and incomes for workers (*West Africa*, May 19-25, 1997; TUC, 1996). The funds for this ambitious enterprise are to be raised through workers' monthly contributions.

A concrete effort by the Mine Workers Union to avert job losses partly induced by the privatization of mines in Ghana led to workers at the Prestea Gold Mine assuming both ownership and control of this mine. After Barnex, a privately-owned company shut the mine down for allegedly being unprofitable, the 1,473 employees pooled their pension funds and formed the Prestea Gold Resources and acquired the mine (*The Ghanaian Worker*, April, 1999:16).

Efforts by the Prestea mine workers to preserve jobs may not only be bold but also run counter to global trends: In different regions of the world, especially the former Soviet bloc, worker-owned enterprises have been collapsing or abandoned since 1990. As it is in Ghana, on the other hand, in the case of the mine workers, they are going to great lengths to preserve good paying jobs and hence workers' economic security by creating worker-owned enterprises. Thus workers seek to fill the huge employment void created by the withdrawal of the state from the economy.

Led by the TUC, Ghanaian unions are devising programs and have expressed interest in improving not only the numbers of women in unions but also importantly, their leadership roles. They are also concerned about women's working conditions and how they are affected by the Structural Adjustment Program (TUC, 1996). The unions seek to accomplish these new objectives at the time when the Structural Adjustment Program makes it extremely complex and difficult to achieve them.

10:9 The Human Resource, Gender and Public Policy Implications of the Structural Adjustment Programs: Some Concluding Observations

A major public policy challenge the Structural Adjustment Program poses is that potentially it could threaten the current relatively equitable distribution of wealth in Ghana. According to the EIU (Ghana Country Profile, 1998 and 1999) Ghana has the least skewed ownership of resources in Sub-Saharan Africa. The Structural Adjustment Program has enormous potential to worsen the distribution of wealth by creating a society of a few and extremely wealthy and technically skilled ("hi-techs") and very cosmopolitan individuals and classes living with a vast majority of illiterate and under employed or unemployed. Ghana could for the first time be on the verge of creating a real and massive underclass of citizens due to the socially inequitable dimensions of the Structural Adjustment Program such as pricing education and health out of the reach of the poor. We might note that the current relatively unskewed distribution of resources in Ghana and the modest gains made by women were the direct consequences of public policies deliberately aimed at enhancing social mobility as part of national development. On the contrary, as Mhone (1995:51) has observed, under the Structural Adjustment Program, schools could "assist in the reproduction of gender differences in human capital formation" to the disadvantage of young girls and women.

Irrespective of the policy objectives officially assigned to the Structural Adjustment Program, it might be observed that in Ghana under the Structural Adjustment Program living and working conditions (from the point of view of the average Ghanaian) have not improved substantially. It has on the other hand engendered economic and social conditions with potentially adverse political repercussions and compounded Ghana's development challenges. For example, the society is grappling with education as a privilege for which the individual is privately and financially responsible. But as fervent advocates of the Structural Adjustment Program would concede, economic growth, even defined narrowly, requires a skilled and healthy work force. This requirement becomes more urgent, when as in the case of contemporary Ghana, attracting foreign private investment is deemed so crucial to the economy.

Notes

[1] Ghana's gross domestic product grew at approximately 5 percent; cocoa, diamond, and timber exports increased and infrastructure, especially roads and seaports for exports were rehabilitated.

238 IMF and World Bank Sponsored Structural Adjustment Programs in Africa

2 As part of the Structural Adjustment Program and to drastically reduce the national budget, since the late 1980s, the Ghana Government has made local governments responsible for 25 percent of the nation's budget.

References

Berry, L. (ed.) (1995). *Ghana: A Country Study.* Washington, D.C., Federal Research Division, Library of Congress.

Department of Labor, Ghana (1995). *Labourscope*, 54.

Department of Labor (1996). *Annual Reports 1975-1990*. Accra: Labor Department.

Dodoo, R. (1994). Brief Notes on the Current Status of the Proposed New Pay and Grading System for Civil Servants. Accra: Public Service Commission.

Economist Intelligence Unit (EU) (1989). *Country Reports*. Ghana. London: EIU.

Economist Intelligence Unit (EU) (1991). *Country Reports*. Ghana. London: EIU.

Economist Intelligence Unit (EU) (1993). *Country Reports*. Ghana. London: EIU.

Economist Intelligence Unit (EU) (1994). *Country Reports*. Ghana. London: EIU.

Economist Intelligence Unit (EU) (1995). *Country Reports*. Ghana. London: EIU.

Economist Intelligence Unit (EU) (1998). *Country Reports*. Ghana. London: EIU.

Economist Intelligence Unit (EU) (1998). *Country Reports*. Ghana. London: EIU.

Economist Intelligence Unit (EU) (1999). *Country Reports*. Ghana. London: EIU.

George, B. S. (1976). *Education in Ghana*. Washington, D.C: U.S. Government Printing Office.

Government of Ghana (Jan. 1995) Ghana-Vision 2020 (The First Step: 1996-2000). Accra: Government Printer.

Greenstreet, M. (1979). "The Woman Wage Earner in Ghana" in U. Damachi and K. Ewusi (eds.) Manpower Supply and Utilization in Ghana, Nigeria and Sierra Leone. Geneva, International Institute for Labor Studies: 18-30.

ILO (1995). Impact of Structural Adjustment in the Public Services (Efficiency, Quality Improvement and Working Conditions). May. Geneva: ILO.

ILO/UNDP (1997). *Jobs for Africa.* Geneva: ILO.

Kraus, J. (1991). "The Struggle over Structural Adjustment in Ghana". *Africa Today.* 4th Quarter, 38:19-37.

Mhone, G. (1995). *The Impact of Structural Adjustment on the Urban Informal Sector in Zimbabwe*. Geneva: ILO.

Ministry of Employment and Social Welfare (1998). *Labor Market Skills*, 1, March. Accra, Ghana.

Obeng-Fosu, P. (1991). *Industrial Relations in Ghana: The Law and Practice*. Accra: Ghana Universities Press.

Panford, K. (1994). "Structural Adjustment, the State and Workers in Ghana". *Africa Development*, xix, 2: 71-95.

Panford, K. (1996). "IMF/World Bank's Structural Adjustment Policies, Labor and Industrial Relations and Employment in Ghana". *African Development Perspectives*

Yearbook 1994/1995, Vol. iv (ed.) Wohlmuth, K. and Messner, F., Lit Verlag, Munster/ Hamburg, 193-205.

Panford, K. (1997). "Ghana: A decade of IMF/World Bank's Policies of Adjustment (1985-1995)". *Scandinavian Journal of Development Alternatives and Area Studies*, 16(2):81-105.

Panford, K. (1999). "The ILO and Employment in Sub-Saharan Africa". *African Development Perspectives Yearbook*, Vol. VII Institute for World Economics and International Management. Bremen, Germany, p. 319-334.

Panford, Kwamina (Forthcoming). "The Ghana TUC, Politics and Democracy in Ghana (1985-1992)", Dakar, Senegal: CODESRIA.

Parry, F.A. (1995). "Institutional Linkages Between Trade Unions Informal Sector in Ghana". (September). ILO/EAMAT, Addis Ababa Ethiopia.

Rothchild, D. ed. (1991). *Ghana: The Political Economy of Recovery.* London, Lynne Reinner.

SAPRIN (1997). "Structural Adjustment Participatory Review International Network". *World Economy*, May, Vol. 20: 285-305.

SAPRIN (1999). "SAPRIN in Ghana". *The Structural Adjustment Participatory Review Initiative.* August 22. <http://www.worldbank.org/research/sapri/ghana.htm>

Schneider, G. (1999). "An institutionalist assessment of structural adjustment programs in Africa". *Journal of Economic Issues.* June XXXIII, 2: 325-334.

Trades Union Congress (1996). *Report of the Executive Board.* 5th Quadrennial Delegates Congress. Accra: TUC.

Tutu, K. (1994). Structural Adjustment Programs and Their Effects on Ghanaian Workers. Accra: Friedrich Ebert Stiftung.

World Bank (1999). "Gender, Growth, and Poverty Reduction". *Technical Paper No. 428: Special Program of Assistance for Africa: 1998 Status Report on Poverty in Sub-Saharan Africa.* Washington, D.C.: World Bank.

11 Structural Adjustment, Policies and Democracy in Ghana*

KWAME BOAFO-ARTHUR

11:1 Introduction

Most African countries have pursued political and economic reform at the same time, frequently under pressure from donors. For many years it was acceptable to donors for governments to be undemocratic, provided they accepted donor conditions on economic policy. However, by the end of the 1980s, donors became less willing to tolerate undemocratic forms of government, even if economic policies were acceptable. With the publication of its 1989 report on adjustment in Africa, the World Bank targeted the political barriers to economic development and raised expectations for political reform, developing a new focus on governance that emphasized transparency and consultation in policy making (World Bank, 1989).

The introduction of political conditionality coincided with, and was in part caused by, the end of the Cold War. The new "political conditionality" adopted by most donors after the democratization of Eastern Europe threatened to cut development assistance flows to countries that failed to show progress in democratization. Consequently, economic liberalization and sustained structural adjustment are seen as being inextricably linked to political institutional reform (Callaghy, 1993).

However, the association of political liberalization and democratization with successful economic reform has been the subject of increasing criticism. The basis of such criticism is that it is difficult to introduce, and even more difficult to sustain, both processes, especially for governments susceptible to popular opinion-as expressed in elections-and to pressure from interest groups allowed to be active as a result of democratization or political liberalization. Similarly, structural adjustment reforms undermine vested interests and may, therefore, be prevented or unduly delayed by pressure from interest groups. Moreover, mac-

---

*An earlier version of this chapter was published in African Studies Review, 42 (2), 1999.

roeconomic balance invariably requires reducing the budget deficit; and this in turn may require major structural reforms such as downsizing the civil service and privatization of state enterprises (Harvey and Robinson, 1995; Herbst, 1993). Delay in such structural reforms, made possible by democratization would, therefore, prevent sound macroeconomic policy from being implemented successfully, as well as obstructing the conditions needed for long-term development. In a nutshell, the fear of advocates of neo-liberal adjustment programs is that with democratic transition democracy will undermine reform programs, as governments cave in to popular demands for public spending, new protections, and rents (Harvey and Robinson, 1995).

It is against this background that this paper examines the factors that in the Ghanaian case explain the pursuit of adjustment programs and democratization without seriously undermining the former. Specifically, the following questions will be addressed.

- What socio-historical factors led Ghana toward the path of market reform and multi-partyism, in spite of the fact that Rawlings's populist rhetoric, in the aftermath of his second coup, initially suggested otherwise?
- How important have donors been in Ghana's political and economic transformation, and what are the possible contradictions of the externally imposed reform?
- Given the characteristics of adjustment programs and of democratic gover nance, to what extent have Ghanaians been able to cope with the former in an era of the latter?
- What factors either promoted or undermined the pursuit of economic and political reforms *pari passu*?

Is it possible for democracy and adjustment to be pursued in tandem?

Structural adjustment in Ghana has had mixed impact on Ghana's development. Nonetheless, international financial institutions (IFIs), the World Bank, and the IMF still underscore adjustment's relevance in a liberalized developing economy. Democratic governance is equally perceived as a sine qua non for national development. The problem seems to be the theoretically assumed antithetical relationship between structural adjustment and democratic governance. I argue, therefore, that if economic crisis and the need for political liberalization have made adjustment and democratization imperative for Ghana's development, there is the need to adjust adjustment" to suit democratic governance.

The chapter is divided into four main parts. Part 1 examines the theoretical debate on the relationship between economic reform and democratization. Part 2 analyzes the background to the introduction of Structural Adjustment Programs. Part 3 discusses the impact of adjustment policies. The final part examines the rationale for the seemingly successful pursuit of adjustment after democratization by the government of the NDC led by Flight Lt. J. J. Rawlings.

11:2 The Theoretical Debate

According to Mkandawire (1992), democracy is incompatible with adjustment packages because invariably countries that implement adjustment programs come under the indirect control of external donors who at times make such controls conditional to the grant of financial assistance. Mkandawire argues further that the International Financial Institutions' (IFIS) insistence on loan repayment will undermine democratic governance. Since most of these countries are poor and depend on external assistance, any rigid insistence on loan repayment may derail the reform program for the simple reason that such countries may not have enough money to build democratic structures. More significantly, democratic governance normally gives a lease on life to civil associational groups whose zeal for survival, through various political and economic modes, might conflict with the pursuit of adjustment policies. For that reason, "adjustment programs will pose severe strains on new democratic governments" (1992: 310).

Stephan Haggard and Robert Kauftnan (1989) offer three main reasons for the perceived incompatibility of adjustment and democracy. First, new democratic leaders are confronted with previously repressed demands, heightened social and economic expectations, and strong pressures to reward supporters and incoming groups. Second, as much as possible, newly installed democratic governments will try to avoid difficult economic policies that might generate both economic and political resistance or unrest. Finally, democratization will give vent to heightened social demands which most probably will lead to the pursuit of policies that will increase the expectations of the electorate, reduce social conflicts in the short run, and thereby garner political support. The concern for public support might compel democratic governments to pursue populist policies at the expense of structural adjustment or strict financial discipline.

Another explanation is that the "medicine of adjustment" is unlikely to thrive in new democracies because of weak political institutions, frequent elections, mobilized populations, and weak party loyalties (Remmer, 1993). Logically, politicians would be inclined more toward the adoption of short-term economic and political policies that would guarantee their popularity and hold on power than the pursuit of adjustment policies that will have the opposite impact.

The foregoing explanations for the assumed incompatibilities fall in line with the reasoning of Sirrowy and Inkeles (1990), who argue that democracy in least developed countries (LDCS) will lead to certain dysfunctional consequences such as political instability. This is because political institutions in such societies tend to be weak and fragile. Democratic governance therefore leads to the exertion of undue pressures on newly created participatory institutions. Pressures are intensified because channels through which impatient groups such as workers and the poor can express their grievances are made available. Ruling elite may accede to such demands in order to win support in elections. Consequently, the democratic institutions become overburdened. Newly democratic regimes under such circumstances shift to the maintenance of internal order.

Sirrowy and Inkeles further contend that because of the political and civil liberties they rest on, democratic regimes only act to inflame social divisions and erode the capacity of the government to act quickly for the expected results to be attained. The assumption is that government officials may be compelled to shift their allegiances among policies based on short-run political expediency rather than focusing exclusively on policies oriented toward national development in the long run.

The fear is that too much pressure in the form of various politico-economic demands may jeopardize the long-term success of adjustment programs. Thus the dictates of democracy will overshadow the rigidities of adjustment, or the strong commitment to adjustment will encourage the flouting of democratic norms. Democratic governments that become too preoccupied with the issues of distribution, whether political or economic, due to pressures from associational groups do so at the expense of fiscal discipline required by adjustment.

One other reason underlying the assumption of incompatible relations between adjustment and democratization is the penchant of democratic governments for the pursuit of populist oriented policies aimed at maximizing electoral gains and support. Such populist policies tend to undermine the stringency of adjustment.

In sum, because of the openness of a democratic system of government and the need for internal legitimacy, such governments tend to become Victim to claims from several segments of the society. These claims invariably engender governmental preoccupation with distributional issues such as expansion of government benefits and welfare policies at the expense of structural adjustment policies of contraction in welfare benefits.

The Ghanaian experiences during the first term of the Fourth Republic from 1993 to 1996 seem to question the assumed incompatibility between adjustment and democratization. Whether adjustment and democratization can be pursued pari passu or not is a question that must be analyzed within the context of a nation's history. As argued by Rudebeck (1992: 268), "social action takes place on a 'stage' structured by class, institutions and culture". Political and economic policies and analyses, for that matter, are informed, shaped, and influenced by a wide array of factors that may make outcomes theoretically anticipated but unrealizable in reality. That is, certain societal factors in particular political settings may tend to sustain adjustment under democracy. In the Ghanaian case, for instance, the prior pursuit of adjustment under a dictatorial government as well as post-transition political dynamics go a long way toward explaining the continued pursuit of adjustment under democratic governance.

11:3 Background to Structural Adjustment Programs

On December 31, 1981, the Provisional National Defense Council (PNDC) under the chairmanship of Flight Lt. Jerry John Rawlings came to power in Ghana after ousting the civilian government of the Peoples' National Party (PNP) led by Dr. Hilla Limann. That was the second time Rawlings intervened in the administration of the country through a military coup. Rawlings' first intervention was on June 4, 1979, when he led junior army officers to oust the Supreme Military Council (SMC 11) from power. Paradoxically, Rawlings overthrew in December 1981 the very government to which he had handed over power after general elections in 1979.[1]

The PNDC espoused socialism and launched what Rawlings termed the "peoples revolution" on December 31, 1981.[2] In April 1983, a landmark policy change occurred. Socialism was abandoned by the PNDC and Ghana embarked on the World Bank/IMF-inspired structural adjustment programs. In fact, between 1983 and 1992, Ghana implemented six IMF reform packages in which the most severe of austerity measures were put in place. In that period, Ghana experienced unprecedented repression under the rule of the PNDC.

For sure, Ghana is not a newcomer to IMF/World Bank-sponsored stabilization and adjustment policies (Hutchful, 1985; Jonah, 1989; Libby, 1976). The Progress Party government of Dr. K. A Busia unsuccessfully implemented economic stabilization measures in 1971. Economic stabilization was also seriously contemplated but not implemented by the Peoples' National Party (PNP) of Dr. Hilla Limann (1979-1981) because it was thought to be a recipe for military intervention. In the early 1960s, Dr. Nkrumah's Convention Peoples' Party (CPP) flirted, albeit abysmally, with stabilization policies. The major question is why the PNDC government that professed socialist leanings implemented IMF/WB-sponsored adjustment policies. This could be answered by examining the historical trajectory that seems to have made the implementation of adjustment policies a kind of a "Hobson's choice" for the PNDC. Such an examination will also help us to appreciate why most of the theoretical predictions have so far not been borne out by the Ghanaian experience.

First and foremost, the economic imperatives that led to the implementation of economic recovery and structural adjustment programs (ERP/SAP) in Africa in general and Ghana in particular have been widely discussed by academics, statesmen, and World Bank officials. For Africa in general, the 1980s ushered in a completely new global economic situation. In the words of Moseley *et al.* (1991:9), serious economic problems "placed a premium on finding ways of bringing down developing countries' payments deficits to the level that could be financed by stagnant aid flows plus rapidly dwindling private commercial lending".

By the early 1980s, Ghana had reached abysmal levels in its socioeconomic development. Only effective and sustainable economic measures could salvage the economy. Because Ghana is a mono-crop agricultural country, which for a long time was dependent on cocoa for a major proportion of its external earning, the lack of incentives for agricultural production, overexploitation of cocoa farmers, and poor agricultural pricing policy were quite destructive. The cumulative effect of the poor agricultural policy over the years was a precipitous decline in cocoa production from 403,000 tons in 1970 to 179,000 tons in 1983 (Kusi, 1991).

Non-performing state-owned enterprises also constituted a huge drain on government finances. Apart from operating deficits of 0.2-3.4 percent of GDP, most of the enterprises depended on government subsidies to the tune of 9 percent of government expenditure (Kusi 1991: 188-89).

Political instability and maladministration over the years had an equally devastating toll on the economy and national development. Of special significance are the frequent changes in government through undemocratic means since independence. Between 1966 and 1981, there were four successful military interventions and one successful palace coup (1978). There were several abortive coups. These unconstitutional regime changes led to grave policy discontinuities that had a deleterious impact on economic planning and development. First, the instability and consequent policy changes discouraged external investors. Second, many domestic entrepreneurs were victimized through unwarranted seizure of properties, especially after the successful military interventions in June 1979 and December 1981. Third, adhocism in policy measures became a distinctive feature of economic planning in Ghana. Irrespective of their merits' it became the norm for the policies of ousted regimes to be jettisoned by the new rulers. National resources were wasted as a result. Typical examples were the Seven Years Development Plan of the First Republic (1957-1966), and the Policy on Rural Development of the Second Republic (1969-

1972). The National Liberation Council (NLC) for instance, abandoned most of the industries established by the Nkrumah government with support from the former Communist countries. Finally, political instability and the accompanying economic stagnation contributed to waves of migration by qualified Ghanaians to other countries in the West African sub-region and other places. It must be underscored that "economic policies, no matter how sound usually do not have instantaneous effect" (Anyemedu, 1993: 18). Thus the political instability and the consequent disruption of economic policies had serious negative implications for the economy.

The consistent decline in cocoa prices on the world market for several years contributed in no small measure to the economic difficulties faced by past regimes. The decline in cocoa prices on the world market went side by side with decline in production in the country. It cannot, however, be gainsaid that the inability of past governments to take bold measures to stabilize the economy was a major contributory factor to the rapid economic deterioration, stagnation, and corruption that became distinct features of both political and economic life. For instance, by the time the PNDC decided to go to the World Bank, the effective exchange rate of the cedi had appreciated by 816 percent (Anyemedu 1993: 14). By the time of the second *coup d'etat* of Rawlings on December 31, 1981, the economy of Ghana had almost been throttled to a halt. Between 1970 and 1982, income per capita fell by 30 percent and real wages by 80 percent; import volume fell by two-thirds; real export earnings fell by one-half, and the ratio of Ghana's exports to GDP dropped from 21 to 4 percent. Furthermore, the domestic savings rate fell from 12 to 3 percent, and the investment rate from 14 to 2 percent of GDP; finally, the government deficit rose from 0.4 to 14.6 percent GDP of total government spending (World Bank 1984: 17). Apart from the poor state of the economy, other factors contributed to the decision of the PNDC to seek IMF/World Bank assistance:

• Between 1978 and 1979 and between 1982 and 1983, severe drought and bush fires adversely affected agriculture and food production. The 1983 drought, for instance, affected the production of hydroelectric power and consequently industrial production.

• In 1983, an estimated two million Ghanaians were expelled from Nigeria. The expulsion came at a time when drought and bush fires had devastated agriculture and food production. The PNDC had to find solutions to the unanticipated socioeconomic as well as political problems created by the "returnees".

• The regime placed much confidence in the Communist bloc's ability to provide needed economic assistance for socialist reconstruction in Ghana. Rawlings was frustrated, however, because little or no assistance from the so-called progressive countries on which much hope had been placed was received. For instance, in April 1982, PNDC delegations to Eastern Europe and Cuba in search of financial support returned empty-handed (Hansen, 1987). In the Soviet Union, the delegation was advised to go to the IMF/WB (Boafo-Arthur, 1989).

Another significant factor was that left-wing elements (who initially dominated the PNDC) were unable to come up with any meaningful suggestions on ways and means for tackling Ghana's economic woes. They constantly repeated their known rhetorical refrain about ensuring "fundamental changes in the structure of the economy with the mobilized mass of the people providing the main dynamo" (Hansen, 1987: 192). The initial policy measures pursued under the heavy influence of the left-wing elements were chaotic and unproductive.

In order to resuscitate the economy, the PNDC launched an IMF/WB sponsored Economic Recovery Program (ERP) in 1983. The major features of the policies pursued included labor retrenchment, trade liberalization and devaluation, subsidy withdrawal, and an increase in user fees (which rendered basic services like health and education unaffordable to the average worker). These measures were seen by many as politically risky because of Ghana's past experiences with the implementation of adjustment policies. For instance, the adoption of an IMF stabilization package in 1971 led to the fall of the Busia regime in 1972 (Libby, 1976). Macroeconomic reforms therefore tend to have very serious political implications in Africa in general and Ghana in particular. Leftwich (1994: 362) argues "no significant change occurs in society without destabilizing some status quo, without decoupling some coalition and building another, [and] without challenging some interests and promoting others". Ghana's experiences with adjustment since the PNDC era demonstrate that a nation's ability to plan and implement adjustment was the outcome, among others, of both political commitment, capacity, and skill, as well as bureaucratic competence, independence, and probity (Healey and Robinson, 1992: 91, 155).

Some argue that with the assistance of the Bretton Wood institutions, the PNDC was able to pursue "Africa's most successful stabilization and structural adjustment program..." (Toye, 1991: 155). In fact, the adjustment programs have proved fairly successful when the current economic situation is compared with the state of the economy before the implementation of the programs. In general terms, however, the outcomes have been mixed. The most important question is the level of success of the programs.

11:4 Impact of Structural Adjustment Programs

For sure, the implementation of Structural Adjustment Programs stalled not only the economic distortion but also its stagnation. Structural Adjustment Programs has proved, however, to be a double-edged sword. In the short term, the economic policies led to appreciable growth in the national economy. GDP in real terms increased by 5.3 percent in 1986. Per capita real income grew by 2.6 percent; agricultural output increased by 4.6 percent, while services expanded by 5.4 percent (Government of Ghana 1987).

Adjustment restored economic growth in various sectors with favorable prospects for industrialization, even though the underutilization of industrial capacity increased alarmingly, falling from 25 percent in 1981 to 18 percent in 1984 (Ewusi, 1987). Similarly, Asenso Okyere (1989) makes the same positive inferences with regard to the response of farmers to adjustment policies. In the view of Donald Rothchild (1991: 10-11), Structural Adjustment Programs "reversed the decline of recent years". Relying on statistical indicators gleaned from the World Bank and other sources (which he acknowledges "lack precision"), Rothchild nonetheless paints a favorable picture of development in Ghana due to the implementation of adjustment policies.

There has been infrastructural development (Boakye-Danquah, 1992). Kwasi Anyemedu (1993) has provided an insightful analysis of the pros of the adjustment program. Apart from justifying the policies on purely macroeconomic grounds, he provides generous statistical indicators to buttress claims of economic growth, reversal of production declines in major sectors (especially agriculture), moderation of inflation, stimulation of exports, improvements in the balance of payments, and increase in domestic savings.

With reference to the level of implementation and macroeconomic outturn, Gibbon (1992) states that adjustment in Ghana has been Africa's most successful. This is based on the fact that as a result of the implementation of structural adjustment, the economy witnessed appreciable macroeconomic outrun. GDP per capita increased by about 2 percent between 1983 and 1987, and cocoa production increased by about one-third between 1982 and 1986, while domestic financing was controlled and revenue collection increased sharply. A World Bank report published in 1993 entitled *Ghana: 2000 and Beyond* also flaunts the positive impact of adjustment on the Ghanaian economy. Among other favorable comments – some of which have been noted by others – is that Ghana's adjustment program is by any yardstick one of the most successful in SSA. A decade of stabilizing policies has yielded, among other results, broad budget balances, strong export growth, and a reasonable external position.

The negative implications for industrial development have equally received attention from various sources. Dharam Ghai (1991) notes that policy measures have deliberately ignored industrialization and rather have given greater incentives to the production of primary commodities. The net result of this bias of adjustment policies, according to Folson (1990), has been the immobilization of industrialization in Ghana. The president of the Association of Ghanaian Industries (AGI) noted the negative impact on industries of import liberalization. In his words, "only limited production can be managed and others are faced with the prospect of closing down their establishments and retrenching their labor" (cited in Anyemedu 1993: 45).

The programs also had deleterious impacts on private companies. The development of such companies has been constrained by high interest on bank loans and inflation. However, there are contradictory figures with regard to the impact of the programs on inflation. This may be due, among other reasons, to different time periods covered by different studies. The fact that some of the periods overlap puts serious question marks on statistical figures on inflation. According to Kusi (1991), studies have found that inflation was drastically reduced from the high 122 percent in 1983 to an average of 32 percent per annum in 1990. Peter Gibbon (1992) also notes that there was over 50 percent decrease in inflation between the early 1980s and 1989. This was a dramatic fall in inflation. The Home Finance Company (HFC), a mortgage company, however, identifies inflation as its major problem. Inflation, according to the managing director of the company, tripled from 23 percent in 1994 to 61 percent in 1995. As a result, interests paid by the company also doubled from 15.5 percent to 35 percent between 1992 and 1994. Inflation and high interest rates have impacted negatively on many companies (Panford 1997). The point to be noted is that inflation is still a problem even though it is no longer in three digit figures.

The benefits from competitive prices paid to farmers have at best been cosmetic. Most are small-scale farmers, so very little is obtained from sales on the basis of economies of scale. The truth, however, is that the gains made on the basis of enhanced pricing for agricultural products have been whittled away by the removal of subsidies and the introduction of user fees in hospitals. Two years after the introduction of substantial increases in user charges in government health centers, Ghana attained the targeted 15 percent savings on the cost of social and health services. However, the cost recovery system "discouraged use of the centers by the rural poor, especially women, turning many of them to less qualified traditional healers, commercial pharmacies, and unlicensed drug sellers" (Sawyerr 1991: 6).

Another negative aspect of adjustment is the heavy reliance on external financial support. Ghana faces a bleak future because of the over-dependence on external financial inflows. The high levels of dependence on foreign aid equally cast a shadow on the impact of structural adjustment. Because of abnormally high levels of assistance from donors there is much difficulty in disentangling the effects of external aid from the effects of adjustment programs on Ghana's development (Killick and Malik 1992). Killick and Malik therefore call into question the resilience of the much-touted economic turnaround.

The financial inflows to Ghana of about 8 percent of GDP in recent years is higher than the 3 percent of annual capital inflows of GDP that went to Malaysia and Thailand in the period 1969-1983. Since the implementation of adjustment programs Ghana has been receiving steady inflows of assistance to the tune of five hundred to six hundred dollars per annum (Ampem 1994). To date, external donors have injected a total of almost $8 billion into the Ghanaian economy (Appiah-Opoku 1998). Clearly the Ghanaian economy is aid-driven. Dependence on external aid has consequently escalated the debt burden with high debt service ratio (see Chapters 2:4.1 in this volume).

It is very clear from the foregoing that the nation's economic future is not so bright due to dependence on foreign aid, mounting debt burden and escalating debt service ratio. To try to ameliorate the social effects of the programs, the government introduced in November 1987 the Program of Action to Mitigate the Social Cost of Adjustment (PAMSCAD). PAMSCAD could not reverse the negative social and economic effects of Structural Adjustment Programs on the people, especially the poor in Ghanaian society. A key government functionary, Captain Kojo Tsikata, the security boss, reiterated this fact in October 1989. In an address to revolutionary cadres, he noted that irrespective of what the IFIs may say in praise of the economic recovery program, "life was tough for the common man who deserves to enjoy more and more, the fruits of recovery" (cited in Anyemedu 1993: 39).

In sum, even though current economic indicators stand in sharp contrast to the dismal picture of 1981 when the PNDC took over, Structural Adjustment Programs in my view have had mixed impacts on economic development and growth. Many macroeconomic indicators show positive effects of structural adjustment on development and growth, many Ghanaians find it difficult to cope with the overall effects of adjustment. The problems posed to many Ghanaians by the high cost of utilities, the withdrawal of subsidies on health, education, and agricultural inputs, the retrenchment of labor and the consequent high unemployment rate, and chronic low salaries for workers, especially in the public sector, continue to engage the attention of many.

A successful imposition of policy measures through undemocratic tactics is completely different from the overall attainment of the set objectives of such policy measures. The implementation of structural adjustment was without citizen participation or input in the formulation and implementation of the various programs. Simply, there was no democratic touch to the whole formulation and implementation. In addition, there existed an illiberal and intimidating political climate that made it difficult for various groups opposed to such programs to rise up against their imposition. There was, therefore, the genuine fear that democratization and its embedded freedoms would lead to a revolt by the masses against further imposition of draconian adjustment policies. Donors were equally concerned that "democratization might make economic reform more difficult or even impossible" (Harvey and Robinson 1995: 2). Ho-Won Jeong (1997) states categorically that the reelection of Rawlings for a second term in December 1996 means that Structural Adjustment Programs has managed to survive popular discontent in a democratic dispensation. That is to say that the conceptualization of an antithetical relationship between adjustment and democratization is not explicitly borne out by the Ghanaian experience. What then accounts for the government's ability to sustain adjustment policies under a constitutional regime?

11:5 Accounting for the Sustenance of Adjustment Under Democracy

The implementation of structural adjustment policies before the transition to democracy without any serious opposition from civil society could be attributed to several factors. These include the mode of governance adopted by the PNDC. In a nutshell, the mode of governance outlawed citizen participation in the formulation and implementation of the programs. The dictatorial policies of the PNDC also made it difficult for those opposed to the reform measures to protest. After the transition, however, the crucial factors that have contributed to the perpetuation of adjustment policies have revolved around the dynamics of Ghanaian politics.

In the Ghanaian situation the perceptions of the most critical sociopolitical forces-the opposition parties and civil society organizations should be the best and foremost yardstick in rationalizing the continued pursuit of adjustment under democratization. First, what have been the reactions of opposition parties to adjustment in an era of democratic dispensation? To what extent does the opposition's agenda differ from the ruling government? How have they approached the issue of the negative impact of adjustment on labor, for instance? Have they been able to exert pressure on the NDC government to change course with regard to the pursuit of adjustment policies? Second, what about the ruling party itself. How strong is its resolve in the implementation of adjustment programs under democratic government? Third, how have the other social forces including the labor movement reacted to adjustment after the transition to democracy? Have the so-called masses been able, at long last, to protest against the pursuit of adjustment policies? Fourth, what has been the role of the international donor community? Has it been supportive or neutral?

To date, the ruling government has been able to manage a fair balance between democracy and structural adjustment because of the inherent dynamics of Ghanaian politics in the Fourth Republic and the continued support of the international community. These domestic dynamics include: the opposition's generally supportive economic agenda; the lack of a viable alternative to adjustment; the commitment of the ruling party to the process and political brinkmanship; and weak social forces. The interplay of the above factors is examined below.

11:6 Opposition Parties

Harvey and Robinson (1995) found in their study of eleven African countries that political debates during the transition process were concerned mainly with political issues. It is instructive to note that the economy did not become an issue during the electioneering campaign of the parties in Ghana. The manifesto of the strongest opposition group (NPP) skirted the market-oriented policies of the erstwhile PNDC. It seemed to have agreed to most of the neo-liberal economic measures being pursued for two main reasons. First, there was a conviction that such programs are efficacious if properly managed. In any case, NPP stalwarts saw such measures to be in line with the bona fide philosophy of their predecessors. The major bone of contention was the role of private enterprise in national development and not the pursuit of adjustment policies per se. Second, top-notch opposition members were not prepared to offend donors for fear that if they came to power on the platform of opposition to Structural Adjustment Programs, external assistance might be curtailed.

The manifesto of the NPP decries in no uncertain terms the controls instituted to curtail efforts by individuals to create wealth and the president's deeply ingrained suspicion and distrust of private entrepreneurs. Their objective, if they had come to power, was to create an enabling climate that would enhance the development and growth of the private sector. But since the ongoing privatization policy appears to focus on this, only a thin line of demarcation exists between what the NPP would have done, if it had come to power, and what the NDC government has been able to do so far and still is pursuing relentlessly. So far as structural adjustment is concerned, it appears that the long-term survival imperatives may have compelled the opposition parties to overlook the issues associated with Structural Adjustment Programs to avoid upsetting the donor community. This perception, most probably, influenced the NPP campaign and overall political strategy.

It is a mark of political objectivity that the manifesto of the NPP endorses the expansion of the mining sector under the adjustment program and acknowledges the higher turnover of trading due to liberalization. The party leadership talks in terms of an "adjustment with a human face". It stands to reason, therefore, that the main opposition groups that have the capabilities and organizational strength to capitalize on the wind of change to oppose adjustment are *ad idem* with the NDC government with regard to the pursuit of adjustment policies. In a sense, and rather naively, the NPP in particular perceives the NDC government as "usurpers", and the Ghanaian elite as "opportunists" who had thrown overboard their own ideology of socialism and appropriated whatever the government had to offer. Such a misconception implied further that the practice of neo-orthodox economic policies – the backbone of liberalism – is the exclusive preserve of the NPP tradition in Ghana and nobody outside the fold could meaningfully implement or practice it.

It was a position that could not be sold to any electorate, especially in Ghana where the majority of the electorate is illiterate. Domestic opposition to the PNDC was premised in most cases on the poor human rights record of the government. Democratization and the improvement in the human lights record of Rawlings removed one of the major planks for organizing the people against the government. Therefore, there appears to be no strong viable platform that will help in mobilizing the people against the NDC.

What further enhanced the political climate for the continuation of the adjustment program was the concept of "doing business" with the government, which was initially broached by the NPP (Boafo-Arthur, 1995). Direct talks with the government that aimed at lessening tensions and resolving outstanding political issues got underway between the NDC and the NPP on November 10, 1993. Issues discussed revolved around the controversial voters' register, citizens' identification cards, and unconditional amnesty for exiled Ghanaians. The government's economic program, namely, structural adjustment, never featured on the agenda. This further underlined the tacit acceptance of the ongoing economic reforms by the opposition. Earlier on May 21, 1993, the NPP presented a memorandum on the 1993 budget to Parliament. In his acceptance speech, the Speaker of Parliament, D. F. Annan, stated, "we need to work together to define our problems and look for solutions that are acceptable to us" (*Daily Graphic*, May 22, 1993). Perhaps the "doing business" policy that later emerged was a direct result of Annan's speech.

Most rank and file supporters of the party feel that the NPP's National

Executive Council erred in not consulting them on the decision to "do business" with the government. In an interview to explain the NPP's position, Dr. Nyaho-Tamakloe (the chairman of NPP's sector committee on health and social services) noted:

> We just want to prove to the western nations who have put so much money into P/NDC ERP-SAP that when we are given the chance we will do better than Rawlings and his people. We want them to have confidence in us (*The Ghanaian Voice* 25/27, Oct. 1993).

Another leading member of the NPP and an economic consultant, Kwame Pianim, noted that the Fourth Republic has brightened the prospects for free enterprise in Ghana. He further commended the government for the congenial environment it has created for liberalization (*Daily Graphic*, Jan.8, 1994).

The position of the NPP with regard to dialogue or doing business with the government was reaffirmed at the party's second annual delegates conference held on December 18, 1993. The objective of the party, as stressed in the resolution, was to ensure that fair and satisfactory conditions exist for holding transparently honest and democratically free elections ... in the future and that a political atmosphere conducive to that and the attainment of this end is created and fostered in this country (*Free Press*, Jan. 7-13, 1994). However, it was added that the national executive could withdraw from the talks if the government did not within a reasonable time exhibit its good faith and willingness to continue with the dialogue. The party did not explicitly attack the adjustment policies being pursued. In the resolution, however, it condemned in no uncertain terms the management of the economy of Ghana that has led to the following: (1) resurgence and acceleration of inflation; (2) rapid depreciation of the cedi exchange rate; (3) stunted growth rate of the national economy and declining per capita incomes; (4) rising unemployment; (5) incidence of increasing mass poverty and falling living standards; and (6) persisting imbalances and inequalities in the development of the regions and between the rural and urban areas.

The NPP's objection to aspects of the adjustment program as discussed further at the conference was only with reference to the mode of implementation of the divestiture or privatization program. Consequently, the resolution condemned the lack of transparency in the manner of implementation of the program and called on the government to suspend the divestiture program pending a review of its manner of implementation (*Free Press*, Jan. 7-13, 1994).

Even though the NPP-NDC dialogue collapsed in no time with accusations and counter-accusations of bad faith, the NPP's stance seems to have augured well for the emergence of some level of political consensus. Such a consensus has, to a large extent, facilitated the continuity of adjustment under democratic governance. The fact is the NPP is the strongest and most well organized political opposition group with the capability to mount effective opposition to the adjustment programs. Once the party shares the philosophy of adjustment, the opposition parties have never considered organizing party supporters against adjustment on the basis of its negative effects a viable strategy.

None of the other opposition groups advocated a contrary economic philosophy during the electioneering campaign. Thus by implication all the other parties, to various degrees, have been supportive of structural adjustment. For instance, J. Frimpong Ansah, who until the eleventh hour was tipped to be the presidential candidate of the People's Heritage Party (PHP), openly commended the PNDC's economic programs and indicated his willingness to build on it rather than to change it (*West Africa*, May 25-31,1992). The other parties lost their bearings after the elections. Attempts made to patch up the differences between the "Nkrumahists"[3] (who splintered into various groups to contest the election) eventually led to the merger of the PHP, National Independence Party (NIP), and a major segment of the National Convention Party (NCP) to form the Peoples' Convention Party (PCP). Thus while the NPP appeared to be in broad agreement with the economic reform program, the other Nkrumahist parties had to contend with effective measures to form one strong party.

Another factor of importance is the absence of a viable alternative to the structural adjustment program. Such an alternative would have been debated or formed the basis of political campaigning by the other contending parties. It appears that any alternative economic blueprint should get the endorsement of international donors. The opposition parties have not been able to float any comparable alternative to structural adjustment. Many leading opposition politicians feel that any contrary economic blueprint that fails to win the support of the donor community will be stillborn.[4] As indicated earlier, the NPP, which seems to have the capability to provide an alternative economic policy, incidentally shares, in a very broad sense, the core strands and philosophy of the current structural adjustment program. As a palliative and a mechanism for winning further support from the opposition, as well as containing criticisms leveled against the implementation process of the adjustment program, the government stepped up its support for the private sector. Thus, until the opposition presents an alternative economic program capable of winning both domestic and external donor support, no strong opposition to the broad outline of economic reform will emerge. The implication, then, is that structural adjustment will continue to coexist with democracy no matter how discomforting such coexistence may prove to be.

11:7 The Political Will of the Government

For the ruling party, there is no doubt about the need to continue with the adjustment programs. The illegal mandate of the PNDC had to be legitimized by the electorate at the polls in late 1992. The NDC's campaign slogan of "continuity" was equally unambiguous, with particular reference to adjustment policies. The initial problem for the NDC, just like the donors, was how to pursue adjustment in a liberalized political environment. Given the demands of democratic governance and the need to pursue adjustment policies, the sheer doggedness or political will demonstrated by the government so far has augured well for the coexistence of the two philosophies. A congenial political and economic climate must of necessity be forged or emerge from the normal internal and external political interactions to provide the symbiotic relations between democracy and adjustment. This is exactly what the government has succeeded in doing to date.

With the democratization process, the government has been constrained to at least give some thought to some of the issues raised by the opposition and the private press.[5] In contrast to the PNDC, which imposed the policy measure, the NDC after winning the elections proceeded to explain most of its programs to the rural people in accordance with democratic norms. The government has been able to appeal successfully to the conscience of the people to garner support for certain policies. By so doing, it has been successful, to some extent, in papering over its unenviable history mired in dictatorship.

The NDC's acceptance of a dialogue with the opposition (though not wholly in good faith) was meant to win the support of the largest section of the opposition party for the continuation of adjustment policies. Although the dialogue failed, such interactions with political opponents seemed to have assured the donor community of the existence of national consensus, a necessary ingredient for the pursuit of adjustment policies under democratic governance.

Ensuring such formal and informal interactions between government and the opposition further underlined the relevance of the opposition in the new political dispensation; more so, when the support of the opposition parties was necessary for the continuation of adjustment under democracy. There is some truth in the words of Eckhart Klein and Thomas Giegerich (1989: 32) that "a democratic state unable to wield repressive power is all the more dependent on the peaceful exchange of ideas, as it cannot exist without a basic consensus".

A major post-transition problem was the transparency of the electoral process. In late 1992, the opposition parties boycotted the parliamentary election on the grounds that the presidential election was rigged. The problem had to be tackled to underscore the existence of a credible electoral system. The government demonstrated its commitment to an efficient electoral system by supporting the formation of the Inter Party Advisory Committee (IPAC). It was under the chairmanship of the National Electoral Commission (NEC). IPAC also had the support from some donor organizations. The main objective of IPAC was to advise on issues of electoral reforms. It replaced the ill-fated dialogue between the NPP and the NDC.

Apart from manifesting the political will of the ruling government, the existence of IPAC symbolized the desire of the political parties to reach consensus on issues of national importance. Given the fact that economic consensus across the political spectrum is a fait accompli, the task revolved around the resolution of divisive political issues, such as the electoral process. IPAC was to play that facilitatory role to foster political bridge building. The NDC, however, withdrew its representatives from IPAC on account of what it alleged to be the imposition of ideas on NEC by the other parties. In his 1995 Sessional Address to Parliament, the president instructed the NDC representatives to return to IPAC. This further underlined the will of the government to thrive on consensus building.

Ison's studies on Costa Rica, the Dominican Republic and Jamaica suggest that democratic governments are capable of adopting different tactics and styles to "manage the political challenges of adjustment" (1992: 208-9). The emphasis is on consultation and persuasion; and consultation and participation have been the hallmarks of Costa Rican and Jamaican politics respectively. She concedes that competitive party systems do not in all cases foster the inclination or the skill to engage in persuasion or consultation or to design partial compensation in order to manage political pressures associated with adjustment in the short term. Nelson further argues that equal distribution of costs and gains as well as economic outcomes may be crucial in combining adjustment and democratization. As a result, the various governments have been relying on extensive compensation for adversely affected groups in addition to the forging of close cooperation between government and major interest groups when it comes to the formulation of policy (1992: 211-12).

Even though it appears the NDC government finds dialogue and consultation very difficult political tools, the existence of the National Tripartite Committee on wage guidelines (made up of the representatives of the government, labor, and employers), which deals with salary adjustments where necessary, has equally contained labor unrest. Thus the predicted assault by labor consequent upon democratic governance has been spasmodic and highly predictable.

Such predictability allows the government to anticipate labor agitation and make plans to contain it. A vivid example was the hurried reconvening of the committee in April 1995 when the government was faced with threats of strike actions by nurses, civil servants, and a section of the Trades Union Congress (TUC). The government engaged in what Panford (1997: 92) terms "procedural wrangling to stall labor negotiations". A typical strategy is the setting up of committees to review salaries of workers.[6] Once the salary review committee is set up, the government keeps appealing to agitating workers to wait for the report of the committee. In most cases, the reports are never implemented. In some instances, by the time the committee completes its work, inflation has rendered the reviewed salaries ridiculous. Meanwhile, the workers have lost the zeal to embark on industrial action. The fact is that the spate of retrenchment due to the privatization of state owned enterprises, as noted earlier, has reduced the strength of the labor front. Workers have therefore become much more concerned about their entitlement after retrenchment than about hitting the streets to agitate for higher salaries.

Another factor of significance is the de facto one-party state created by virtue of the 1992 boycott of the parliamentary elections by the opposition parties. With the exception of two independent parliamentarians, during the first term of the Fourth Republic, all two hundred-members of the Parliament were part of the ruling Progressive Alliance made up of the NDC, the Every Ghanaian Living Everywhere (EGLE), and the National Convention Party (NCP). Such a parliament should not be expected to be overly against policies that are in furtherance of adjustment simply because almost all the members believe in the indispensability of adjustment policies. The passing of controversial bills under a "certificate of urgency" by the NDC-dominated Parliament also became a standard practice. Under this practice, the first Parliament of the Fourth Republic delayed important legislation until it was about to recess. Then, as if under intense pressure to finish pending work before the recess, the Parliament rushed through certain bills. Invariably, the government managed to get its way without any serious opposition. This explains why not even a single economic bill was debated and rejected by Parliament during the first term. The government further adopted the tactic of censuring "errant" NDC parliamentarians. Those deemed critical of government policies during debates ended up being summoned privately, to the Castle (seat of government) for verbal assault by the president and other party leaders.[7]

During the second term of the Fourth Republic, which started in 1996 with the opposition parties fully represented in Parliament, the NDC has utilized its numerical advantage to pass adjustment-related bills in spite of strong opposition. A typical example is the reintroduction of the Value Added Tax (VAT) bill, which after intense debate was passed on the wings of 107 to 55 votes in February 1997.

11:8 Other Associational Groups

In spite of democratization, civil society and other associational groups remain politically weak. Those groups, which originally were active have not been able to make any meaningful impact by way of agitation for changes in adjustment policies for a number of reasons.

First is the failure of the forces opposed to Structural Adjustment Programs to capitalize on the democratic changes to organize the rural people and the urban poor who are hard hit by adjustment. Even though medical and educational costs are prohibitive since the introduction of user fees, these people, who are in the majority, only grumble without articulating their grievances. Invariably, the high level of illiteracy makes it difficult for such groups to make their problems known all by themselves or to demonstrate as a way of voicing their views. They need to be well organized for such purposes. However, one needs to note that the government had a head start in the organization and mobilization of such peoples. Through the introduction of the Non-Formal Education (NFED), the government has been able to explain various policies and assign reasons for the harsh economic climate to the illiterate citizenry who register for mass adult education. The mobilization unit of the Ministry of Employment and Social Welfare has also been utilized effectively to explain various government measures.

In Ghana, rural farmers and fishermen are potent forces in any social set-up. They form the majority of the population. The interests of these critical segments of the society should have been well championed by the political parties – especially those in opposition – and by civil society organizations. Unfortunately, this has not been the case. The level of mobilization of these people by social forces is very low. They are almost always left alone to fend for themselves when the government implements policies that affect their livelihoods and, ipso facto, their standard of living. For instance, fishermen in the Central Region grounded their canoes when the subsidy on premix fuel (which is for the exclusive use of fishing motors) was removed in the 1994 budget (*Daily Graphic*, Jan. 24, 1994). No associational group, whether a political party or civil associational group, openly supported the fishermen (beyond rhetorical condemnation of the withdrawal) or agitated for the restoration of the subsidy. The general attitude of these social forces seemed to be supportive of the prevailing economic rationality under whose rubric all subsidies have been withdrawn.

The rural poor were left to their fate.

Second, some members of the other associational groups, such as the Association of Recognized Professional Bodies (ARPB) and the Ghana Bar Association (GBA), appear to have compromised their stand by virtue of being direct beneficiaries of the adjustment policies. One can categorize the beneficiaries of Structural Adjustment Programs as follows:

• Top executives in the public services who, owing to the technocratic demands of Structural Adjustment Programs junket around the globe attending conferences, seminars, and workshops and also as consultants and advisors;

• Local agents of foreign business partners and consultants to such businesses and institutions providing specialized services such as hotels and advertising agencies;

• Top executives in private business who deal with foreign capital and have benefited from the injection of foreign exchange into the economy;

• Beneficiaries of divestiture measures who are able to buy up or buy into state owned enterprises; and

• Large landowners and big-time commercial farmers who benefit from higher producer prices, and other incentive packages (Boakye-Danquah 1992: 246).

Most leading members of the opposition and of other social groups who have been commending the government for the economic liberalization fall under one or more of the categories of beneficiaries. Nothing therefore compels such influential people to finance, organize, or support moves aimed at undermining the pursuit of adjustment policies.

One would have thought that the favorable Supreme Court verdict obtained by the NPP on the democratic right to demonstrate without police permit would signal sustained pressure on the government for changes in economic direction.[8] This has not been the case because the capability of social forces to mount a credible opposition to incumbent elites hinges in part on "organizational resources such as the strength and independence of associations in civil society" (Bratton and van de Walle, 1992:50). Where these are lacking for any reason, it becomes difficult for social forces to react to certain political and economic situations.

Third, the TUC appears to be confused about how best to handle the issue of the negative impact of Structural Adjustment Programs on workers. It more or less acquiesced to the retrenchment of labor unprecedented in the nation's history. Now that many people are outside rather than inside the labor market, the leadership has become extra cautious. It is on record that the Secretary General of TUC exhorted workers to adopt new measures to tackle labor disputes instead of the usual industrial actions.[9] Demands by labor since the inception of the Fourth Republic have not necessarily been geared toward a call for changes in adjustment programs. For instance, between April 1, 1993, and November 30, 1993, the TUC recorded eighteen industrial actions by various categories of workers. In almost all cases causes for such work stoppages revolved around demands for salary adjustment. The wrong impression seems to have been created. First, the government only has to make certain concessions on wages. Second, efforts should be made at making workers appreciate the economic problems facing the nation. Once these measures are taken there would be no agitation for the abandonment of adjustment policies, not even from the labor front whose members are hard hit by the policies.

Fourth, the immediate past reminds some Ghanaians of the ruthlessness of the regime from which the NDC takes its roots. It may be instructive to add that the traumatic psychological impact on the national psyche of the brutal repression during the revolutionary period cannot be wished away. More important, the dramatis personae of the PNDC era have all donned democratic clothing. The refusal by the NDC government to carry out genuine disbandment of paramilitary groups equally puts fear into civil society. Again, the faceless members of the Bureau of National Investigations (BNI) are still at large. Thus the lingering past of the brutal measures adopted earlier seems indirectly to have dealt a mortal blow to the collective will of most Ghanaians. The Commonwealth Observer Group (1992: 68), in spite of its favorable report on the presidential election of 1992, could not gloss over this, noting that

> The events of the first years of the revolution cast a long shadow, and memories of traumatic events were still fresh in people's minds ... The unpleasant experiences of the past were clearly among the underlying reasons for the acute polarization of society and for the bitterness that was so apparent ...

The situation has changed little. It explains further why many Ghanaians express genuine trepidation, even under democratic governance, when there is the need for massive demonstrations against some government policies. The organized pro-government demonstration in December 1994 to counteract a protest march organized largely by the opposition parties against the seizure of the broadcasting equipment of a private radio station known as the *"Radio Eye"* represented what many had feared for a long time. The police confronted the protesters and beat them up.

Again, on March 1, 1995, parliament promulgated Act 486 which imposed a 17.5 percent Value Added Tax (VAT) in place of a 15.5 percent sales tax. The successful antigovernment demonstrations organized by a loose coalition called the Alliance for Change (AFC) held on May 11 and 25, 1995, in Accra and Kumasi respectively, and later in Takoradi and other regional capitals, were met with brutal force by agents of the NDC. Four of the demonstrators in Accra were killed by gunshots allegedly fired by government agents. Even though the VAT was repealed as a result of the opposition on June 14, 1995, it appears many Ghanaians would not like a repeat of such confrontations that might culminate in the death of innocent citizens. This may also explain the docility of social forces in the face of Structural Adjustment Program-induced economic problems.

11:9 The Donor Community

The activities of the donor community as well as utterances of major functionaries of the BWls also seem to send signals that adjustment policies are the best for the economy and must be pursued at all cost. The implicit faith of the donor community in the efficacy of adjustment and the regime's capabilities has been exhibited in several ways. In June 1993, barely six months after the transition to democracy, Ghana received a commitment of $2.1 billion in assistance for 1993-1994 at a donors' meeting in Paris chaired by the World Bank. The bank had earlier indicated that Ghana needed $1.7 billion (*Ghana News*, June 24,1993). The commitment was an absolute vote of confidence in the program and how it had been handled by the PNDC now parading as the NDC political party.

Without doubt, the donor community supports democratic changes that will not change adjustment policies. This seems to contradict the emerging World Bank view which tends to cast doubt on the vaunted positive impact of adjustment policies on African economies (Elbadawi *et al.*, 1992), and by implication, the necessity for its continuation.

The insistence by the World Bank on governance is equally revealing. The narrow meaning of governance calls for countervailing powers, which will foster accountability, increased honesty, and efficiency in the public sector. In the broader sense, however, governance is to promote a free media, an autonomous judiciary, active interest groups, and so on. The preference, however, seems to be for effective economic measures that will facilitate the rapid repayment of contracted loans. The total support for adjusting countries is premised on an increasing and palpable fear that reform programs currently being pursued in the developing countries would fail for lack of indigenous support in the wake of democratization (Harvey and Robinson, 1995).

The price of failure for this experiment (that is, the pursuit of adjustment and democratization *pari passu*) is projected to be devastating to the credibility of the Bretton Wood institutions especially. It is therefore in the interest of donors to ensure the pursuit of adjustment side by side with democratic governance. In a very frank and blunt statement, Edward Jaycox, the World Bank Vice President for Africa stressed:

> The alternative–a series of failed programs in Africa–is not worth basic idea of moving to a market economy, shifting policies out of grandiosity to step-by-step solid progress will be discredited. If they fail in a series of countries... then it is a failure of our approach to the economy, a failure of our institution, a failure of our political will, and there is no way that we'll be able to say that it is just the failure of Africa! So we have a very big stake in this (Cited in Callaghy, 1993).

It is very clear that the BWIs are desperately in need of a success story to advertise the potency, efficacy, and the need for the continuation of adjustment policies alongside democratization. In their estimation, Ghana's program is a model, so assistance ought to be given, especially in a constitutional era in which opposition to such programs was expected to mount.

The donors cannot afford to leave Ghana to her fate for two main reasons: (1) Since the launching of the ERP/SAP in 1983 donors have injected over $8 billion in aid to underwrite the program. Ghana has thus become one of the largest per capita aid recipients in the world; and (2) the donors need a success story, as Edward Jaycox openly admits, to justify such large infusions of external capital. It is most probably in pursuance of these objectives that Jaycox, while addressing the annual conference of the African-American Institute, mentioned certain basic changes being contemplated by the World Bank. These include an end to the imposition of development plans and foreign expertise on reluctant African governments and the intensification of efforts to build Africa's capability to ensure development (Hultman, 1993). The projected changes are premised on the realization that the bank's technical assistance to Africa has proved to be "a systematic destructive force" (Hultman, 1993).

Probably as a result of such anticipated changes, a Structural Adjustment Participatory Review Initiative (SAPRI) was inaugurated in Accra in November 1997 with the aim of reviewing Structural Adjustment Programs and facilitating the incorporation of local inputs in the formulation of reform programs. Earlier, in September 1997, the World Bank approved a new Country Assistance Strategy (CAS) for Ghana. Under this strategy, the economic and social development goals of the government for the medium term and the bank's strategy for assisting the government to achieve them have been spelled out (EIU, 4th quarter, 1997). It is significant to note that the World Bank accepted some blame for not retooling "assistance to suit the changed political environment since the return to democracy in 1992" (EIU 1997:15). This explains the inauguration of SAPRI. One can only hope that SAPRI will not follow in the footsteps of PAMSCAD. In light of the political and economic dynamics of the Ghanaian situation, successful retooling of assistance to suit current political circumstances is likely to have a very positive impact on the coexistence of structural adjustment and democratization.

11:10 Conclusion

The aim of this chapter has been to examine the factors that in the Ghanaian context have assisted in the continued pursuit of adjustment policies after transition to democracy. The main issue is the assumed inherent contradiction between the two processes. Democratic principles, which stress competition for political power, open political process, and respect for minority interests are deemed to contradict adjustment policies which in the African context are seen to be best pursued by an undemocratic and ruthless regime. My argument in this chapter has been that contrary to the theoretical postulations, specific conditions in a particular country and other externalities may make it possible for the simultaneous pursuit of structural adjustment and democracy. In the Ghanaian case, I need to emphasize that the government's commitment to adjustment policies was initially strengthened by the nature of the economic crisis that made Structural Adjustment Programs an indispensable economic tool. Its continuation under democratic governance called for compromises, reaching out to the people and statecraft by the NDC government.

The Ghanaian experience has further shown that sufficient and propitious conditions existed which were well taken advantage of by the government. These include:

- The prior pursuit of adjustment before democratization;
- the good will of the donor community;
- the weak nature and apparent lack of autonomy of civil society and other associational groups;
- the belief of the strongest opposition party (NPP) in the core strands of adjustment policies and democracy;
- the dogged commitment of the government;
- the high level of mobilization of support through patronage and other incen-

tives to the rural population, such as electrification projects;
- the lessons of earlier implementation of the most politically ruinous of the adjustment program by the PNDC; and
- the willingness on the part of the government to pursue dialogue with opposition groups coupled with serious attempts to govern democratically.

In spite of its mixed outcomes, my position is that both adjustment and democratization have become indispensable tools for Ghana's development. As such, there is the need to "adjust adjustment" to suit democratic governance. This may call for changes in some aspects of the reform package so as to ensure the following:

- Some level of state participation in the economy is necessary. This could be done through state management of viable state enterprises. Note ought to be taken of the fact that the vaunted economic miracles of the Newly Industrializing Countries (NICS) were not attained on the altar of unmitigated liberalization. While not becoming overly interventionist, the state can still participate in selected areas of economic development. The World Bank's and the Monetary Fund's policy toward Ghana and many African countries contradicts their policies toward authoritarian Asian states such as Indonesia, China, and Malaysia (Jeong, 1997). One has to agree with Botchwey (1994: 4) "a basic minimal level of government regulation and intervention is absolutely essential if markets are to function properly". The plain truth is that there cannot be said to be a natural compatibility of democracy and the market as exemplified by the 1995 VAT fiasco. Nonetheless, the NDC government has been successful thus far. It seems, however, that the long-term coexistence of the two would require a minimum of state intervention. On this, Hutchful (1992: 14) aptly notes, "with the exception of the U.S., all mature capitalist democracies are highly managed economies; none corresponds to the 'free market' envisaged in the World Bank literature". It is common knowledge, however, that the U.S. government heavily subsidizes sugar production. Meanwhile, the Washington-influenced IMF/WB almost always insists on the withdrawal of subsidies by developing countries as one of the conditions for financial assistance.
- Some degree of protectionism for local industries is necessary. This will safeguard the interest of domestic entrepreneurs who could then be encouraged to enter into viable partnership with foreign counterparts.
- On the political front, the commitment of the government to democracy should graduate from the realms of rhetoric to practical reality. Government officials must realize that the days the PNDC legislated, adjudicated, and implemented policies with careless abandon are over with the inauguration of the

Fourth Republic on January 7, 1993.

- The spirit of dialogue and consultation with political stakeholders must replace that of mutual suspicion. Such an approach will lead to peace, consensus building, and development.
- The donor community must of necessity appreciate the complexities inherent in the joint pursuit of the two processes and show pragmatism in its demands for strict adherence to the spirit and letter of adjustment policies. In this connection, SAPRI should not just be in theory.

In sum, it is difficult to predict the future with regard to the continued coexistence of structural adjustment and democratization. What is certain is that political and economic developments cannot take place in an environment of uncontrolled inflation, adverse balance of payment problems, and the absence of international capital investment. Furthermore, there is no future for any economic policy that favors the political elite and their international cronies at the expense of the masses. Whether structural adjustment can be sustained under democratization in the years ahead will depend on its overall long-term impact on the standards of living of the people in general. Sustained negative effects could breach the tolerance level of the masses and burst the bubble. That is why the Ghanaian situation calls for cautious optimism.

Notes

[1] Rawlings came to power in 1979 when the SMC II had put in place modalities for the return of the country to civilian rule. Rawlings had very little choice but to promise to hand over power to the winning party in the pending general elections. The Peoples' National Party emerged victorious in the elections. Jerry Rawlings fulfilled his promise and handed over power to the victorious party.

[2] Inquisitorial strategies were adopted. These included phony trials before Citizens Vetting Committees (CVCs), arrests by the Bureau of National Investigation without cause, and so on. For instance, Kwesi Pratt, a political activist, was arrested and imprisoned by the BNI over a dozen times.

[3] The Nkrumahist parties trace their roots to the Convention Peoples' Party of Dr. Kwame Nkrumah, the first leader of independent Ghana.

[4] In my view, Dr. Nyaho Tamakloe, the chairman of the NPP's Sector Committee on Health and Social Services, meant precisely this when he stated that the opposition wanted to prove to Western donors, if given the chance, that they would have performed more creditably than the P/NDC.

[5] The opposition called for the revision of the voters' register which was bloated, the

provision of voter identity cards to check multiple voting by individuals, the provision of transparent boxes to preempt pre-stuffing of boxes with ballots, and the disbandment of the paramilitary forces who continued to harass citizens.

[6] These were the Gyampoh Salary Review and the Greenstreet Salary Review Committees. In 1997, the Pricewater House Salary Rationalization Commission was also set up.

[7] Critical party parliamentarians who are summoned are always reminded that they are jeopardizing their chances of being appointed a minister of state or deputy minister of state. Such a member may even lose his position on a board if he/ she does not show enough commitment to the beliefs and goals of the party.

[8] Article 21 (1) (d) of the Ghanaian Constitution guarantees the right to freedom of assembly, demonstrations, and processions. However, the NDC government and the police still wanted people to obtain a police permit before embarking on any demonstration or procession. In the case of *AIPP v the Inspector* General *of Police (IGP)*, the Supreme Court ruled as unconstitutional all existing laws that required permission of the executive or one of its agencies before the right of assembly is exercised

[9] The author was one of the resource persons for the workshop held at the St. Charles Beach Resort in Winneba in the Central Region of Ghana.

References

African Business (1990). March, 10-11.

Ampem, Nana Wereko (1994). *Ghana's Economic Progress: The Dreams, the Realities and the Prospects*. Accra: Accra Academy.

Anyemedu, K. (1993). "The Economic Policies of the PNDC" in Gyimah Boadi, ed., *Ghana Under PADC Rule*. Dakar: CODESRIA, 13-47.

Appiah-Opoku, S. (1998). "The World Bank's Structural Adjustment Programs in Africa: The Case of Ghana". *Africa Notes*. Ithaca, NY: Cornell University, 6-10.

Arthiabah, P. (1994). *Trade Unions and Economic Structural Adjustment in Ghana*. Accra: Fredrick Ebert Foundation.

Asenso-Okyere, W. (1989). *The Response of Farmers to Ghana's Adjustment Policies* (mimeo).

Bangura, Y. (1992). "Authoritarian Rule and Democracy in Africa: A Theoretical Discourse" in Peter Gibbon, Yusuf Bangura and Arve Ofslad, eds., *Authoritarianism, Democracy and Adjustment: The Politics of Reforms in Africa*. Uppsala: Scandinavian Institute of African Studies.

Boafo-Arthur, K. (1989). "Prelude to Ghana's Transition: An Examination of the Process" in Kwame A. Ninsin and Kofi Drah, eds., *Ghana's Transition to Authoritarian Rule*.

Accra: University of Ghana Press, 34-49.

Bangura, Yusuf (1995). "Managing Inter Party Conflict in Ghanaian Politics: Lessons from the NPP-NDC Dialogue" in Mike Oquaye, ed., *Democracy and Conflict Resolution in Ghana*. Accra: Gold Type Publishers.

Boakye-Danquah, Y. (1992). "Structural Adjustment Programs and Welfare Interventions: The Case of Ghana". *Africa Insight*, 22/4: 240-52.

Botchwey, K. (1993). "The Political Economy of Structural Adjustment: What Role Planning?". *DPMN Bulletin*, 1/1: 4-6.

Bratton, N. and van de Walle (1992). "Towards Governance in Africa: Popular Demands and State Responses" in Goran Hyden and Michael Bratton, eds., *Governance and Politics in Africa*. Boulder, CO: Lynne Rienner.

Callaghy, T. M. (1990). "Lost Between State and Market: The Politics of Economic Adjustment in Ghana, Zambia and Nigeria" in J. Nelson, ed., *Economic Crisis and Policy Choice in Africa*. Boulder, CO: Lynne Rienner.

Callaghy, T. (1993). "Political Passions and Economic Interests: Economic Reform and Political Structure in Africa" in T. M. Callaghy and J. Ravenhill, eds., *Hemmed in: Responses to Africa's Economic Decline*. New York: Columbia University Press, 463-519.

Commonwealth Secretariat. (1992). *Report on the Presidential Elections in Ghana*. London: Marlborough House.

Daily Graphic (1993). May 22.

Daily Graphic (1994). January 8, 24.

Elbadawi, I. A., Dhaneshwar, G. and Uwujaren, G. (1992). *How Structual Adjustment Has Not Succeeded in Sub-Saharan Africa*. World Bank Policy Research Working Paper. Washington, DC: World Bank.

Ewusi, K. (1987). *Structural Adjustment and Stabilization Policies in Developing Countries: The Case Study of Ghana's Experience, 1983-86*. Tema: Ghana Publishing Corporation.

Free Press (1993). January 7-13.

Ghai, D (1991). *The INW and the South: The Social Impact*. London: Zed Books

Ghana News Bulletin (1993). June 24. *The of Crisis and Adjustment*.

Gibbon, P. (1992). "Structural Adjustment and Pressure Towards Sub-Saharan Africa" in Gibbon, P and Bangura, S. eds., *Authoritarianism, Democracy and Adjustment: The Politics of Economic Reforms in Africa*. Uppsala: Scandinavian Institute of African Affairs, 156-57.

Government of Ghana (1987). *Budget Statement*. Tema: State Publishing Corporation.

Haggard, S. and R. R. Kaufman. (1989). "Economic Adjustment in New Democracies" in Joan Nelson, ed., *Fragile Coalitions: The Politics of Economic Adjustment*. Oxford: Transaction Books, 1989, 59-60.

Healey, J. and Robinson, M. (1992). *Democracy, Governance and Economic Policy: Sub-Saharan Africa in Comparative Perspective*. London: ODI.

Healey, J. and Robinson, M. (1995). *The Design of Economic Reforms in the Context of Political Liberalization.* London: ODI.

Herbst, J. (1993). *The Politics of Reforms in Ghana, 1982-1991*. Berkeley: University of California Press.

Hultman, T. (1993). "World Bank Takes Turn: Help Africa Help Itself". *Herald Tribune*. May 22-23.

Hutchful, E. (1985). "IMF Adjustment in Ghana Since 1966". *Africa Development 10/12*.

Hutchful, E. (1992). "The International Dimensions of the Democratization Process in Africa" (Paper presented at the 7th General Assembly on Democratization Processes in Africa). Dakar: CODESRIA.

Jeong, Ho-Won (1997). "The Role of African States in Economic Development". *Africa Insight* 27/2: 85-90.

Jonah, K. (1989). "Changing Relations between IMF and the Government of *Ghana: 1960-1987*" in E. Hanson and Y.A. Ninsin, eds., *The State, Development and Politics in Ghana*. London: CODESRIA, 94-115.

Klein, E. and T. Giegerich. (1989). "Parliamentary Democracy under the Basic Law: Characteristics and Status". *Universitas 31*.

Kusi, N. K. (1991). "Ghana: Can Adjustment Reforms be Sustained?". *Africa Development* 16/3-4.

Leftwich, A. (1994). "Governance, the State and Politics of Development". *Development and Change* 25.

Libby, R. L. (1976). "External Cooptation of a Less Developed Country's Policy Making: The Case of Ghana". *World Politics* 1/10, 67-89.

Mkandawire, T. (1992). "The Political Economy of Development with a Democratic Face" in G. A. Cornia, der Hoeven and T. Mkandawire, eds., *Africa's Recovery in the 1990s*. London: St. Martin's Press.

Moseley, P., J. Harrigan and J. Toye (1991). *Aid and Power. The World Bank and Policy-Based Lending in the 1980s*. London: Routledge.

Nelson, Joan. (1990). "The Politics of Economic Adjustment in Developing Countries" in Joan Nelson, ed., *Fragile Coalitions: The Politics of Economic Adjustment*. Princeton, NJ: Princeton University Press.

Panford, Kwarnina. (1997). "Ghana: A Decade of IMF/World Bank's Policies of Adjustment (1985-1995)". *Scandinavian Journal of Development Alternatives and Area Studies*, 16/2 (June).

Remmer, Karen L. (1993). "The Political Economy of Elections in Latin America". *African Political Science Review*, 87/2.

Rothchild, D. (1991). *Ghana: The Political Economy of Recovery*. Boulder, CO: Lynne Rienner.

Rudebeck, L. (1992). "Politics and Structural Adjustment in a West African Village" in Lars Rudebeck, ed., *When Democracy Makes Sense*. AKUT: Uppsala University Press.

Sawyerr, A. (1991). *Marginalization of Africa and Human Development* (mimeo).

Schmitz, G. and E. Hutchful (1992). *Democratization and Popular Participation in Africa*. Ottawa: North-South Institute.

Sirrowy, L. and A. Inkeles (1990). "The Effects of Democracy on Economic Growth and

Inequality: A Review". *Studies in Comparative International Development, 25/1* (Spring).

Toye, J. (1991). "Ghana" in P. Moseley, J. Harrigan and J. Toye, eds., *Aid and Power. The World Bank and Policy-Based Lending in the 1980s.* Vol. 2, New York: Routledge.

West Africa (1993) (July 19-25).

World Bank (1984). *Ghana: Policies and Program for Adjustment.* Washington, D.C: World Bank.

World Bank (1993). *Ghana: 2000 and Beyond.* Washington, D.C: World Bank.

12 Migration and Remittances: Rural Household Strategies for Coping with Structural Adjustment Programs in Ghana

SIAW AKWAWUA

12:1 Introduction

The economic hardships and political instability that faced Ghanaians in the 1970s and the early 1980s stimulated considerable migrations of the population. It is estimated that millions of Ghanaians, both skilled and unskilled, who could not endure the poor economic conditions, moved to foreign countries especially to Nigeria in the 1970s (Loxley, 1991; Konadu-Agyemang, 1999). Konadu-Agyemang (1999) points out that several more Ghanaians moved out of the country in the early 1980s after the introduction of International Monetary Fund (IMF) and World Bank sponsored Structural Adjustment Programs (SAPs). Although the Structural Adjustment Programs were intended to solve the economic problems facing the country, they rather created more economic and social hardships (Loxley, 1991; Riddell, 1992; Konadu-Agyeman, 1999) thus inducing further outmigrations from the country.

One positive consequence of migration that forms an important component in the local economies of poor countries of the Third World is the flow of cash remittances and other resources from migrants to their home regions (Jones, 1998; Conway and Cohen, 1998). Ramsay (1999) for instance, points out that ".... hundreds of thousands of professionals who began their schooling under Nkrumah work overseas, annually remitting an estimated $1 billion to the Ghanaian economy" (Ramsay, 1999: 37). It is argued that migration and remittances are survival strategies that families and households adopt in times of economic adversity (Conway and Cohen, 1998; Rubenstein, 1992; Poirine, 1997). In Ghana, as elsewhere in poor Third World countries, the implementation of Structural Adjustment Programs and their attendant socio-economic hardships, have created situations in which migration and remittances have become critical for household sustenance and survival, especially for poor rural families. The objective of this chapter is to examine the extent to which rural households employed migration and remittances as adaptation to national economic stress that was engendered by the implementation of Structural Adjustment Programs. The exercise contributes to current literature on the social and economic impacts of migration and remittances on rural households in poor regions of the Third World.

This chapter is divided into four sections. Following this introduction, an overview of Ghana's economic crisis and the government's solution is provided in section 2. Section 3 contains a description of the rural household survey and reports the main findings on migration and remittances as important survival strategies. Some conclusions are offered in section 4.

12:2 Economic Crisis in Ghana and Government Solutions

The economic crisis that faced Ghana in the early 1980s can be blamed on both internal and external factors. Political instability, introduced by the involvement of the military in politics starting from the 1966 worsened the internal factors of corruption, nepotism and economic mismanagement. External factors include unfavorable terms of international trade for Ghana's primary exports products particularly cocoa and gold, the international economic recession in the mid-1980s and increases in OPEC oil prices since 1979. The combined effect of these negative factors was the virtual economic collapse that the country suffered in the 1970s and early 1980s. Between 1970 and 1983, real GDP fell 36 percent; the country suffered acute food shortages and foreign exchange; inflation rates were over 100 percent, unofficial black markets flourished, and the government budget registered a deficit of 14 percent of GDP (World Bank, 1984). These problems were worsened by a severe drought in 1982-83 and the expulsion of over a million Ghanaians from Nigeria. It is interesting to note that the government of Ghana displayed a characteristic lack of initiative and insensitivity to fellow Ghanaians by closing the borders of the country on those expelled from Nigeria.

The economic crisis negatively affected large sections of Ghanaians who suffered declining living standards. For both rural dwellers and urban workers, real incomes and consumption levels fell and poverty increased. For instance, the real incomes of cocoa farmers fell by 80 percent from 1970 to 1983; the minimum wage of urban workers was estimated to cover only 2.6 percent of household budget for a family of five (Loxley, 1991). Transport and basic infrastructure such as schools and hospitals collapsed and this seriously affected the welfare of many Ghanaians. Epidemic diseases such as yaws and yellow fever broke out in the Northern and Upper Regions of the country; infant mortality rose to as high as 110-20 per thousand in 1983-84; and malnutrition and under-nourishment increased in all regions of the country (Loxley, 1991).

Unable to find any viable solution to the possible economic bankruptcy facing the country, the government turned to the IMF and the World Bank for assistance. These Bretton Woods institutions suggested the implementation of structural adjustment programs, known collectively as Ghana's Economic Recovery Program (ERP), which was launched in 1983. The program involved cutting government spending, devaluing the local currency (the cedi), liberalizing trade, investing in natural resource exporting industries, selling state enterprises, and creating incentives to attract foreign investors (Government of Ghana, 1987a, b, c; Loxley, 1991; Danaher, 1994). All these policy objectives were the conditionalities imposed by the IMF, and accepted by the government of Ghana.

The government of Ghana was unaware that IMF conditionalities are uniformly applied to all countries regardless of the specific circumstances (Cheru, 1989). Ghana's peculiar circumstances of official corruption, mismanagement of public funds, unnecessarily huge military expenditure, a politicized military elite and a rich upper class with consumption habits modeled on Western standards etc., were not taken into account by the IMF and the World Bank. Ghana is described as the World Bank's African showcase because since 1983 real GDP has been raised 3 percent in excess of its 1980 level. What is not considered is the high population growth (about 3.1 percent per year) and the massive inflows of foreign exchange from donors. Under ERP, IMF funding totaled over $1.35 billion and over $8 billion was obtained from bilateral and multilateral sources (Danaher, 1994).

The human suffering resulting from the implementation of the structural adjustment programs is not adequately assessed. Few Ghanaians benefit from the program since it ignores the plight of those not involved in the export sector as IMF reforms emphasize primary resource extraction for export. The government shifted resources toward cocoa rehabilitation and other export sectors, not toward food production. As part of civil service reforms, many government employees, especially those in state enterprises, lost their jobs and this has contributed to rising unemployment. Farmers suffered as the percentage of the total budget devoted to agriculture fell from 10 percent in 1983 to 4.2 percent in 1986 and to 3.5 percent in 1988. Although cocoa contributed less to Ghana's GDP than food crops, cocoa nonetheless received 9 percent of capital expenditures in the late 1980s; at the same time it received roughly 67 percent of recurrent agricultural expenditures because of its export value (Danaher, 1994).

This increased investment in the cocoa sector did not produce the desired positive effects. For one thing the world market price of cocoa has been dropping since the 1980s. It's estimated that while the supply of cocoa has increased by 6 to 7 percent, world consumption has increased by only 2 percent annually (Danaher, 1994). Concentration of ownership in the cocoa sector has occurred since the 1970s, with about 7 percent of cocoa producers owning almost half of the land under cocoa cultivation, the greater percentage of farmers possessing only very small landholdings. In this sense, the benefits accruing from increases in producer prices of cocoa do not spread to the masses of small cocoa farmers. As Stewart (1991) argues, emphasis on export crop production means that rural producers of subsistence crops do not benefit from structural adjustment, and in fact the poor suffer during the adjustment process. How people manage to survive these economic hardships will be addressed later in the chapter.

12:2.2 Structural Adjustment Programs and Poverty Alleviation

One effect of Ghana's structural adjustment program is the increasing polarization of the Ghanaian society. Rich people have fared quite well under adjustment and it is felt that the rich are getting richer and the poor becoming poorer. For instance the 1987 Living Standards Measurement Survey showed an increase in income inequality in the 1980s compared to the 1970s. Some concerned citizens naturally criticized the structural adjustment program's effect on the poorer segments of the population. And in response to the criticisms, the government initiated a US$85 million Program of Action to Mitigate the Social Costs of Adjustment (PAMSCAD) in 1988. This program sought to create 40,000 jobs over a two-year period. It was aimed at the poorest individuals, small-scale miners and artisans in particular, and communities were to be helped to implement labor intensive self-help projects.

Under PAMSCAD, 10 billion cedis was allocated in the 1993 budget for the rehabilitation and development of rural and urban social infrastructure. The program was designed to focus on improving water supply, sanitation, primary education, and health care. An additional 51 billion cedis was set aside for redeployment and end-of-service benefits for those who had lost their jobs in civil service and parastatal reorganizations. The success of PAMSCAD is variously debated; however, the important idea here is that by instituting the program, the Ghana government acknowledged the immense havoc that structural adjustment was wrecking on the poor population in the country.

12:3 Coping Strategies of Rural Households under Structural Adjustment Programs

12:3.1 The Rural Household Survey

This section provides an overview of the strategies that rural households adopt to cope with structural adjustment programs. The analysis is based on data collected from a survey of households conducted at Dwenase, a rural village in the southeastern region of Ghana during the summer of 1999. Dwenase is located about 15 kilometers from the diamond mining town of Akwatia, and about 150 kilometers from the capital city of Accra. The total population of the village, according to the latest census of 1984 was 1,468. A random sample of 100 households was taken and interviewers were assigned to administer questionnaires in each household. A sampling frame was prepared by enumeration and listing of occupied dwelling units and random numbers were used for the selection of specific households to be home-interviewed on a variety of demographic, economic, social and migration characteristics. The interview schedule contained detailed questions on migration histories of household members. Data was collected on household income, expenditures, and remittances from migrant household members. Other questions related to the channels and flow of information, home visitations, rural-urban links and reasons for migration.

The introduction of fee-free compulsory education by the far-sighted government of Kwame Nkrumah impacted upon the population. About 48 percent of the heads of households had received primary education; 42 percent had secondary, technical and other forms of post-secondary education; only about 10 percent of the heads of households could not read or write. About 44 percent of the sample had lived outside the survey village before; while 65 percent of all household members had lived and worked in another town before. This characteristic return migration phenomenon among the survey population is explained by the disparity in socio-economic development between rural and urban areas of the country. While there is a concentration of social and economic infrastructure in the urban areas, the rural areas are invariably neglected in terms of the provision of social infrastructure.

In view of this, those people who want to rise up the socio-economic ladder must necessarily move to the towns and cities. At Dwenase most of the educated had opted for urban jobs and had migrated to the cities, especially the capital, Accra; the regional capital, Koforidua; Kumasi, the second largest city in the country and Takoradi, the third largest city. These initial migration streams set in train a certain amount of chain migration, in which families and relatives of the early migrants moved to join their kith and kin and to take up residence in the cities.

At the time of the survey, the structural adjustment program had markedly affected socio-economic conditions at Dwenase. The immediate effect of the introduction of Structural Adjustment Program resulted in changes in the rural economy: first to be felt was a sharp increase in rural transportation fares and fuel, which resulted in a rise in food prices (all of which were the direct result of the devaluation of the local currency) and, second, increased commodity prices for basic household required requirements (soap, salt, kerosene, sugar etc.). What compound the problems for the rural dwellers are the frequent increases in the price of kerosene. The much-publicized rural electrification program has not penetrated most rural areas of the country, including Dwenase. Majority of the people who depend on kerosene as a source of lighting are rural dwellers and any increase in the price therefore affect them. Even more significant for the survey village is the fact that as a result of the government's civil sector retrenchment policy, several household members (45 percent) had returned to the village after being declared unemployed. The addition of these returnees to the village population strained the local food economy further although their presence was a blessing in disguise. It is known that migrants generally return to their home origins in an improved financial state; they are often more cosmo-politan or better educated, and occasionally are concerned with changing exist-ing social, economic, and political conditions (Rubenstein, 1992). Such were the attributes of the return migrants who were encountered at Dwenase. Most of them (85 percent) were completely dissatisfied with the prevailing socio-eco-nomic conditions in the country generally. About 95 percent expressed a desire to migrate again or support a household member to leave the country.

12:3.2 Migration and Remittances as Coping Strategies

Migration has a positive impact on household income, in that out of altrusic concerns or an implicit family loan arrangement (Poirine, 1997), the migrant remits money home to support the family left behind. The survey at Dwenase revealed that over 80 percent of the sample households had one or two absent members either in a city in the country, or in a foreign country. It was observed that most migrants were either encouraged to leave the village or were actually sponsored to migrate. About 62 percent of the sample stated that household members contribute money to sponsor a potential migrant to leave for the city or a foreign country, on the understanding that the migrant will remit money back home to support the family. European countries and North America are the preferred destinations of most of these migrants, with the majority (45 percent) moving to Britain, West Germany (20 percent), Italy (14 percent), Holland (10 percent). Other destinations include Denmark, Norway and the US and Canada.

12:3.3 Remittances from Migrants

The survey at Dwenase revealed that 90 percent the sample households received remittance income from their relatives in cities and foreign countries. Only 5 percent felt that their migrant relatives have not been sending money back home on a regular basis. 85 percentof the sample households stated that they would be worse off without remittance money from their migrant relatives abroad. Most return migrants (80 percent) claimed that they used to send remittances home when they lived abroad. Substantial sums of money are received by the survey population on a regular basis. About 45 percent of the sample households receive about $200 per month, while another 10 percent receive nearly $300 about every three months. Only 5 percent of the households receive less than $100 per month.

The respondents were asked to indicate the channels through which they receive their money remittances. 90 percent of the sample stated that they receive remittances through informal channels such as by mail or by friends and occasionally, by personal delivery. A few, about 15 percent receive money through the formal banking system. Most respondents convert their remittances in foreign currency into local currency through the forex bureaus. Now the need to convert remittances to local currency is obviated by the acceptance of foreign currency in local transactions. The most preferred currencies are the British pound sterling and the US dollar. Others are the German mark, the French CFA and the Japanese yen.

12:3.4 Uses of Remittances

The majority of the sample population, about 60 percent, are food farmers and cocoa producers. Other cash crops such as oil palm, kola nuts, and sugar cane are also produced. Illicit diamond and gold mining are part of the informal economic activities in the survey village. An expansion of informal economic activities occurred with the introduction of ERP, as rural dwellers strained themselves to adapt to the harsh economic conditions. About 55 percent of sample household members engage in petty trading in food products, textile and clothing, soap and cosmetics, craft goods, fish and vegetables.

In an attempt to quantify the use of remittances, it became obvious that rural households depend heavily on the earnings of migrants to support themselves and their informal economic activities. A very large proportion – 75 percent – of the remittances goes into the upkeep (especially feeding) of household members. About 12 percent is used to assist the education of the migrants' children and other relatives at home. Ten percent is used for the purchase of land, home improvements, construction of new houses and other productive investments in local craft industries, in agriculture, shops and bars etc. Only a very small proportion of remittances -3 percent is spent on marriages, funeral ceremonies and other social celebrations.

Some previous studies of migrant-remittance systems found that much remittance income is wasted on conspicuous consumption (Rubenstein, 1992). This conclusion is not corroborated by the findings of this study since not much is spent on luxury commodities. The main concern of rural households is with sheer survival in an environment that is neglected by the national government. Results of the survey point to the fact that migration and remittances must be seen as safety nets (Jones, 1998) and strategies for adaptation to economic stress (Conway and Cohen, 1998).

12:4 Conclusions

In this chapter, attention has been focused on migration and remittance impacts on rural households at Dwenase, southeastern Ghana. Migration and remittances are conceptualized as rural household survival strategies for coping with the adverse economic effects of Structural Adjustment Programs. In Ghana, as elsewhere in the Third World, the Structural Adjustment Programs have been especially harsh for rural households who have come to depend on remittances from migrants for supplementary income. Remittances are very important in the lives rural dwellers because of the contraction of the rural economy and government neglect. The findings of the research reported in this paper indicate that remittances are invested in rural household social economies. This corroborates the findings of Massey and Parrado (1994) that "migradollars" play a significant role in the local and regional economy of Mexico.

Improvements in rural living conditions cannot be attributable to any positive impacts of Structural Adjustment Programs. On the contrary, it is the rural households who have managed to maintain the rural economy by utilizing their own human capital resource, in the form of investing in a household member to migrate. The evidence suggests that rural dwellers are completely dissatisfied with current socio-economic conditions in the country. The increasing polarization of the society, the marginalization of large sections of the population, particularly the poor, should not be overlooked in government policy instruments.

A final point of note is the feeling among rural households that it is the Ghanaian elites who formulate policies such as Structural Adjustment Programs, which invariably benefit themselves. The elites are blamed for not demonstrating any willingness to champion the cause of the poor and underprivileged, especially the rural dwellers. Nor is the government prepared to address the growing pauperization of the rural dweller and the extreme social differentiation that is currently taking place in the country. Perhaps the situation can be reversed only with the appearance of another over-ambitious trigger-happy "Junior Jesus"![1]

Note

[1] J.J. Rawlings, Ghana's Head of State, was hailed as "Junior Jesus" by his supporters when he usurped power through a *coup d'etat* in 1979.

References

Cheru, F. (1989). *The Silent Revolution in Africa: Debt, Development and Democracy*. London: Zed Books Ltd.

Conway, D. and Cohen, J.H. (1998). "Consequences of Migration and Remittances for Mexican Transnational Communities". *Economic Geography*, 74(1): 26-44.

Danaher, K. (1994). *50 Years is Enough: The Case Against the World Bank and IMF*. Boston, MA: South End Press.

Government of Ghana (1987a). *The PNDC Budget Statement and Economic Policy for 1987*. Accra: February 20.

Government of Ghana (1987b). "A Programme of Structural Adjustment". Report prepared for the Fourth Meeting of the Consultative Group for Ghana, Paris. Accra: May.

Government of Ghana (1987c). *National Programme for Economic Development* (Revised). Accra: July 1st.

Jones, R.C. (1998). "Remittances and Inequality: A Question of Migration Stage and Geographic Scale". *Economic Geography*, 74(1) 8-25.

Konadu-Agyeman, K. (1999). "Characteristics and Migration Experience of Africans in Canada with Specific Reference to Ghanaians in Greater Toronto". *Canadian Geographer*, 43(4): 400-414.

Massey, D.S. and Parrado, E. (1994). "Migradollars: The remittances and savings of Mexican migrants to the USA". *Population Research and Policy Review*, 13: 3-30.

Poirine, B. (1997). "A Theory of Remittances as an Implicit Family Loan Arrangement". *World Development*, 25(4): 589-611.

Ramsay, J.F. (1999). *Global Studies: Africa*. 8th ed. Guilford, CT: Dushkin/McGraw-Hill.

Rubenstein, H. (1992). "Migration, Development and Remittances in Rural Mexico". *International Migration*, 30(2): 127-151.

Stewart, F. (1991). "The Many Faces of Adjustment". *World Development*, 19(12): 1847-1864.

World Bank.(1984) *World Development Report 1983*. Oxford University Press.

13 Rural Banking and Credit Inter-mediation in an Era of Structural Adjustments

ANDY C.Y. KWAWUKUME

13:1 Introduction

Although the Economic Recovery Programme (ERP) began in 1983, the Financial Sector Adjustment Program 1 (FINSAP 1) was not introduced until 1987. By then, it was clear that the financial sector had become a major bottleneck in the implementation process; and remarks were made that the reform policies should have been applied to the financial sector first. This viewpoint is buttressed by the importance of savings and investment in generating economic growth, which the SAP was designed to achieve.

The need to reform the financial sector is crucial in the case of Ghana, because the country is noted for its below-average level of gross national savings of not more than 15 percent of GDP at best. This compares unfavorably to other countries with similar per capita, even in sub-Saharan Africa (Boehmer, et al., 1994). Gross national savings was at its lowest level (below 5 percent) in 1983 due to a near total loss of confidence in Ghana's formal financial institutions, the result of some unpopular measures taken by government. In 1982, the Provisional National Defense Council (PNDC) government decided to freeze and vet accounts over 50,000 cedis, demonetized the 50 cedi notes, restricted bank loans to the trading sector, and required business transactions over 1,000 cedis to be conducted using checks. These requirements exacerbated the preference of Ghanaians to hold their savings in largely unproductive assets rather than in deposits with financial institutions. Consequently by 1994, after many years of liberalization in the financial sector, the national savings rate stood at below 10 percent, compared to savings rates of 30 percent for East Asian countries (Boehmer, et al., 1994:1).

Ghana's rural sector, with 63 percent of Ghana's population, continues to provide the largest source of employment. Agriculture alone employs about 70 percent of the working population and provides about 50 percent of the GDP presently. Moreover, Ghana's mineral wealth is extracted from the rural districts. In spite of the importance of the rural areas in the country's economy, it was not until July 1989 that FINSAP 2 initiated adjustment policies for the rural credit system under the Rural Finance Project (RFP). This was part of an agreement signed between the Bank of Ghana (BoG) and the International Development Agency (IDA) of the World Bank. The agreement called for a credit line of US$20 million *"to help increase the volume of production, improve efficiency and ensure sustainability of rural financial intermediation"* (RFD Workshop Paper: 1, quoted in Kwawukume, 1995:107).

There is apparent disagreement about the contribution of the formal sector to the credit needs of Ghana's rural farmers. According to Huq (1989:100), by 1984 only 2 percent of the estimated credit needs of small-scale farmers were being met by formal sector institutions. Rural informal credit sources responsible for as much as 90 percent of agricultural investment. This latter figure, observed after several years of operation of the rural banks, appears to be rather unimpressive. On the other hand, The BoG reported in 1984 that 40 percent of the credit needs of the small-scale farmers was being met by the formal institutions. Sarris and Sham (1991:128) also observed that in 1975, *"only about 7 percent of small farmers had access to institutional credit"*. But because of the *"aggressive lending policies pursued by the banking system, over 400,000 small farmers"*, constituting 40 percent of the about 1.6 million farmers in the country, had access to formal institutional credit by 1983.

However, the 40 percent of farmers reportedly benefiting from credit facilities in 1983 had fallen to a very low level by 1986 on the eve of the introduction of FINSAP 1, as banks had decreased lending to the agricultural sector due to the crisis of confidence in the formal financial institutions, the unstable economy and the conditionalities of SAP. Sarris and Sham (1991:127-8) pointed out that between 1983 and 1986, agricultural sector credit *"rose from 2.0 billion cedis to 7.5 billion cedis, or roughly 3.7 times ... in nominal terms"*, which amounted to 94 percent increase in real terms. The banks that provided *"31 percent of total credit to the agricultural sector in 1983 reduced this to 18 percent by end of 1986"*. Reportedly, poor recovery of loans and high interest rates partly accounted for this decline. It is perhaps no wonder that agricultural growth has been constantly low, just over 3.1 percent during the SAP period.

Even though the rural banks had already been playing very important roles in savings mobilization and intermediation of credit to the all-important rural economy, FINSAP 1 did not cover them. FINSAP 1 covered only the BoG and the big financial institutions. Its implementation, however, affected the rural areas and the rural banks, as we shall observe.

13:2 The Rural Banks of Ghana

The rural banks are formed, owned and managed by the people of designated "rural" areas, with the initial financial and technical assistance of the BoG. Their main objectives are mobilizing rural savings and *"using the resources for the extension of credit to small peasant farmers and other small entrepreneurs as a means of promoting rural development"*, and national development (Addo, 1984:5; Addeah, 1989:14).

The rural banks were set up because of the failure of the formal and informal financial institutions to meet the credit needs of the rural producers. A former Governor of BoG during whose time the program was developed observed that:

> It was the realization that the existing institutional credit system was not well suited for rural development that led to the search for a credit institution devoid of the disabilities of the existing banking institutions but possessing the advantages of the non-institutional credit agencies (Addo, 1984:6; Kwawukume, 1995:72-5).

Huq reiterated the governor's observation (1989:190):

> ...The need for the rural banks became apparent in the closing years of the 1960s, when it was discovered that the ADB was not meeting all the credit needs of the small-scale rural farmers and industrialists.... Moreover, funds mobilized in the rural areas by the commercial banks were transferred to the cities and larger towns for investment in sectors like commerce and housing, thus leaving the rural communities short of investment funds.

Rural development is thus a principal rationale for setting up the rural banks. Furthermore, the need to bypass informal moneylenders perceived as charging usurious interest rates was also paramount. A former Deputy Governor of BoG lamented in 1991:

The much detested money-lender is still in the system and thriving. His success is due to his style of operation and to the absence of a suitable financial institution [...]. The main purpose of the rural banking system is to supplant the traditional money-lender and to save the rural borrower from atrocious interest charges (Owiredu, 1991).

The first rural bank was opened at Agona Nyakrom in the Central Region on July 9, 1976, with 146,000 cedis initial capital invested by 125 shareholders. By 1983, there were 59 rural banks reporting total assets of 636.6 million cedis, with about 318.8 million cedis in loans and advances. By 1986, there were 106 rural banks with accumulated assets worth 1.2 billion cedis and 686.3 million cedis in loans and advances. By September 30, 1992, the 123 rural banks in operation had total deposits of 7.021 billion cedis with 3.393 billion cedis in loans and advances. BoG bonds and government securities purchased by them in 1986 stood at 345.8 million cedis. This increased to 2.216 billion cedis at September 1992, representing 2.2 percent of government internal debt. The comparable figure for the second quarter of 1999 was 59.2 billion cedis bought by 110 rural banks.

The rural banks, where they have been operating satisfactorily, have indeed been making an impact on the lives of many rural dwellers through the advances they make for all kinds of productive and consumption spending and the assistance they have been offering towards community projects. For example, grants have been made for the building and roofing of school blocks, clinics, and the erection of electric poles.

Many rural banks have opened shops through which all kinds of commodities ranging from batteries, cutlasses, axes, radios, wax prints, irrigation equipment, outboard motors, fishing nets, building materials and corn-mills are sold to the people. Credit sales and high purchase terms are often available to cover the more expensive durable items.

An important element in the early 1980s, when general shortages plagued the country, was the channeling of inputs financed under the EU/ACP Lomé Convention Pacts through the rural banks to farmers and fishermen. Lomé 1 provided in 1982 a grant of $2.5 million to eleven rural banks, while Lomé II provided grants totaling $7.93 million to 60 rural banks for the acquisition of inputs for farmers and fishermen who were customers of the rural banks in 1986. Proceeds realized were put into the Cedi Counterpart Fund for financing projects. It is therefore undisputed that the rural banks have been playing an important role in improving the quality of life of people in rural Ghana.

This brief overview of the rural banks highlights the very large growth in their assets and mobilized funds and the important role they have been playing in bringing credit and both consumer goods and productive inputs to hitherto marginalized areas. But their early rapid expansion had been halted by 1986. The BoG actually suspended the opening of new rural banks in 1986 when the FINSAP 1 policy was being drawn. Accordingly, there were 228 unprocessed applications with 44 approved. Only one opened in 1987, bringing the total to 107 (Obben, 1991:95; Kwawukume, 1995:97).

However, we must note that all rural banks have been opening branches known as "agencies" and "mobilization centers", and thereby extending the spread and availability of rural banks. By so doing, they effectively nullified their initial design as "unit banks without branches". More importantly, they have expanded beyond their originally designated "catchment areas" into areas unable to open their own rural banks, thereby stymieing prospective plans to open independent banks in those areas.

13:3 FINSAP and Rural Credit Intermediation

According to an Agricultural Development Bank (ADB) report for the period 1984-88, FINSAP 1 has as main components management and financial audits, as well as detailed portfolio reviews of key banks in order to determine the extent of their problems and recommend remedial measures; restructuring the capital base of the commercial/development banks to ease the financial strain on them; the strengthening of the BoG in monitoring and supervising the banking system; and measures to develop capital markets in the country.

Some commercial banks such as the Barclays Bank and the Co-operative Bank responded to the implementation of FINSAP 1 by closing down some of their rural branches, while all banks started reducing lending to the rural sector. The Co-operative Bank, for instance, had been run inefficiently and was thus plagued by threats of bankruptcy on many occasions since the late 1970s, with direct consequences for the rural banks and many of their clients. It was thus noted by the then Rural Banking Department in a seminar presentation on 13 September, 1986 that:

> The collapse of the Co-operative Bank which some rural dwellers
> patronized had adversely affected saving mobilization in the rural
> areas. Most depositors, especially in the Brong Ahafo and the Ashanti
> Regions, lost money when the Co-operative Bank folded up. This had
> to some extent eroded confidence in the rural banks. It is hoped that
> with the re-organization of the Co-operative Bank, the confidence in
> the rural banking sector would be restored (Source: RFD Document
> 1985).

The loss of deposits and shares owned actually spread all over the operational areas of the Co-operative Bank. With the introduction of FINSAP 1 and 2, more branch closures occurred. The rural customers of the Co-operative Bank were handed over to rural banks where the Co-operative Bank closed branches. With these developments, the Co-operative Bank, by the end of 1993, had become a mainly urban-based bank, with the bulk of its loans going to salaried workers and the commercial sector.

As events turned out, the Co-operative Bank had not performed creditably in its new role area either, and was slated for liquidation in 1999. With this chequered history, we can safely conclude that co-operative banking had failed in providing the necessary credit to the rural producers it was set up to serve.

The fall in the agricultural sector's share in the loans of the big financial institutions is depicted in Table 13.1.

As demonstrated in Table 13.1, the situation got worse in the late 1980s and in the 1990s, underpinning the sluggish growth in agriculture and fishing during the SAP period. The common belief under-riding these changes in loans portfolio is that the agricultural sector, in the absence of any form of insurance or government guarantees against loans may default due to the vagaries of the weather or poor markets, is therefore is a credit risk. This is not entirely true, as the BoG has a compensatory scheme.

Since it is now clear that the extension of commercial banks to the rural areas, as usually recommended (FAO, 1986, a and b), has proven ineffective, largely due to the high transaction and other operational costs in rural operations and the inability to match effectiveness of the informal sector moneylenders. Bell (1988) states, *"using commercial banks to challenge the market power of the informal sector money lenders may be expensive and inefficient"* (Frimpong-Ansah and Ingham, 1992:47). Some alternatives therefore must be found for the operation of commercial banks in the rural areas.

13:4 FINSAP 2 and the Rural Finance Project, 1989-95

Following the passage of the Banking Law 1989, teams of chartered accountants were hired in 1989 by the newly created RFD to evaluate and restructure the rural banks in order to bring the deserving ones into the RFP. This was done under the institutional building component of the RFP.

The objectives of this component were to assess the "capital adequacy, management capabilities and overall operational performance of the Banks". with the intention to "ensure the liquidity, profitability and solvency of the Banks in line with the provisions of the Banking Law 1989, PNDC Law 225" (RFD

Table 13.1 Percentage of Loans and Advances to the Agricultural Sector by Commercial and Secondary Banks

YEAR	COMMERCIAL BANKS	SECONDARY BANKS*	
1984	31.9	29.7	
1985	22.9	23.5	
1986	18.0	18.7	
1987	21.3	17.9	
1988	16.6	13.2	
1989	15.5	13.4	
1994+	8.3	COCOA MARKETING++	
1995	9.7	1994	2.6
1996	10.5	1995	4.8
1997**	11.9	1996	1.4
1998	12.2	1997	6.8
		1998	2.7

Source: *Min. of Agric, 1991: 19; BoG Annual Reports, 1996: 16; 1998:21)*

*Secondary banks include the African Development Bank, National Investment Bank, National Savings & Credit Bank, Social Security Bank and Merchant Bank (Ghana) Ltd.

+Figures amalgamated from 1994 and comprise 11 banks

**Coverage extended from 11 to 17 banks from Dec. 1997

++Cocoa marketing has its own special financing account

Workshop Paper, 1993:6).

The conditionality of the "preparation and restructuring plans for some participating institutions including the Agricultural Development Bank and the Rural Banks" meant a delay in disbursement until April 1990 (RFD Workshop Paper, 1993:6). The exercise was thus carried out between June 1990 and January 1992.

The institutional building plan also included strengthening the financial management and loan appraisal capacity of the of the rural banks through staff training; strengthening of the Association of Rural Banks to provide training, advisory and promotional services previously provided by the Rural Banking section of the BoG; and restructuring the Rural Finance Department (RFD) through the provision of technical assistance and logistic support. The RFD was later renamed Rural Finance Inspection Department (RFID) to assume the roles of supervising the activities of rural and community banks, and regulating the establishment of new ones.

Restructuring Action Programs (RAPs) were submitted in stages to each rural bank by the consultants indicating which actions to implement in order to bring the operations of the banks in line with the provisions of the Banking Law 1989 by December 31, 1992. This was conditional to having their licenses renewed but as things turned out, the forbearance was extended into 1993 and eventually relaxed until 1999.

The reports of the consultants formed the basis for the classification of the rural banks in July 1992 into four groups, subject to subsequent reviews. Twenty-three were classified as performing satisfactorily as they met the required capital adequacy of 6 percent, and thus qualified to participate in the On-lending Component. They were required to maintain a cash reserve ratio of at least 10 percent of total deposits. They were also seen as eligible to participate in a "proposed networking of strong rural banks via an intermediary rural bank or national bank".

Twenty-six rural banks, "showing capital inadequacy, illiquidity and negative profitability", were classified in Mediocre Group A, and expected to meet a series of conditions which included loans and advances to customers not to exceed 25 percent of total deposits; cash reserves ratio of at least 20 percent of total deposits; 55 percent of total deposits to be invested in first class risk free securities; and reduction of staff to four personnel (BoG Notice No.BG/RB/92/1, July 23rd, 1992).

Fifty-three banks classified in Mediocre Group B, were found to be having high capital inadequacy and on the verge of insolvency, and were expected to maintain cash reserve ratio of at least 20 percent of total deposits; invest 80 percent of total deposits in first class risk free securities; cease all lending activity subject to review after six months; and reduce staff to only four personnel. Both Mediocre Groups A and B were to be assisted with "re-capitalization loans to improve upon their capital base" (RFD Workshop Paper, 1993:7).

Finally, twenty-one were declared insolvent; distressed, and unable to "meet the withdrawal claims of depositors and/or honor payment of workers' salaries routed through [them]…by the Accountant General". Initially, they were to be assisted to pay off their depositors prior to closure but this did not happen. They were to cease all lending activities and invest all mobilized funds and recovered loans into BoG bonds and government securities.

Needless to say, some of the recommendations such cutting staff and cessation of lending activities were found unrealistic to the rural banks concerned. Workers of many rural banks were already much over-burdened and often worked very late hours. Many therefore retained more than four staff. Some distressed rural banks visited in 1993 continued to make "safe loans" to trusted customers and salaried workers, as they claimed they needed the interest on such loans to pay staff salaries and overhead expenses.

Elsewhere, I have classified the period up to December 1984 as the Period of Optimism and Rapid Growth, and the Structural Adjustment Program era as the Period of Crisis and Doubt for the rural banks (Kwawukume, 1995:76-7,99,104,191). The Structural Adjustment Program period derailed the plans for the further development of the rural banking system along the objectives originally conceived, and rapidly realized during the first period. Rural banks were seen as having made "remarkable progress" in their goal of mobilizing rural savings. As put by the then Governor of BoG:

> The rural banks have been able to mobilize the resources; it now remains to apply the resources to the full to develop the rural areas. This will be the next stage in the development of the rural banking scheme, i.e., to develop strategy to utilize the excess liquidity of the rural banks. The approach will be to conduct baseline studies in all the areas where rural banks are situated in order to identify the resources and the potential projects that can be carried out in the area (Addo, 1984:11).

The Centre for Development Studies of the University of Cape Coast carried out a baseline study in the catchments area of the Biriwa Rural Bank Ltd. on behalf of BoG. The study was intended to help to correct the perceived "misplaced tendency" of most rural banks to invest excess liquidity in "sterile assets" such as buildings, vehicles and "government stocks", instead of investing in "primary economic activities" (CDS, 1983:1). This is a clear indication that the rural banks were encountering problems even then. A former Director of the National Investment Bank (NIB) observed that:

> [...] Of the total increase in resources of 203 million cedis between June 1981 and June 1982, only 5.9 percent went to loans and advances; 37.7 percent to Government Securities and 20.8 percent to other assets. The remainder of 35.6 percent was kept in idle cash and balances with banks. It is glaring that a large proportion of the total resources of the banks is either used to finance the Government securities or is in the form of idle cash and balances with the banks. This implies that the resources of the rural banks are not being effectively directed towards investment that will benefit the rural folks (Nyonator, 1983).

It is evident that even the commercial banks were getting more mobilized funds than all the rural borrowers, with the central government already taking the lion's share. As we shall see later, what was considered an anomaly in 1983 had become the accepted norm by 1992, with official mandate.

The rapid growth during the first period had exposed a lot of organizational, managerial, staff and generic problems, leading to a ban on the further opening of rural banks, already noted. As the Governor of BoG put it at the opening of the 105th rural bank at Abokobi near Accra:

> We are not interested in only the numbers of banks but good quality personnel who will manage the banks efficiently ... we therefore have to slow down the rate of growth ... in order to be able to select, train and give the right orientation to the prospective management personnel (ARB Briefing Notes No.3, April 1985:2).

The problem with personnel was spelt out by the Secretariat of the ARB: "grievous events and mishaps have befallen some rural banks". It went on to lament:

> The cruel deeds of some managers and staff and the behavior of some chairmen, board of directors either individually or collectively, are creating concern. In general, the Bank of Ghana guidelines and regulations are being flouted; no satisfactory liquidity ratios are maintained, the mandatory purchase of stocks is ignored and returns are not submitted on time. Bills Payable checks are drawn on Bank of Ghana when funds are not available to meet them. Loans are given haphazardly by directors often to themselves and their relatives or friends with little or no effort to retrieve those that become overdue. Managers collude with customers to defraud the very banks, which they are supposed to manage. Wherein lies their loyalty! Indeed, those acts are soiling the name of the Rural Banks and gradually eroding confidence in the system (ARB Briefing Notes, April 1985:3-4).

Other anomalies revealed included the high default rates already being observed, and collusion with corrupt staff of the Produce Buying Company (PBC) of the Cocoa Marketing Board to embezzle funds meant for the purchase of Akuafo checks. This problem has been traced to the recruitment of so-called "experienced staff", which turned out to be "rejects of their former employers, due to involvement in fraud, negligence of duty or allied offences. Others were retired personnel of the established banks whose energies have already been sapped by their previous employers" (West Africa, April 6, 1987:662; Kwawukume, 1995:243). But the overwhelming reason why these people had free rein to engage in their mismanagement and malpractice was the very inadequate supervision of the rural banks by the BoG.

Concerning loan default, it cannot be denied that the culture of non-repayment perceived elsewhere (Ahmad, 1983:7,11-12), whereby the borrowers perceive loans as a form of relief aid, or a kind of national cake-sharing not to be repaid, has been established in Ghana too. The CPP government's credit policies of the early 1950s and 1960s share the blame for establishing this culture. Subsequent governments' lukewarm attitude in not putting pressure on the state-owned banks to enforce repayment has served to reinforce this attitude of defaulting on loans in Ghana. Experience has shown that sanctions are rarely invoked to secure repayment.

13:5 The Structural Adjustment Programs Period: Crisis and Doubt

Rural finance under the Structural Adjustment Program period can be divided into three phases: January 1985 to the inception of the Rural Finance Project in July 1989; July 1989 to July 1992, when new directives confirmed the decision to "commercialize" the rural banks after undergoing restructuring; and July 1992 to the present, marking the period when bureaucratic imperatives have fully taken over the operations of the rural banks.

The first phase saw the initiation of FINSAP 1 directed at BoG and the main banks. Dominated by public policy, it was associated with crisis as well as doubt for many banks. All banks and some of their customers were still reeling from the unpopular actions taken by the PNDC in 1982 in freezing bank accounts.

There was also deepening financial crisis relating to inadequate capital liquidity in the face of the fast deteriorating value of the national currency thereby causing the decapitalization of banks; increasing default rate with accumulated huge portfolios of non-performing assets; and damaging publicity surrounding management lapses and rampant abuses by some management and staff of some rural banks as well as the big banks. Accordingly, in 1986, a high level of money supply (48 percent) was being held outside the banking system.

Accordingly, as part of the corrective measures adopted under FINSAP II in Phase two, the government assumed responsibilities for the liabilities of the commercial and secondary banks. It released in 1992 an amount totaling $100 million (about 340 billion cedis then) to re-capitalize these banks (People's Daily Graphic, May 29, 1992). The Non-performing Assets Recovery Trust (NPART) was established under the management of BoG, and entrusted with the recovery or liquidation of these defaulted loans. Rural banks in difficulties were not initially extended this support until well into 1993. According to a Director of NPART, as the rural banks extend credit without demand for collateral securities, there were no identifiable assets to take over and liquidate (Kwawukume, 1995:105). The BoG later paid 524.81 million cedis to 12,574 customers of 18 out of the 19 distressed rural banks by December 1993, a tiny amount compared to the amount being spent on the big banks (BoG Annual Report, 1993:20).

13:6 The Rural Finance Project

The Phase two era saw the initiation of the RFP under FINSAP 2. The RFP was also carried out under the agreement signed by the BoG and the IDA.

The RFP had two components, comprising an On-lending component of $15 million, which only banks meeting certain conditions could participate in, and the Institutional Building component of $5 million. The main objectives of the RFP were specified as followed:

- Expansion of productive capacity and employment in the rural sector through provision of finance for viable projects in agriculture, livestock, fisheries, agro-processing and input supplies through lending approach.
- Implementation of a financial restructuring program for participating institutions and organizations to enable them become more efficient in deposit mobilization and the delivery of credit, especially to smallholders.
- Build an enhanced capacity for the appraisal of rural credits.
- Strengthen the rural banks examination unit in the Bank of Ghana.
  (RFD Workshop Paper, 1993:4-5)

Notable was that the project "was drawn to be consistent with the IDA's lending strategy in the financial sector and emphasizes restructuring of financial institutions, especially the Rural Banks and the Credit Unions, to enable them mobilize more deposits and thus increase the flow of credit to the rural sector" (RFD Workshop Paper, 1993: 3). We can therefore concluder that the premises of external actors have become predominant in determining the organization of the rural banks and rural development in Ghana.

The main conditions to meet in order to participate in the On-lending component "were having an updated audited accounts, compliance with the BoG specified liquidity reserves, undergoing of financial restructuring plan if found necessary, and finally for the rural banks as well, compliance with the 6 percent capital adequacy requirements under PNDC Law 225" (RFD Workshop Paper, 1993; Kwawukume, 1995:107). Other conditions were imposed. For instance, all projects in the rural sector with the exception of industry, real estate and speculative trading, could be financed, provided the estimated gross margin was not less than 20 percent after debt service; the beneficiary provided 10 percent of the cost; and the sub-loan financing would not exceed 70 percent of total investment cost or 80 percent of loan amount approved by the participating financial institution (RFD Workshop Paper, 1993:4).

The level of disbursement as at December 31, 1992 is shown in Table 13.2.

Only nine banks out of the number found qualified to participate received US$1.1m (21 percent) of the total funds from the On-lending component. Clearly, the project was performing below projections. When conceived, it was expected to last for three years, ending in December 1992, but it carried on till December 1995, well behind schedule.

The need to meet the new bureaucratic imperatives now operative in Phase 3, which found 99 rural banks ineligible to participate, contributed to the delay. Also, weak human resources in both management and board of directors, and the high lending ceiling of 50 million cedis on individual sub-projects became major contributory factors for the low performance and the delay (RFD Workshop Paper, 1993).

Also, to many managers, the stringent conditions attached to the loans and the perceived high cost made it unattractive, as the RFD also testified. Besides,

Table 13.2 Disbursement Under Rural Finance Project as of December 31, 1999 ($'000)

CATEGORY	ALLOCATION	TOTAL DISBURSEMENT	BALANCE
1. Sub-loans	15,000.0	5,351.0	9,648.9
2. Consultancy & Training	2,700.0	2,931.2	(231.2)
3. Equipment & Vehicles, etc.	800.0	125.0	675.0
4. Operating Costs (Institutional building)			
RFD	300.0	25.2	274.8
ARB	150.0	6.9	143.1
5. Refunding of PPF	2.7	--	2.7
6. Unallocated	1,047.3	--	1,047.3
TOTAL	$20.000.0	$8,439.4	$11,560.6

Source: *Kwawukume, 1995:109; based on RFD sources*

due to the prevailing high-risks in lending to agricultural projects, and the need for making safe loans, the eligible rural banks were already having a problem of "excess liquidity", which they were investing in government securities, or keeping with the commercial banks. In 1993 about 60.69 billion cedis in shareholders funds were utilized in this fashion.

The national level banks expected "to be prime movers of the on-lending component" did not show particular interest. By 1993, only five had signed up, and only three (the Agricultural Development Bank, Ghana Commercial Bank and Standard Chartered Bank Ltd.) actually participated (RFD Workshop Paper, 1993:9). Barclays Bank and the Social Security Bank failed to participate. It is not surprising as these banks were not much willing to extend credit to the rural sector anyway, and were actually withdrawing loans and closing rural branches.

Some proposed solutions to address these problems included the amendment of the formula for calculating the interest chargeable on the sub-loans and reducing the interest rate from 14.4 percent to 12.75 percent; intensification of the ARB's training programs for directors and staff; removal of the ceiling for financing projects; and re-capitalization scheme by the BoG to allow more potential financial institutions to participate – the activities of NPART and BoG in this regard have already been mentioned.

At the end of 1993, nineteen rural banks had participated in the re-financing facility "to provide funds for lending to agriculture and agriculture-related rural enterprises". These rural banks received 1.57 billion cedis, representing 16.2 percent of total funds disbursed (BoG Annual Report 1993:20). This represents a modest turn towards increasing advances to the agricultural sector as reflected in the proportion of loans received in subsequent years.

The rural banks were also reclassified into three categories: satisfactory, mediocre and distressed (BoG Annual Report 1993:19). The highly enhanced supervision of the RFID and the Banking Supervision Department, coupled with the financial support the rural banks were receiving, were clearly showing positive effects on some banks.

In 1994, twenty-five mediocre rural banks were reclassified into the satisfactory category due to their improved financial indicators, thus bringing this category of rural banks to fifty-five. There were fifty-one mediocre rural banks, with nineteen still distressed. Two new rural banks were opened during the year, heralding a cautious relaxation of the suspension order to open rural banks in much disadvantaged communities not taken care of by branches of other rural banks. But it was not until 1996 that four new community banks were opened, bringing the total of rural/community banks to 129. Three more were formed in 1997 and one in 1999, while twenty-three were eventually closed.

The Akuafo checks system was revised to take account of a new checks system by which rural banks now officially clear their checks through the commercial banks. These were attempts to streamline the Akuafo checks system and redress the problems faced by the participating rural banks since the introduction of the Akuafo checks in 1982 to pay for the purchase of cocoa, coffee and shea-nuts. At the end of the RFP in 1995, 67 rural banks purchased a total of 46.054 billion cedis worth of Akuafo checks as against 15.603 billion cedis purchased in 1993/94, a very important role they continue to play in financing the export of Ghana's main export crops.

Most rural banks finance the purchase of Akuafo checks with their own funds before seeking reimbursement from the BoG. As far back as 1985, it was observed that some chairmen and managers had to be "nightmarishly chasing funds. An added problem was the sharing of the meager 1 percent margin with the Social Security Bank, not to mention the unhealthy competitive attitude of some commercial banks" (ARB Briefing Notes, April 1985:3). Disillusionment led to some rural banks threatening to withdraw from the scheme but they were reminded that, "they are doing a service to the people for whose benefit their bank was established, and that with the assurance of the Bank of Ghana, their woes will be turned into cheers" (ARB Briefing Notes 3, April 1985:4). This apparently had not happened by 1993, as witnessed to by the Manager of a rural bank earlier interviewed seen on a corridor of the RFID in Accra. Asked what he was doing there, he responded almost in tears: "I just want my money! I want my money! I did not plan to come to Accra. I just went to Kumasi to collect money and here I am!" (Kwawukume, 1995:102-3). Some managers of rural banks interviewed claimed that they could make more money lending mobilized funds to salaried workers instead of pre-financing purchases of Akuafo Checks for a meager commission of 1 percent, from which expenses such as mobile banking services have to be deducted, apart from a commission of 30 percent taken by the Social Security Bank for clearing their checks. But it cannot be denied that these commissions were the major sources of income to some rural banks before the advent of salaried workers taking the bulk of loans.

In 1995 rural banks also paid public sector workers salaries amounting to 51.8 million cedis per month on behalf of the government (compare to 4.4 million cedis per month in 1994). This is another public policy role assumed which eventually led to the salaried workers becoming their main shareholders and recipients of loans and advances. In the same year the RFID managed to achieve complete inspection of all the 125 rural banks for the first time. Financial indicators were reported as making "significant improvements" (BoG Annual Report, 1995:14). In recognition of their improved performances, forty rural banks were considered qualified to be re-licensed as community banks.

Total deposits increased from16.01 billion cedis by December 1994 to 33.51 billion cedis by December 1995, that is, an increase of 109.31 percent. Outstanding loans and advances also saw an increase of 85.59 percent, that is, increasing by 6.65 billion cedis in December1994 from 7.77 billion cedis to 14.42 billion cedis by the end of December 1995. The average overdue ratio decreased from 41 percent in 1994 to 29.37 percent, and by 1998 stood at 12 percent, a significant improvement.

Meanwhile, there was a hefty increase of 175.6 percent in investments in BoG bonds, treasury bills and other government securities of 14.51 billion cedis, an amount bigger than loans and advances in 1995 (BoG Annual Report, 1995:14). This is truly a large part of the prudent portfolio management that accounts for the declining overdue ratio.

Moreover, in response to measures introduced under FINSAP 2, the rural banks, already accustomed to disregarding BoG directives on the sectoral lending when it existed; keeping mandatory reserves and submitting returns, with equal disregard to sanctions imposed on them for not doing so (Obben, 1991:103; Kwawukume, 1995:113, 221-4), refused to lend to the agricultural sector until they got RFP on-lending funds. A large number of them, emboldened by the lifting of the sectoral controls on the allocation of loans in January 1990, with only 20 percent reserved for agriculture, and the BoG directives of July 1992, had loans portfolios with over 80 percent of loans going to salaried workers at the end of 1992. As much as 60.69 billion cedis of mobilized funds were also kept in shareholders' funds by December 1993.

It appears that a sure way to meet the new guidelines and to ensure being listed among the satisfactorily performing rural banks now depended upon the number of salaried workers a rural bank has as customer-shareholders.

By April 1993, a typical satisfactorily performing rural bank had a shares distribution as displayed in Table 13.3. Workers, predominantly male, thus constituted 80.47 percent and farmers and others only 19.53 percent, respectively, of this rural bank by April 1993. These salaried workers used to be among the miscellaneous borrowers who were expected to take only 5 percent of loans. The distribution of loans for some reporting rural banks is presented in Table 13.4.

Together with trading, salaried workers took 74.37 percent of loans in 1996, 72.70 percent in 1998, and 71.50 percent for the second quarter of 1999. As if to hide this fact of goal displacement, while publicly espousing the original goals,

Table 13.3 Distribution of Shareholders of Amanano Rural Bank Ltd.

Head Office, Nyinahin According to Occupation and Gender as of April 30,1993						
OCCUPATION	MALE	%	FEMALE	%	TOTAL	%
Workers	525	86.07	85	13.93	610	80.47
Farmers, etc.	120	81.08	28	18.92	148	19.53
TOTAL	645	85.09	113	14.91	758	100

Source: *Kwawukume 1995:201*
Based on data collected from Amanono Rural Bank Ltd, Nyinahin-Ash, 1993

salaried workers (the major recipient of loans and advances), are grouped under the category "Others" in the BoG's Annual Reports.

Some salaried workers, mostly teachers, had organized to take over some banks, as the reported case of the troubled Biriwa Rural Bank (Ghanaian Times, March 22, 1993:1). This shift to giving the bulk of the loans to salaried workers was happening in spite of no proclaimed changes in the objectives of the rural banks, except for the declared commercialization policy by the bureaucrats at BoG, which seems to contradict the government's declared public policy of supporting agriculture and cottage industries in order to promote rural development. This state of affairs has led to the situation whereby by December 1999, the "rural sector's share of total formal credit remains a mere 8 percent" (InfoShop, Jan. 7, 2000). This share of formal credit is clearly no higher than the 7 percent the rural sector was supposedly receiving in 1975 before the first rural bank was formed.

Some managers and directors cited the same reasons as the commercial banks for this state of affairs, claiming that the risky nature of agriculture meant farmers and fishermen had defaulted heavily on loans, and they could no longer be expected to approve bigger loans for them. As aptly expressed by one manager:

> If you give all your money to these fishermen and farmers, and you failed, you'll be classified as 'distressed' and be closed down. The person who puts money in your bank wants interest, and the shareholders want dividends. Why do you give it to people who can't pay you? [...] This is why I'm saying there's conflict between the initial policy [of the rural banks], what rural bank was supposed to do and what's happening now (Kwawukume, 1995:174).

Table 13.4 Sectoral Breakdown of Loans and Advances for the years 1996, 1998 and the 2nd Quarter 1999 (billion cedis)

SECTOR	1996		1998		2nd Qtr. 1999	
	Loans & Advances	% of Total Loans	Loans & Advances	% of Total Loans	Loans & Advances	% Of Total Loans+
Agriculture	1.48	13.41	5.57	15.41	7.2	18.00
Cottage Industry	0.56	5.07	1.49	4.15	1.7	4.25
Transport	0.79	7.16	2.77	7.74	2.5	6.25
Trading	2.86	25.91	9.32	25.59	8.9	22.25
*Others**	*5.35*	*48.46*	*16.65*	*47.11*	*19.7*	*49.25*
No.of Banks	43		79		67	
TOTAL	**11.04**	**100**	**35.50**	**100**	**40.0**	**100**

Source: *BoG Annual Reports 1996; 18;24; Quarterly Economic Bulletin Apr-June 1999:12*
Others comprise salaried workers.

The indications are, the bureaucrats in the RFID and ARB who were then facing severe, negative criticisms in the national media, gave tacit approval to the changes in the shares ownership and loans recipients in the rural banks (Aidoo, 1992:2, and field data).

The shift in loans portfolios saw loans recovery rate rising from 44.7 percent as at December 1989 to 52.1 percent as at December 1990 (RFD Document, April 1992). But then it worsened again from the 52.1 percent recovery rate to 46.5 percent as at 30 September 1992. It appears the rural banks were about to discover for themselves what the commercial banks and researchers already knew: that the large-scale farmers, the commercial and transport sectors were not, contrary to popular expectation, better at repaying loans. A study conducted by the ADB on its Credit Commodity Scheme revealed that between 1971 and 1977, the recovery rate among small-scale producers had a mean of 43 percent, whereas among large-scale producers the rate was 23 percent (Hansen, 1989:207).

Experiences elsewhere, for example, the Comilla Cooperative Programme in Bangladesh (Ahmad, 1983:32), and the "People's Bank" in Nigeria (Moghalu,

1990:6), had also shown similar tendencies for the richer elements in society to be both the main beneficiaries and defaulters in credit schemes in the developing countries. Gran (1983:285) referred to Blair's (1978) demonstration of how the focus of bureaucrats on default in rural Bangladesh led to loans going to bigger farmers, in spite of the fact that they defaulted more. As Blair (1978:73-74) explains:

> ...This is irrelevant. What is relevant is that the bureaucracy be able to defend itself against a charge of fiscal irresponsibility in its distribution policy by showing that it lends only to the "best" credit risks.... The wise bureaucracy as a whole, in seeking to show objective measure of progress, tries to minimize default, but in so doing forces officials to pursue policies that in fact tend to maximize default. Too much supervision, just like too little supervision, results in the benefits going to the local rich.

For the rural banks, Obben (1991:128) had already noticed an increase in trading loans' share of total loans resulting in decreased profitability. The stage was thus set for making more secured and safer loans and other investment. A brief examination of this is follows.

It is clear from Table 13.5 that, even though the rural/community banks have succeeded in boosting investment funds through mobilization campaigns and attracting deposits from salaried workers, most of the mobilized funds have been going into purchasing BoG bonds, government securities and the higher than required primary reserves they have been keeping in cash and balances with other banks.

By December 1998, primary reserves increased by 87.2 percent (14.4 billion cedis) to 30.9 billion cedis representing 27.2 percent of total deposit liabilities instead of the required ratio of 10 percent. Secondary reserves increased by 58.3 percent (22.1 billion cedis) to 59.9 billion cedis, representing 52.6 percent of total deposit liabilities. Outstanding loans and advances by the 108 reporting rural banks however increased by 41.8 percent (15.9 billion cedis) to 53.9 billion cedis.

The government's policy of using fiscal measures to finance the fiscal deficit instead of using them as tools to smooth out fluctuations in the money supply has been noted as an impediment to private investment and economic growth. The private sector's previous equal share of investment funds started falling, beginning 1992 (Boehmer, et al.1994). With a very low ratio of private credit to GDP of 5.3 percent, far below even African standards, the credit squeeze on

Table 13.5 Selected Financial Indicators of Rural Banks

(Billion Cedis)

	1984	1987	1989	1993	1995	1996	1997	1998	1999*
ASSETS									
Total Assets	0.92	2.50	5.55	n.a.	45.25	72.54	113.2	163.2	182.0
Primary Reserves	0.09	0.26	0.47	3.88	8.94	12.37	16.5	30.9	29.4
Secondary Reserves+	0.15	0.52	0.88	5.54	14.51	25.33	37.8	59.9	60.1
Loans & Advances	0.51	1.02	2.13	6.82	14.42	24.34	38.1	53.9	66.8
Total Liabilities	0.92	2.50	5.55	n.a.	45.25	72.54	113.2	163.2	182.0
Total Deposits	0.67	1.87	3.66	13.24	33.51	50.13	79.9	113.8	130.4
Savings	0.53	1.40	2.60	n.a.	21.48	31.85	51.1	70.0	75.8
Demand	0.16	0.47	1.06	n.a.	10.46	10.46	24.6	33.9	43.1
Time	n.a.	n.a.	n.a.	n.a.	1.56	2.00	4.2	9.8	11.5
Shareholders' Funds	0.04	0.09	0.13	60.69	3.23	4.68	7.9	11.3	16.2

Source: *various BoG and RFD Reports*
* *1999 End of second quarter*
+*Secondary Reserves actually represent government securities and BoG bonds*

rural producers presents a problem given the key role played by rural areas in Ghana's development. For the rural dwellers, the problem of equity in the intermediation of credit in rural development under Structural Adjustment Programs and accusations of favoring the rural elites are two are two major complaints. Mikell noted when discussing rural development during the Structural Adjustment Program period in Ghana that it was the educated men and well-connected men and women, usually *"large-scale farmer or farmer-bureaucrat"*, not well integrated into the local society, who are leading this local development. Thus, these "new development procedures effectively lock out older farmers, stranger farmers, small-scale farmers, and women farmers from equal participation in local development; they create a young, privileged and relatively educated group that is tied to the government" (Mikell, 1991:93).

Interviews conducted in 1993 confirmed the fact that many of the salaried workers, not necessarily tied to the government but certainly appreciative of its new liberalized attitude towards mobilized funds for the rural banks, were investing either their second or third loans into cash crops; having often used the first or second to acquire durable household items. Some bank managers felt this was a positive development in replacing the supposedly "old, illiterate and conservative farmers".

It is interesting how either too little or too much supervision of the rural banks had had the same effect of elite dominance during the pre-RFP years and post-RFP era respectively. In the pre-RFP period, apart from the cases of gross anomalies, the bigger loans were found to going to the local elites and directors of the rural banks.

Such was the extent of the negative effects of poor supervision in the pre-RFP era in Phase 1 that Ashe and Cosslett (1989) noted that, the relevance to the UNDP of the experiences of Ghana's rural banks:

> Is the importance of proper supervision of the unit offices? Problems faced by the Rural Banks are largely due to inadequate supervision. A majority of the local banks do not have Bank of Ghana supervision; training is inadequate and local staff have little incentive to be productive (Ashe and Cosslett, 1989:29).

This poor supervision by BoG was largely due to the reduction of supervisory and monitoring staff as a result of the embargo on recruitment introduced by Structural Adjustment Programs, and later the restructuring exercise in the BoG itself under FINSAP 1, focusing on the curtailment of staff.

In a report dated April 1992 on the operation of the RFD for the period July 1990 to December 1991, a previous Director of the RFD recognized the need for the RFD "to step up considerably its inspection role to ensure that the banks stay on course". Instead of this happening, he claims that owing to massive transfer of officers of the RFD to other Departments over the past few years, the Department has become grossly understaffed. For instance, the Monitoring, Appraisal and Inspection Office of the RFD is now down to 4 officers from a previous strength of twelve (Kwawukume, 1995:190).

BoG representatives to rural banks, already noted for their irregular attendance at boards meetings, were officially withdrawn in January 1992 amidst protests from the rural banks (Kumah, 1992). The BoG handed over its training role to the ARB (BoG letter No. RFD/N/4/92/342), and prepared to return to its traditional role of supervision and monitoring of the whole banking system.

These contingent developments exemplify how certain assumptions about Ghana's rural areas do not work out in practice (Obben, 1991:102-104, 116; Brown, 1993; Kwawukume, 1994:198-207; 1995:172-192,230-45). There are many vicious socio-economic, cultural and demographic constraints, which are not well understood and properly taken care of before the setting up of micro-credit organizations projects.

A solution suggested by Bell (1988) to these generic problems, which has found favor with the World Bank under FINSAP 2, is using private lenders as agents for rural banks. However, room is allowed for credit unions and Rotating

Savings and Credits Associations (ROSCAs) such as *susu* to be agents, in spite of problems of inefficiency due to possible barriers to entry. Bell's conclusion (1988:791) that, "the establishment of rural banks should be motivated by the desire to encourage thrift and mobilize rural savings rather than challenge traditional lenders directly at the outset", was found to be relevant to African rural savings mobilization (Frimpong-Ansah and Ingham, eds. 1992:48). But the rural banks were established to challenge the traditional informal operators such as moneylenders, whose command of local knowledge, ironically, the rural banks were expected to emulate (Owiredu, 1991).

13:7 Lessons from the Rural Finance Project

Valuable lessons were learned from the RFP, which formed the basis for measures incorporated into a new Rural Financial Services Project (RFSP) that was planned in 1999 with the support of IDA, Africa Development Bank, IFAD and the Ghana government (The InfoShop, Jan. 7, 2000). These were:

• Bank staff/Management Training. This emerged as "the most important factor contributing to the development of a viable rural banking sector with improved transparency and accountability".

• Need for an Apex Institution. It emerged that the lack of an apex bank made it difficult to sustainably deal with common constraints, e.g., check clearing, and specie supply.

• Need for stronger oversight in order to enhance depositors' confidence in the overall rural banking sector (The InfoShop, Jan. 7, 2000).

These findings were not new, as we have noticed. As far back as the middle of the 1980s, similar proposals were made to resolve those problems (ARB Briefing Notes 3, April 1985:2,5; West Africa, April, 1987:663; Kwawukume, 1995:101-4, 106). The RFID was not unaware of these problems and solutions, as they exist in published forms (for example Aidoo, 1992:2-5). It may therefore not be a hasty conjecture to note that the introduction of FINSAP disrupted these correctives measures previously specified.

The consequences of these developments for agriculture and rural development cannot be underestimated. The sluggish growth in agricultural output, and the increasing import bill for food items led to the introduction of the Medium-Term Agricultural Development Strategy (MTADS) in 1990. It was formulated by the Government with the assistance of IDA, and implemented at a total cost of $26 million (Public Information Center, July 8, 1994). But this project and others have not made much impact as the essential credit and other inputs, services and infrastructure such as storage, marketing and transportation were not made readily available. It is clear a lot more need to be done if the objectives of Ghana's Vision 2020 are to be met in this regard.

New government initiatives supported by multi-lateral donors such as the IDA, IFAD, African Development Bank, the World Bank and NGOs such as Technoserve and Women in Development are responding by setting up new rural credit schemes to fill the gap created by the withdrawal from lending to the rural sector. It is hoped a big boost in investment in agriculture and secondary industries in the rural areas will be realized.

But for rural development to take place, there is the need to undertake long-term investment in capital and social infrastructure that may not have immediate or direct benefits. The commercial and rural banks are not equipped to shoulder these responsibilities. As pointed out by the Manager of a rural bank, which has contributed a lot to community projects in his area of operation, the rural banks cannot be expected to invest their deposits and all their profits into community projects, as their shareholders expect dividends. This is a role for development banks (Kwawukume, 1995:171).

13:8 Conclusion

In conclusion, it is clear that Structural Adjustment Programs have had adverse effects on credit intermediation to the rural economy with agriculture and cottage industries suffering the worst consequences. Even though FINSAP 2 has brought a major increase in the mobilization of funds by the rural banks and restructured many to perform more satisfactorily, this has been achieved through goal displacement. As the commercial banks either withdrew from, or reduced lending to the rural dwellers, the rural banks with tacit BoG support, also began diverting the bulk of their mobilized funds into government securities and loans to salaried workers and the commercial sector. Accordingly, the share of the rural sector in formal institutional finance has declined since the inception of Structural Adjustment Programs.

Indications are that Structural Adjustment Programs halted the expansion of the rural banks in their objectives of lending to the rural productive sector and promoting rural development. In order to meet the strict financial controls under Structural Adjustment Programs, they chose to invest in high interest risk-free securities and lend salaried workers, instead of in risky agriculture and shaky cottage industries. This means that the funds which would have entered the important informal investment chain were diverted.

Even though it can be argued that the government has been incurring huge fiscal deficits partly in pursuance of its policy of developing the rural areas through the provision of such facilities as electricity and roads, district capitals and the major towns have been the main beneficiaries. The majority of the rural dwellers live in smaller settlements not yet easily accessible or benefiting from such amenities. Many rural people cannot afford the cost of installing electricity in their houses as even their basic needs of adequate housing, food and clothing remain largely unmet due to widespread poverty.

One way mobilized rural savings can be made to benefit the rural people, apart from the donations some rural banks make towards community projects many of which often remain unfulfilled, is for mortgages to be given to local and district councils to undertake such projects to completion. Payment derived from levies and taxes on the citizens can then be spread over many years, giving the rural banks regular income. It is only then that we may be able to conclude that meaningful rural development is reaching the mass of Ghanaian rural dwellers through the rural bank projects and activities.

References

ADB Report (1975). *Agricultural Credit Programmes in Ghana*. A country paper first presented at FAO/Finland Regional Credit Seminar for Africa, December 1973. Accra: The Research and Planning Division: Loans Department. Accra: ADB.

ADB Report (1989). *Performance of the Ghanaian Economy: 1984-88*. Accra: ADB.

Addeah, K. (1989). *An Introduction to the Law of Rural Banking in Ghana*. Accra: Amantah Publications.

Addo, J.S. (1984). Governor of Bank of Ghana. Speech delivered at the Annual Dinner of the Institute of Bankers (January 28) Accra.

Ahmad, Razia S. (1983). *Financing the Rural Poor: Obstacles and Realities*. Dhaka: University Press.

Aidoo, Joe (1992). "Operational, Manpower and Management Problems of Rural Banks". Address by Manager of the Lower Pra Rural Bank Ltd., Shama, at the seminar Rural Banking in Ghana - Strategies for Survival (August 29) University of Ghana. Legon.

Association of Rural Banks (1985). *Briefing Notes*, No. 3. April. Accra: ARB.

Atsu, S.Y., Owusu, P.M. (1982). *Food Production and Resource Use on the Traditional Food Farms in the Eastern Region of Ghana*. Accra: ISSER. University of Ghana. Legon.

Bell, C. (1988). "Credit markets and interlinked transactions" in H. Chenery and T. Srinivasan, eds. *Handbook of Development Economics*. Vol. 1,

Amsterdam: North-Holland

Blair, H. (1978). "Rural Development, Class Structure and Bureaucracy in Bangladesh". *World Development.* Vol. 6, No. 1: 65-82.

Boehmer, Hans-Martin, Wetzel, D., Gupta, A. (1994). "Ghana Financial Sector Review: Bringing Savers and Investors Together". *Report No 13423 – Ghana.* Washington D.C.: West Central Department, World Bank.

BoG (1992). "New operating procedures for Rural Banks". Notice to all Rural Banks, Letter No. BG/RB/92/1 (July 23). Accra: Bank of Ghana.

BoG (1992). Letter No. RFD/NA/4/92/342. Accra: Bank of Ghana.

BoG Annual Reports for 1993 to 1999.

BoG (1999). Quarterly Economic Bulletin, April-June.

Bose, R.S. (1974). *World Development.* Vol. 2, No. 8, August. p. 21-28.

Brown, C.K. (1993). "Socio-cultural Aspects of Rural Credit in Ghana". Paper presented at the Workshop on Financing Micro-Agricultural Enterprise (February 24-27). Sponsored by Ministry of Finance and Economic Planning. Bank of Ghana and the World Bank. GIMPA.

Bureau of Statistical Service (1994). Population Census: Regional Demographic Data. Accra.

CDS (1983). Socio-economic Baseline Survey of the Biriwa Rural Bank Ltd. Centre for Development Studies. University of Cape Coast.

Crook, R.C. (1991). "Decentralisation and participation in Ghana and Côte d'Ivoire" in R.C. Crook and M. Jerve, eds. *Government and Participation: Institutional Development, Decentralisation and Democracy in the Third World.* Bergen: Chr. Michelsen Institute.

Daily Graphic (May 29, 1992). Front page news: "$100 million Released to Recapitalise banks".

Duncan, K.A. (1987). "Growth of Banking". *West Africa.* December 14.

FAO (1986a). *Schemes for Agricultural Credit Development – Approach and Methods.* Rome: SACRED.

FAO (1986b). "Savings mobilization for agricultural and rural development in Africa". Paper presented at The Third International Symposium on the Mobilization of Personal Savings in Developing Countries. Yaounde, Cameroon (December 10-14, 1984).

Frimpong-Ansah, J.H. and Ingham, B. eds. (1992). *Saving for Economic Recovery in Africa.* New Hampshire: Heinemann Educational Books Inc.

Ghanaian Times (1993). March 22.

Gran, G. (1983). *Development By People: Citizen Construction of a Just World.* NY: Praeger Publishers.

Hansen, E. (1989). "The State and Food Agriculture" in E. Hansen and K. A.

Ninsin, eds. *The State, Development and Politics in Ghana*. London: CODESRIA Book Series.

Huq, M.M. (1989). *The Economy of Ghana: The First 25 Years Since Independence*. London: The Macmillan Press Ltd.

InfoShop (2000). *Ghana Rural Financial Services Project* (January 7). Washington D.C: The World Bank.

Kohlhoff, H. and The Rural Banking Department (1985). *Operational Manual for Rural Banks*. BoG. Accra: Jupiter Printing Press.

Kumah, D.C. (1992). Vice-President of the Ashanti Regional ARB. Letter dated March 4.

Kwawukume, Andy C.Y. (1994). "Constraints to Realisation of Local Rationality: The Case of the Rural Banks of Ghana" in Inge Amundsen, ed. *Knowledge and Development*. Proceedings of the NFU Annual International Conference. May 27-29, 1994:181-211. University of Tromsö. Tromsö: Center for Environmental and Development Studies (SEMUT).

Kwawukume, Andy C.Y. (1995). "Rural Banks in Ghana: Participatory Credit Organisations for Rural Development?". M.Phil. Dissertation submitted to the Department of Administration and Organisation Science, Faculty of Social Sciences. University of Bergen, partial fulfilment of the requirements for the higher degree of Candidatus rerum politicarum.

Mikell, G. (1991). "Equity Issues in Ghana's Rural Development" in D. Rothchild, ed. *Ghana: The Political Economy of Recovery*. Boulder and London: Lynne Rienner Pub.

Ministry of Agriculture (1991). *Agriculture in Ghana: Facts and Figures* (November). Accra: Policy, Planning, Monitoring and Evaluation, and MOA.

Moghalu, K.(1990). "Nigeria: Poor People's Bank Takes Root". *Development Forum* No. 5, Sept-October.

Nyonator, C.Y. (1983)."Mobilization of Financing Resources for Regional Development". Paper presented to the Volta Regional Economic Planning Committee Conference. Ho: Ministry of Finance and Economic Planning.

Obben, J. (1991). "Evaluating and Modelling the Rural Banks of Ghana". Thesis submitted for the Degree of Doctor of Philosophy (November). University of New England. Faculty of Economics Studies. Armidale, Australia.

Owiredu, Frank (1991). "Whither the Rural Banks"in *The Ghanaian Chronicle*. November 18-24, 1991:8.

Public Information Center (1994). *Ghana Agricultural Sector Investment Project*. Washington, D.C: The World Bank.

Public Information Center (1997). "Ghana Non-Bank Financial Institutions". Washington, D.C: The World Bank.

RFD (1993). *Rural Finance Project.* Presented by the Rural Finance Inspection Department of BoG at the Workshop on Financing Micro-Agricultural Enterprises, February 24-26, 1993. GIMPA, Accra.

Sarris, A. and Shams, H. (1991). *Ghana Under Structural Adjustment: The Impact on Agriculture and the Rural Poor.* Published for the International Fund for Agricultural Development. New York: University Press.

Sowa, K. N. (1989). "Financial Intermediation and Economic Development" in Emmanuel Hansen and Kwame A. Ninsin, eds. *The State, Development and Politics in Ghana.* London: CODESRIA Book Series.

Wilson, F.A. (1980). "Commercial Banks and Rural Credit". University of Bradford Project Planning Centre for Developing Countries. Reprint Series No. 4. Paper originally appeared in Borrowers and Lenders: Rural Financial Markets and Institutions in Developing Countries, John Howell ed. London: Overseas Development Institute.

14 Structural Adjustment Programs and Emerging Urban Forms

IAN E.A. YEBOAH

14:1 Introduction

At its core, the relationship between structural adjustment and urban spatial form is also a relationship between economic growth and urbanization. As Simon (1997) argues for Africa as a whole, this relationship is not clear-cut. For Sub-Saharan African, the relationship between structural adjustment programs and urban form has hardly been studied. This is despite the fact that 42 countries in Africa have embarked upon structural adjustment programs. Even though research on this relationship exists for Latin America (Gilbert, 1993; Riofrio, 1996) and Asia (McGee, 1991; Thong, 1995; Ocampo, 1995; Lo and Yeung, 1996), the limited research on cities of Sub-Saharan Africa is rather recent and contradictory in some respects. An early assessment of the effects of structural adjustment programs on African cities (not just their spatial form) was that cities of the region will move from positions of leadership in national economies which attract people from the countryside to become the focal point of national depression (Riddell, 1997: 1303). Recent evidence from cities such as Accra and Dar es Salaam suggests that cities in the region have rather become the focal points of national development, or better, the focal points of economic growth, especially the consumption of cultural attributes of Western culture (Briggs and Mwamfupe, 2000; Briggs and Yeboah, 2000; Yeboah, 2000). The initial assessments on the relationship between structural adjustment and Sub-Saharan African cities may have been based on insufficient data, and with time these effects have become clearer than they were in the mid-1990s (Briggs and Yeboah, 2000).

The dearth of research on the relationship between structural and urban form of Sub-Saharan African cities means that urban planners and managers have not been able to capture the implications of the growth and expansion of cities that have occurred under structural adjustment. This chapter raises and assesses questions that relate to planning, urbanization, and social stratification

311

implications of the relationship between Ghana's structural adjustment program and emerging urban form of Accra. City planners and urban managers should be aware of the implications of urban change of Accra if they are to be both efficient in their decision making and avert potentially expensive problems (both in economic and political terms) in the near future. Furthermore, the specific case of Accra may have heuristic potential for other Sub-Saharan African cities.

The first part of the chapter sets the stage for assessing the urban system, planning, management, and social stratification implications of Accra's recent expansion under structural adjustment. This is achieved by describing what has happened to Accra's urban form since the inception of structural adjustment programs, and addressing why such changes have occurred. Based on this, the chapter outlines specific urbanization and urban system implications of Accra's emerging urban form. This is followed by an assessment of planning and management questions of the city's expansion. Before concluding, an assessment of the potential social stratification and segregation effects of Accra's emerging urban form are addressed. It must be stressed that this chapter is not about specific planning programs that have to be considered for Accra. Those are dealt with in the next chapter by Aryeetey-Attoh. This chapter is about the kinds of issues that planners and managers of the city, and possibly of other Sub-Saharan African cities that have undergone similar changes as Accra, in the last two decades, will have to consider in their decision-making.

14:2 What has Happened in Accra?

A continually declining economy and the need to survive as a usurpal leader seemed to have compelled J.J. Rawlings to abandon his populist rhetoric in favor of World Bank and International Monetary Fund (IMF) sponsored structural adjustment program (Ninsin, 1993) in 1983 (Adepoju, 1993). The Ghanaian economy since independence had been characterized by high inflation, increasing balance of payment deficits, increasing debt, and a shortage of basic or essential commodities (Ewusi, 1984; Frimpong-Ansah, 1991). This unchanging state of affairs resulted in a weakened capacity of successive governments to maintain political and economic order in the country (Ninsin, 1991; Rothchild, 1991). Rawlings realized that to entrench his power, he had to look to international capital, thus his embarking upon a structural adjustment program. Adepoju (1993) argues that Ghana's structural adjustment program was started after the state had undertaken an economic recovery program from 1983-1986. Irrespective of when the program was started, there is debate about the effects of structural adjustment in Ghana. Whereas the World Bank classifies Ghana as its success story in Africa, evidence from various researchers suggest that structural adjustment has had detrimental effets on education (Daddieh, 1995), health (Sowah, 1993), employment, and food entitlement (Gyimah-Boadi, 1991; McCarthy-Arnolds, 1994). Kofi (1994) concludes that the effects of structural adjustment are rather sector specific. A blanket statement will therefore not capture the effects of structural adjustment on Ghana and Ghanaians.

Despite the debate on the effects of structural adjustment, there is no doubt that the Ghanaian economy has grown under structural adjustment (ISSER, 1995). Unfortunately, this growth of the national economy does not necessarily mean development. Structural adjustment in Ghana has increasingly tied the Ghanaian economy to the global one. Thus, by embarking upon structural adjustment, Ghana has become part of the new global production, distribution and consumption network of goods, services, political systems and cultures (after Corbridge, 1992). Unfortunately, what structural adjustment has done is that it has strengthened Ghana's traditional role as a producer of raw materials (such as cocoa, timber, and minerals) to feed industries in the West and a consumer of manufactured goods from the Western factories. An assessment of funding of projects under structural adjustment suggests that the development of infrastructure for the extraction of raw materials has been the main emphasis of Ghana's structural adjustment program rather than the provision of social services. Thus, Klak's (1998) assessment that globalization for the Caribbean is not new is applicable to Ghana, since structural adjustment, as a part of globalization, has strengthened Ghana's primary producer role. The effect of colonial polices on the internal structure of cities and urban systems of African countries that Drakakis-Smith (1991) describes under the world systems approach is worth noting here. The creation of ethnic enclaves and land use associated with resources extraction, colonial administration, and European residence in Cantonments and Ridges seems to be repeating itself under Ghana's structural adjustment, albeit along the basis of class (not ethnicity and race). Also the primacy of African capital cities, such as Accra, seems to be strengthened by global forces resulting from structural adjustment.

It is therefore safe to argue that even though Ghana has been increasingly tied to the global system through structural adjustment, its peripheral status in the global system (Rodney, 1982; Howard, 1978) has not really changed. Ghana's peripheral globalization is borne out by external trade statistics of the country. Ghana's trade statistics show that its dependence on Western markets for both imports and exports has intensified even under the new globalization. From 1981 to 1985, between 72 percent and 86 percent of Ghana's exports (by value) went to the West each year, and increased to between 78-91 percent from 1992 to 1995. The pattern of Ghana's imports, also demonstrate an entrenched dependence on Western markets. From 1981 to 1985, between 18 percent and 40 percent of Ghana's imports were from other African countries, but from 1992 to 1995, this value dropped to between 0.5 percent and 20 percent per year. The proportion of imports from the West, on the other hand, went up from between 52 percent and 64 percent per year in the first period to between 65 percent and 88 percent per year in the second period (Computed from Trade Tables in Quarterly Digest of Statistics, June 1990 and December, 1996). Such a profile is typical of the core-periphery relationship between Ghana and the West.

Ghana's peripheral globalization (even though the phrase sounds contradictory), should not be surprising since the neoliberal ideology of international capitalist expansion is geared toward the state providing enabling circumstances for international capital, and to a lesser extent local capital, to intensify their extraction of resources (Corbridge, 1992). Mining companies from all over the Western World are now stationed in the gold mining formation of the Tarkawian gold ore deposit of southwestern Ghana. The Ghanaian state has acquired loans from international sources that have been used to improve roads, telecommunications, and railways to ensure that extraction of raw materials is efficient. The state has thus cut back on its involvement in social sectors of the economy and in the process international capital now punishes the Ghanaian poor. The literature that suggests that the poor have suffered in terms of access and affordability to health care and education, alluded to above, is buttressed by the recent protests against the World Bank and IMF in Washington D.C. at the annual meeting of Finance Ministers.

As expected, in the case of Ghana, the expanding demand for consumer goods produced in the West has also been harnessed by international capital. Thus, telecommunications firms from Japan and South Africa, and automobile manufacturers from Newly Industrializing Countries (NICs) such as South Korea have targeted the Ghanaian middle-class consumer with their new goods and services. The availability of new and used vehicles in Ghana today (new cars could not be purchased in Ghana before structural adjustment) attest to the view that the Ghanaian middle class has been targeted by both international and local capital in indirect ways in some cases. This peripheral globalization has produced certain contradictions for Ghana's economy and its urbanization, especially that of its capital city. These contradictions center around the fact that even though the economies of both Ghana and its capital Accra have grown, this growth does not necessarily result from development. After all, the growth of Accra and the Ghanaian economy in general, have not translated into better access to education, employment, health care and other social services for the working class and the poor of Ghana. Rather, it is the middle-class that has benefited from the economic growth evident in Accra and parts of Ghana.

The specific effects of structural adjustment on Accra's urban form are phenomenal. Since structural adjustment began in Ghana, Accra has remarkably expanded in area. Even though demographic data to measure this expansion does not exist, Figure 14.1 maps the city's built-up area based on the 1975 topographic map and aerial photographs from 1986, 1992 and 1997.

During the eleven years before structural adjustment (1975-86), the city grew in area by approximately 29 square miles from a 1975 area of 63 square miles. Yet, in the fist eleven years since structural adjustment (1986-97) the city grew in area from 92 square miles 1986 to 172 square miles in 1997. Most of the city's growth has been in the late structural adjustment era (i.e. post 1992). This growth of the city has not just been phenomenal, it has also changed the urban form of Accra.

What has emerged in Accra since structural adjustment is a new urban form that is best described as a quality residential sprawl of the city into peri-urban areas, with the city exhibiting unicentric tendencies. Seven attributes of today's Accra support such a characterization. First, there is very little commercial, office, or industrial building associated with this peri-urban sprawl. It is mostly residential in nature. The only commercial activities associated with the peri-urban areas deal with further expansion of they city and they include block manufacture, building material sales, and "chop bar" (local restaurant) operations. Second, most of the development is of low density both in terms of population and building intensity, most buildings are one or two stories and there are large tracts of green spaces (often land that is held in speculation) between buildings. Third, high quality residential buildings owned mostly by the new rich, an emerging middle class, and Ghanaians resident abroad is the norm. This is what makes it quality. The poor have been excluded from building in peri-urban Accra, as has been the case in many Sub-Saharan African (SSA) capitals (Stren and White, 1989). Fourth, most of the buildings are built incrementally by individuals as is practiced in Abidjan (Dubersson, 1997). It is only in the last seven years that development companies (American-style) have become important builders in Accra. Despite this recent trend, the majority of houses built in Accra today are still built incrementally by individuals. Fifth, most of the building in peri-urban Accra has been spontaneous and unplanned (Odame Larbi, 1996). Builders often ignore basic issues of land delivery, regulation, and building permits. Sixth, most houses are built in anticipation of service and infra-

Figure 14.1 Expansion of Accra, 1975-1997

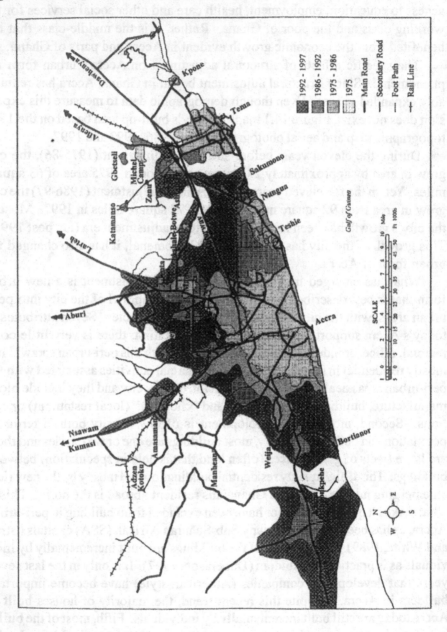

structure such as telephones, electricity, and roads. The expectation is that in future such services will be provided by the state at one point in time. The fifth and sixth attributes make it a sprawl and may lead to questions about quality, but quality here refers to the quality of structures rather than of the built environment. Seventh, a strong link between peri-urban Accra and Accra central exists in terms of journey-to-work and shopping trips. Most residents of peri-urban developments work and shop in Accra central. Thus Accra's expansion is not just a quality residential sprawl but it has unicentric tendencies just as Riofrio (1996) describes for Lima, Peru.

There is no doubt that Accra's expansion under structural adjustment has been phenomenal. The argument here is that the relationship between Accra's growth and structural adjustment is not coincidental. The question is whether structural adjustment and its attendant economic growth have caused Accra's growth, or whether structural adjustment has only facilitated this growth. The specific attributes of the city's growth identified above partially help in answering this question. To get a more complete answer to this question, an understanding of why Accra has grown the way it has is necessary. Global and local forces have interacted to produce Accra's quality residential sprawl with unicentric tendencies. Structural adjustment has only facilitated this residential sprawl. Two attributes of structural adjustment, namely export and import trade liberalization, and foreign currency liberalization have created circumstance for preexisting local factors to manifest themselves in the increasing quality residential sprawl of Accra. These local factors can be summarized as demand-side economic conditions, supply-side innovations in Ghana's housing industry, Ghanaian cultural imperatives, and institutional factors. The interaction of global and local forces has manifest itself spatially in the quality residential sprawl of Accra. Structural adjustment has not directly caused Accra's growth neither is it coincidental to the city's growth. It has been a facilitator of the emerging urban form. It is therefore necessary to tease out the specific local and global forces and how they have interacted to influence Accra's expansion.

One of the objectives of structural adjustment in Ghana is to ensure that foreign capital can flow into extractive sectors of the economy (such as mining and lumbering) with very few bottlenecks. Profits earned by foreign companies in Ghana can also be repatriated to their source countries. To ensure this, the Ghanaian state has liberalized the flow of foreign currency into Ghana. Even though it was illegal to own and trade in foreign currency, such as the US dollar in the 1970s, such activities have now been legalized through the establishment of licensed forex bureaus. The forex bureaus were designed to facilitate the movement of foreign capital in and out of Ghana yet, their operations have been to the advantage of Ghanaians in a number of ways. Ghanaians resident abroad have been able to remit large sums of foreign currency into Ghana for a variety of purposes. One of these is investment in the built environment. Estimates are that US$276 million was remitted to Ghana in 1996 (*Daily Graphic*, 1998).

Apart from such remittances, foreign trade liberalization has ensured that petty international traders have been able to take foreign currency out of Ghana to purchase a variety of foreign inputs, both used (what I call Eurocarcass) and new. Such petty international traders have made great profits from their trading. A casual observer of Ghana will be impressed with the plethora of goods that are now available on the Ghanaian market. These include automobiles (cars and trucks), building materials (such as floor tiles, sinks, toilet bowls, baths, and air conditioners), household appliances and other consumer goods. These two processes (trade liberalization and foreign currency liberalization) have therefore facilitated or made it possible to both import foreign building materials and bring foreign currency into the country. Such foreign injections have been directed toward the expansion of Accra. Why have such injections of foreign input gone toward the built environment of Accra and not to manufacturing investment or the built environment of other cities in Ghana such as Kumasi or Takoradi or Tamale? The answer is that these global forces have enabled preexisting local forces to manifest themselves in the emerging urban form of Accra, so it is necessary to identify and tease out the specific ways in which the local forces have acted.

The first of these local forces is the acute primacy or importance of Accra in the Ghanaian urban system. Despite the projection by Riddell (1997) that African capitals will diminish in importance relative to their national urban systems, Accra seems to have increased its domination of the Ghanaian urban system. Even though the state launched a decentralization program in 1987 (Ayee, 1997), head offices of government ministries, major public and private establishments outside agriculture and mining are all still located in Accra. Even international agencies (Non Governmental Organizations, NGOS, and those for profit) are head quartered in Accra. Thus, a high demand for houses in Accra (by expatriates and, to a lesser extent, Ghanaians working with such foreign agencies) is to be expected. The typical house rented to expatriates may have three or four bedrooms with a garage, "boys quarters" and a swimming pool. The owners rent them for US dollars and take a three-year rent advance. Often the three-year rent advance could provide the owner with seed capital of about thirty thousand dollars ($30,000) which is then reinvested in further housing construction. Most builders in Accra believe they can benefit from renting to expatriates for US dollars, this is why they build quality houses.

The second set of pre-existing local factors is related to the primacy of Accra but it has more to do with the paucity of the Ghanaian economy and the high demand for houses. The Ghanaian economy has been characterized by high inflation rates since the 1970s (ISSER, 1995). In the absence of a strong stock market, residential real estate in Accra has become an attractive investment option (Dickerman, 1988; Diko and Tipple, 1992), since quality built houses in Accra can be rented out to expatriates to earn foreign exchange. Houses are only sold under conditions of extreme hardship. They are held for their investment value. Accra, as the capital city of Ghana, has attracted a disproportionate share of housing built for investment purposes. A still small but expanding middle-class has realized the safety of real estate as a form of investment. This middle class is made up of first, international petty traders (described above) who have gained financially from international import trade. They build houses either to live in or to rent out to expatriates. The second is a group of university educated graduates in the technical fields who can sell their services to establishments that are geared toward profit and are willing to pay for skilled workers. These people often do not build houses but their employers guarantee their request for mortgage loans for them. Thus, in places such as Sakumono flats on the periphery of Tema, many up-and-coming well-educated Ghanaians have mortgaged flats through the Home finance Company (HFC).

The third set of factors is on the supply-side of the Ghanaian economy. As with most of Ghana, there is no doubt that there is a need for houses in Accra. Acquaye and Associates (1989) predict that Ghana will need 227,460 houses by the year 2010. Accra will definitely need a lion's share of these houses. Historically, building a house in urban Ghana was done by individuals on an incremental basis. In some circumstances, a small contractor will be employed to undertake certain parts of the construction but more often, masons, carpenters, electricians and plumbers will be retained on a piecemeal basis. Whenever funds run out the project is halted and can be continued at a later date. There was basically no market for loans to finance home building and ownership (such as mortgages). The implications of this for scheduling are that a three-bedroom house may be completed five years later as a five-bedroom house. Thus, the supply of housing in Accra has been piecemeal but this may be changing.

Under structural adjustment, two innovations in Ghana's housing industry have emerged. These are the creation of the HFC in 1993 by the state as a mortgage granting agency, and retooling by many of the small contractors (who used to be involved in home building) and many larger ones to establish Residential Development companies (RDCs). Examples of these are Regimanuel-Gray (the industry leader), Parakou Estates, African Concrete Products, and

Rabshold Ltd. Some of the RDCs are partly foreign owned. Even government agencies (such as the State Housing Corporation and Tema Development Corporation) that used to build houses for sale to the public have retooled and have started operating like the RDCs. All RDCs are under the umbrella of a new trade group known as Ghana Real Estate Developers Association (GREDA). By mid 1997, Regimanuel-Gray had built over 8000 houses in Accra alone. Increasingly, the HFC has become a popular source of mortgage lending. Between 1992 and 1996, the HFC disbursed 2,807 mortgage loans. Loans are often given by the HFC at a 12.5 percent interest rate (HFC, 1995), over ten years, and repayment is in US dollars and British Pound Sterling. Even though the HFC and RDCs do not dominate the housing industry, their potential to do so is great and increasingly they are convenient means to home ownership by middle-class Ghanaians resident both at home and abroad. There is however no doubt that these innovations in Ghana's housing industry have contributed toward the expansion of Accra. Examples of some of the estates RDCs have developed are Adenta Estates, Ashongman Estates, East Airport, and the massive building boom that has developed on the Spintex Road between Teshie, Nungua, and Tema.

The fourth pre-existing local factor has to do with Ghanaian cultural imperatives. There is both a desire and aspiration on the part of most Ghanaians to build a house in Ghana. A measure of a person's success in life is often gauged by whether they have built a house or not (Malpezzi and Tipple, 1990). Increasingly, Ghanaians resident in the country as well as others resident abroad see building a house as their goal. The extended family system is such that up-and-coming members are encouraged and even pressured to build a house in Ghana, be it in their hometown or in other cities. Diko and Tipple (1992) describe ways in which Ghanaians resident in Britain had remitted funds home for building houses. Under structural adjustment the specific means of such remittance has changed from petty traders carrying money back to Ghana, or shopping for their wares in Britain with money given to them by Ghanaians and paying the Cedi equivalent in Ghana, once their goods have been sold.

Today, Ghanaians living abroad legally transfer funds to Ghana either by Western Union or other commercial banks, such as the Ghana Commercial Bank. Increasingly, the new methods are sophisticated and safer than the reliance on petty international traders. Ghanaians of all socio-economic classes have embarked on remitting money home from working either menial or professional jobs in countries such as USA, Canada, Germany, Saudi Arabia, South Africa, and Libya. Even though some of these international builders may never live in their houses in Ghana (or at best will live in them when they retire), they choose to build in Accra rather than other cities because of its primacy. It is in Accra that they can potentially rent their houses to expatriates so that they can recoup some of the foreign currency they have invested. So the building of houses in Accra is not only by Ghanaians who live in the country. Neither is it for immediate use. It is often as an investment (as rental property) that is used by the builder at retirement and it is influenced by both encouragement and pressure from the extended family (Owusu, 1998).

The fifth and final set of local factors is related to Ghanaian culture, especially land tenure. However, it is more institutional in nature and it has to do with the inefficient operation of planning and land delivery institutions and the conflict between traditional systems of land tenure and modern systems of land delivery. Historically, land in Accra belongs to the Ga. Either Chiefs (stools) or family heads hold it for members of the community (Acquaye and Associates, 1989). Land cannot be sold but it can be leased for 99 years for building a house. Even though stools and families hold land, the modern state apparatus does its delivery, registration, and planning. Odame Larbi (1996) argues that the inability of these two sets of institutions to work together has resulted in a fragmented land delivery system in Accra. The outdated planning framework (Town and Country Planning Act, 1945) that exists in Ghana today worsens such fragmentation of land delivery.

Representatives of stools and families allocate land for building so far as customary rites (a presentation of drinks and money) have been performed. Increasingly, different families and stools claim ownership to the same land or different members of the same family sell the same land to different persons. Litigation is therefore rampant and by the time land has been subdivided by planning agencies, buildings may have been erected in the middle of streets. Thus, the fragmented delivery of land has made it easier for stools and families to sell unplanned and unserviced land. In the context of such litigation and a lack of regulation such land is very cheap (Acquaye and Associates, 1989; Asiama, 1989). It has therefore been relatively easy for people who wish to build in Accra's peri-urban area to find cheap unserviced land.

All these local factors existed prior to structural adjustment. What structural adjustment has done is that it has provided enabling circumstances for all these factors to manifest themselves in Accra's quality residential sprawl with unicentric tendencies. Thus, the nexus of global and local circumstances is expressing itself in the space and economy of Accra but its effects or implications are not limited to Accra alone. There are specific urbanization and urban system implications, planning and management implications, as well as and social stratification implications resulting from these developments. The next section will examine some of these issues.

14:3 Urbanization and Urban System Implications

Increasingly Accra has become primate in the Ghanaian urban system. There is no doubt that there has been building in other Ghanaian cities but the rate of Accra's growth under structural adjustment has not been matched in any way by the growth of other cities. As Ghanaians often say, Accra is the center of everything since it accounts for the majority of industrial establishment, government head office, and head offices of international agencies. In fact, RDCs build almost exclusively in Accra and, the majority (95 percent) of the HFC's loans have been for buildings in Accra. Thus, the domination of Accra of the Ghanaian urban system is an issue that requires attention. The question that need to be addressed relate to the extent to which Accra, as the focus of Ghana, can stimulate development. This is a question of whether development can trickle-down from Accra. Contrary to Riddell's (1997) suggestion for African cities, Accra has become the center of growth in Ghana. For a country that has embarked upon a decentralization program, such a trend needs to be reversed. Accra's increased primacy reinforces the view that Ghana's decentralization program is a political one without spatial considerations.

Also, the massive expansion of the Accra coupled with the unicentric pattern that is emerging means that these changes in the city's internal structure will impact its management. Increasingly, Accra is becoming an unmanageable city with the potential of experiencing a crisis in transportation, provision of services, and infrastructure. Demographic data on Accra's growth does not exist but it is estimated that Accra has a population of over 1.7 million people. In the mornings most workers head toward central Accra and Tema and in the evenings they head out. The transportation bottlenecks of this need to be studied and planned for. Also, how will services such as water, sewage and electricity be provided for residents of Accra and its peri-urban area?

14:4 Planning and Management Implications

Related to the urbanization effects of Accra's growth and its emerging urban form under structural adjustment is a set of effects that can be classified as planning implications. These issues need further attention by planners and city managers. First, because of the remarkable expansion of Accra into its peri-urban area, the city now sprawls over four administrative districts in an almost contiguous manner. These are the Accra Metropolitan Assembly itself, Tema

Metropolitan Assembly, Ga Rural District Assembly and Ewutu Efutu District Assembly. The first two are urban whereas the last two are rural. Even though the first three are part of the Greater Accra Region, the last is part of the Central Region. The implication is that planning and managing the present and future growth of Accra now cuts across a variety of jurisdictions, each with its own planning and management traditions as well as revenue generating strategies. The question is whether a legal and administrative framework can be established to synchronize planning and management of the city. Is there a need for a centralized system? Can a decentralized system (such as the Metropolitan and District Assemblies) of planning and management work efficiently for Accra? How can these be established? What can be done about issues of sovereignty amongst the various administrative bodies? Sovereignty is often an issue when it comes to merging administrative responsibilities. Such turf battles are similar to those of the development of local government structures that Johannesburg experienced immediately after the transition in South Africa (Beavon, 1997).

The second planning and management problem is that of service provision. How best can water, electricity and telephone services be provided for the ever-expanding city? At present, centralized government agencies (Water and Sewage Corporation, Electricity Corporation and Ghana Telecom) provide these services to all parts of Ghana and for that matter, Accra. Apart from being erratic, quality of service provision differs from location to location in Accra itself. Often, well-to-do neighborhoods receive better service. What model can ensure relatively uniform spatial distribution in Accra? What can be done to strengthen the centralized system of service provision? Or, is there a need to decentralize service provision and at what administrative level should it be decentralized? What balance should be established between private and public provision of service? These issues are germane to most SSA cities such as Abidjan (Dubresson, 1997).

A third planning problem deals with transportation. Just like Lima, Peru (Riofrio, 1996), Accra has increasingly become unicentric. Most jobs are located in the central part of the city (from Kwame Nkrumah Circle, in the north, to Accra Central, in the south, and from the Ministries, in the east, to Timber Market, in the west), as well as the industrial settlement of Tema. Each morning, traffic flows from the peri-urban areas and other residential areas to the city center and Tema. In the evenings, the pattern reverses. It is so frustrating to get around in Accra (Quaye and Badoe, 1994). On weekdays, between 8 a.m. and noon, it takes about two hours to travel from Alajo Junction to Kwame Nkrumah Circle, a distance of no more than three miles. The growth of the city under neoliberal conditions has therefore contributed toward the creation of major traffic congestion in the city. This is the result of a poorly planned street network. What options exist to solve the city's transportation problem? Will a third ring road help? Can the city and country afford a light rail or subway system? Or will it be possible to restrict the daily movement of vehicles as has been done in Mexico City (Friedrich, 1984), for example?

Despite the problems highlighted above, there are positive planning impacts or effects of Accra's emerging urban form. One positive effect is that the RDCs provide infrastructure and services for the houses they build. Thus, the quality of the housing environment is a lot more pleasant than those built independently either by the poor or the middle-class. There are, therefore, differences in the quality of middle-class areas that are self-built and those that are built by RDCs. Middle-class persons who rely on RDCs often find that houses are well serviced and the infrastructure is adequate. Those which are self-built often lack roads, drains, recreational spaces, electricity, water and telephones. Even though some of the houses that are self-built are mansions (compared to the often small and standardized houses of RDCs), they border on the verge of slums because of the poor level of services and infrastructure. In a sense, the quality of housing stock provided by RDCs is a lot more attractive than that built by individuals. With an intensification of RDC activities, the general quality of the built environment of Accra may improve.

14:5 Social Stratification and Segregation Implications

A result of Accra's expansion and emerging urban form is the emergence of a gap between the middle-class and the poor or vulnerable class. This is manifest in terms of residential segregation, socio-economic stratification and differences in quality of housing infrastructure and services. The differentiation under Structural Adjustment Programs is however, tampered by the uniquely African (Ghanaian) social attribute of the extended family.

RDCs build houses in the American suburban development fashion. Even with the fragmented land delivery system, RDCs are able to consolidate large tracts of land (over 50 acres at a time) by negotiating with stools and families who own such lands. Unlike individuals who seek land for building in Accra, most RDCs employ legal experts who write out tenure agreements with the stool or family involved, register the title of such land with the Lands Commission, and seek building permits from the appropriate district or municipal assembly. Often, the development company walls such land, and employs security companies to patrol them, if building is not eminent. Once such land has been consolidated and the company is ready to build, it provides infrastructure and services such as roads, drains, water, electricity and telephones. Thus, the land that is used by RDCs has appreciated in value. ACP's estates at Pokuase and Regimanuel-Gray's at East Airport have both been developed along this line.

The result of such a process is that the cost of houses built by RDCs is high and the poor are effectively excluded from living in the same area as the middle-class. As Table 14.1 shows, the average price for different sizes of houses built by RDCs far exceeds what the minimum wage worker in Ghana can afford. The minimum daily wage in Ghana (1999) is two thousand Cedis or about eighty cents US. Assuming that all a minimum wage worker does is to save his/her earnings without spending any of it on basic necessities, it will take over 300 years of savings to purchase the least priced one-room unit at $9,100 (RDCs only provide the one room only as an attachment to a main house).

It can be argued that most Ghanaians do not live on the minimum wage. A schoolteacher with 10 years experience earns about two hundred and fifty thousand Cedis (about $100) a month. If all their earnings are saved for a one-room unit, it will take over seven and a half years to purchase such a unit (that is if inflation is held constant). A university lecturer who makes about five hundred thousand Cedis a month will require close to a total of four year's saved income to purchase the same lowest priced one-room unit. So far as RDCs are concerned, most Ghanaians are excluded from owning even a single room in Accra, let alone a three-bedroom house.

Two other factors help exclude the poor from enjoying the effects of Accra's

Table 14.1 Cost of Development of Company Houses

Cost of Houses by Number of Bedrooms in '000 US $ (1997 Prices)			
Number of Bedrooms	Regimanuel Gray	ACPLtd.	Parakou Estates
1	9.1-19.8		
2	15.2-27.8		17.3-23.8
3	28.8-58.6	41.1-85.2	27.8-80.0
4	52.6-99.0		

Price differences are due to area, quality of finish etc.

expansion. First, the payment terms that RDCs offer their prospective clients favor the middle-class rather than the poor. Often, 50 percent of payment is expected before construction begins, the other 50 percent payment is required before the keys are handed over to the owner (often six months after breaking ground). It is the middle-class, not the poor, who can afford such funds in such a short period of time. Second, it can be argued that the poor can depend on the HFC for mortgages to buy a house. Assuming that the poor qualify and are granted the 12.5 percent per annum loan (over a ten year period), the HFC stipulates that it will only grant loans for property that is serviced by roads, electricity, and water. This means that land that has been serviced and therefore has appreciated in value. The poor cannot afford such land.

Even in terms of self-built housing, the prices of houses are not different from those provided by RDCs. The major difference is that most of the self-built houses are built in anticipation of services and infrastructure. So even though the houses may be of excellent quality, they are hardly serviced. Thus even though the middle class is the group that is building and buying houses in Accra, there is a difference in the appearance and availability of services and infrastructure between those built by RDCs and those self-built. It can be predicted that in the near future, houses built by RDCs will be worth more than similar houses that are self-built.

Under Ghana's neoliberal Structural Adjustment Programs, therefore, the implication of Accra's expansion and the city's emerging urban form is that the poor and vulnerable will be segregated from the middle-class. And even amongst the middle class, there will be a schism between those who buy from RDCs and those who self-build. It is not surprising that parts of Accra like East Legon, Haatso-Papao, Sakumono, Kasoa and Madina reflect differences in the socio-economic profiles of their owners. There is a decreasing level of socio-economic status from East Legon to Madina. Yet, the extended family concept and the responsibilities it imposes on members who are relatively well to do means that this is not always the case. Often, members of the extended family who build in Accra, whether they rely on RDCs or they self build, have responsibility for poorer members such as aunts, uncles, nephews and nieces who live with them in their house. Apart from providing a place to live, they also support such members of their family through the payment of school fees, payments for learning a trade, and even providing small capital for the establishment of trades or businesses. The poor members of the extended family often provide houseboy and maid services for their wealthier relative. Thus, in an area where you will expect to find only middle-class residents, you also find those who, based on income and education, will be considered poor. Thus, even though neoliberalism has widened the differential between the classes, a uniquely Ghanaian attribute of the extended family has made it rather difficult to identify the poor and vulnerable in particular territories (squatter settlements) of cities such as Accra. Contrary to what Konadu-Agyemang (1991) argues, it is not just religious values that prevent the emergence of squatter settlements in Ghana. The extended family concept has also contributed toward the absence of squatter settlements in Accra.

14:6 Conclusion

World systems analysis has historically argued that cities of the Third World have been shaped and impacted by global circumstances (Armstrong and McGee, 1985; Drakakis-Smith, 1991). The peripheral status of Sub-Saharan African cities to the global community resulted in their internal structure and the interaction between cities of an urban system being determined by imperial and colonial forces beyond the control of indigenes of cities such as Accra, Lagos, Dar es Salaam and Nairobi. In today's era of globalization, this view is only partly true for Sub-Saharan African cities since they are still peripheral to the global system of production, distribution and consumption (Simon, 1997). Increasingly, the effects of local forces have become important sources of urban growth and expansion.

For the most part, Sub-Saharan African countries and cities have been tied to the globalization project of international capital and neoliberal agenda because these countries have embarked upon structural adjustment programs. Despite their historically peripheral status to the global system, though, local forces seem to manifest themselves in spatial patterns within the enabling framework provided by forces of globalization. Increasingly, the internal structure of Accra is being shaped along the basis of class rather than race, colonial administration, and ethnicity. Thus a nexus of local and global forces seem to be at work in shaping Sub-Saharan African cities in the era of globalization and cities like Accra are beginning to exhibit changes in their urban spatial form and structure. The effects of this nexus on cities of the region will increasingly become apparent in the near future. Some of these will be positive and others will be negative.

Since Ghana was one of the first countries to embark upon structural adjustment in the region and the World Bank has labeled it a success story (Alderman, 1994), the experience of its capital city could provide some insight for other cities in the region. These insights relate to issues of urban system interactions and urban primacy of capital cities, planning, management, and livability of cities and, the social stratification or segregation issues. Specifically for Accra, though city leaders, managers, administrators, planners, as well as residents have to deal with these effects of the city's emerging urban form by devising specific programs and projects to tackle these problems head on. The next chapter by Aryeetey-Attoh provides some specific examples of such programs and projects.

References

Acquaye, E. and Associates (1989). *Study of Institutional/Legal Problems Associated with Land Delivery in Accra.* Report prepared for Accra Planning and Development Programme Government of Ghana.

Adepoju, A. (1993). *Introduction* in *The Impact of Structural Adjustment on the Population of Africa: The Implications for Education, Health and Employment.* Edited by A. Adepoju. Portsmouth: UNFPA: 1-6.

Alderman, H. (1994). "Ghana: Adjustment's Star Pupil?" in *Adjusting to Policy Failure in African Economies.* Edited by David E. Sahn. Ithaca: Cornell University Press: 23-52.

Armstrong, W. and McGee, T.G. (1985). *Theatres of Accumulation: Studies in Asian and Latin American Urbanization.* London: Methuen.

Asiama, S.O. (1989). *Land Management in Kumasi, Ghana.* Kumasi: Report to the World Bank.

Ayee, J.R.A. (1997). The Adjustment of Central Bodies to Decentralization: The Case of the Ghanaian Bureaucracy. *African Studies Review,* 40(2): 37-57.

Beavon, K.S. (1997). "Johannesburg: A City and Metropolitan Area in Transformation". in *The Urban Challenge in Africa: Growth and Management of its Largest Cities.* Edited by Carol Rakodi. Tokyo: United Nations University Press: 150-191.

Briggs, J. and Mwamfupe, D. (2000). "Peri-urban Development in an Era of Globalization and Structural Adjustment: The City of Dar es Salaam, Tanzania". *Urban Studies* 37(4) 79-89.

Briggs, J. and Yeboah, I.E.A. (2000). "Structural Adjustment and the Contemporary Sub-Saharan African City" (forthcoming in *Area*).

Corbridge, S. (1992). "Discipline and Punishment: the New Right and the Policing of International Debt Crisis". *Geoforum,* 23(4): 285-301.

Daddieh, C. (1995). "Education Adjustment under Severe Recessionary Pressures: The Case of Ghana" in *Beyond Economic Liberalization in Af-*

rica: *Structural Adjustment and the Alternatives*. Edited by Kidane Megisteab and B. Ikubolajeh Logan. London: Zed Books: 23-55.

Dickerman, C. (1988). "Urban Land Concentration" in *Land and Society in Contemporary Africa*. Edited by R.E. Downs and S. P. Reyna. Hanover: University of New England Press.

Diko, J. and Tipple, A.G. (1992). "Migrants Build at Home; Long Distance Housing Development by Ghanaians in London". *Cities*, November: 288-294.

Drakakis-Smith, D. (1991). "Colonial Urbanization in Asia and Africa: A Structural Analysis" in W.K.D. Davies (ed.). *Human Geography from Wales: Proceedings of the E.G. Bowen Memorial Conference*, 16: 123-150.

Dubersson, A. (1997). "Abidjan: From the Public Making of a Modern City to Urban Management of a Metropolis" in *The Urban Challenge in Africa: Growth and Management of its Largest Cities*, edited by Carol Rakodi. Tokyo: United Nations University Press: 252-291.

Ewusi, K. (1984). *The Political Economy of Ghana in the Post Independence Period: Description and Analysis of the Decadence of the Political Economy of Ghana and the Survival Techniques of Her Citizens*. Institute of Statistical, Social and Economic Research. Unpublished Discussion Paper No. 14, Accra.

Frimpong-Ansah, J. H. (1991). *The Vampire State in Africa. The Political Economy of Decline in Ghana*. London: James Currey.

Gilbert, A. (1993). "Third World Cities: The Changing National Settlement System". *Urban Studies*, 30, (4/5): 721-740.

Government of Ghana (1990). *Quarterly Digest of Statistics*, V111 (2). Accra: Statistical Services.

Government of Ghana (1996). *Quarterly Digest of Statistics*, XIII (4). Accra: Statistical Services (diskette version).

Gyimah-Boadi, E. (1991). "State Enterprises Divestiture: Recent Ghanaian Experiences" in Donald Rothchild (ed). *Ghana: The Political Economy of Reform*. Boulder: Lynne Rienner Publishers: 193-208.

HFC (Home Finance Company) (1995). *Own a Home at Home Scheme. Information Brochure Specifically for Non-resident Ghanaians* (unpublished brochure).

Howard, R. (1978). *Colonialism and Underdevelopment in Ghana*. London: Croom Helm.

ISSER (Institute of Statistical, Social and Economic Research) (1995). *The State of the Ghanaian Economy*. Accra, Legon: Institute of Statistical, Social and Economic Research.

Klak, T. (1998). "Thirteen Theses on Globalization and Neoliberalism" in *Globalization and Neoliberalism, The Caribbean Context*, edited by T. Klak. Lanham: Rowman and Little.

Konadu-Agyemang, K.O. (1991). "Reflections on the Absence of Squatter Settlements in West Africa: The Case of Kumasi, Ghana". *Urban Studies*, 28(1): 139-51.

Lo, Fu-chen and Yeung Yue-man (1996). Introduction in *Emerging World Cities in Pacific Asia*. Tokyo: United Nations University Press.

Malpezzi, S.J. and Tipple, A.G. (1990). *Costs and Benefits of Rent Control: A Case Study of Kumasi, Ghana*. World Bank Discussion Paper no. 74. Washington D.C.: The World Bank.

McCarthy-Arnolds, E. (1994). "The Right to Food: Questions of Entitlement Under Structural Adjustment Policies" in *Economic Justice in Africa: Adjustment and Sustainable Development*, edited by George W. Shepherd, JR. and Karamo N.M. Sonko. Westport, CT: Greenwood Press: 117-135.

McGee, Terry G. (1991). "Eurocentrism in Geography: The Case of Asian Urbanization".*The Canadian Geographer*, 35 (4): 332-44.

Ninsin, K. A. (1991). "The PNDC and the Problem of Legitimacy" in *Ghana: The Political Economy of Recovery*, edited by Donald Rothchild. Boulder: Lynne Rienner Publishers: 49-67.

Ocampo, R. B. (1995). "The Metro Manila Mega-Region" in *The Mega-Urban Regions of Southeast Asia*, edited by Terry G. McGee and Ira M. Robinson. Vancouver: UBC Press: 282-295.

Odame Larbi, W. (1996). "Spatial Planning and Urban Fragmentation in Accra". *Third World Planning Review* 18 (2): 193-215.

Owusu, T. Y. (1998). "To Buy or Not to Buy: Determinants of Home Ownership Among Ghanaian Immigrants in Toronto". *The Canadian Geographer* 42 (1): 40-52.

Quaye, K. and Badoe, D. (1994). "Delivery of Urban Transport in Sub-Saharan Africa Case Study of Accra, Ghana". *Journal of Advanced Transportation* 30 (1): 75-94.

Riddell, B. (1997). "Structural Adjustment Programmes and the City in Tropical Africa". *Urban Studies* 34 (8): 1297-1307.

Riofrio, G. (1996). "Lima: Mega-City and Mega-Problem" in *The Mega-City in Latin America*, edited by Alan Gilbert. Tokyo: United Nations University Press: 155-172.

Rodney, W. (1982). *How Europe Underdeveloped Africa*. Washington, D.C.: Howard University Press.

Rothchild, D. (1991). "Ghana and Structural Adjustment: An Overview" in

Ghana: The Political Economy of Reform, edited by Donald Rothchild. Boulder: Lynne Rienner Publishers: 3-17.

Simon, D. (1997). "Urbanization, Globaization and Economic Crisis in Africa" in *The Urban Challenge in Africa: Growth and Management of its Largest Cities*, edited by Carol Rakodi. Tokyo: United Nations University Press: 74-108.

Sowah, N. K. (1993). "Ghana" in *The Impact of Structural Adjustment on the Population of Africa: The Implications for Education, Health and Employment*, edited by Aderanti Adepoju. Portsmouth: UNFPA, pp. 7-24.

Stren, R. E. and White, R. R. (1989). *African Cities in Crisis: Managing Rapid Urban Growth*. Boulder: Westview Press.

Thong, L. B. (1995). "Challenges of Superinduced Development: The Mega-Urban Region of Kuala Lumpur-Klang Valley" in *The Mega-Urban Regions of Southeast Asia*, edited by Terry G. McGee and Ira M. Robinson. Vancouver: UBC Press: 315-327.

Yeboah, I.E.A. (2000). "Structural adjustment and Emerging Urban Form of Accra, Ghana". *Africa Today* 47(2): 61-90.

15 Urban Planning and Management under Structural Adjustment

SAMUEL ARYEETEY-ATTOH

15:1 Introduction

Since the 1980s, the World Bank and International Monetary Fund's Structural Adjustment Programs (SAPs) have conditioned financial and technical assistance to African countries. The SAPs were initiated to adjust malfunctioning economies, assist in debt restructuring, and promote greater economic efficiency and economic growth in order to position Sub-Saharan economies to be more competitive in today's global economy. By 1990, 32 African countries had launched structural adjustment programs or borrowed from the International Monetary Fund (IMF) to support policy reforms.

Candidates for structural adjustment are countries with budget deficits, balance-of-payment problems, high inflationary rates, ineffective state bureaucracies, inefficient agricultural and industrial production sectors, overvalued currencies, and inefficient credit institutions. As a result, countries that buy into the program are required to: devalue their currencies to reduce expenditures on imports and release resources for exports; restructure the public sector and state-owned enterprises by trimming overstaffed bureaucracies and padded payrolls, thereby improving institutional management, and encouraging privatization; eliminate price controls and subsidies such as artificially reducing the price of food in urban areas, thus lowering farm incomes. They also have to restructure the productive sectors of the economy by liberalizing trade and removing import quotas and high tariffs that protect uncompetitive firms, and by providing export incentives to promote export growth particularly in agriculture.

A problem with such macroeconomic policies is their tendency to rely heavily on a sectoral approach that sets up targets for the country and for various sectors, and formulates investment plans and restructuring policies to reach these targets. What is usually lacking, for the most part, is a complementary spatial strategy that examines the impact of such macroeconomic policies on key constituents and productive sectors of the urban economy, including the formal and informal sectors. Spatial and place-specific strategies have always occupied a marginal position in Ghana's arsenal of economic reform policies. This is because urban and community-based strategies run counter to a neo-classical economic development strategy that presumes that people are moti-vated by rational self interest in making life decisions. The norms and precepts of classical neoclassical theory inform us that place-specific, and thus urban-specific policies, are inefficient in comparison with policies that emphasize mac-roeconomic growth, human capital, and individual mobility.

One area that has received little attention in the scholarly literature on struc-tural adjustment programs is the planning and management of African cities. The impacts of structural adjustment programs on forestry (Owusu, 1998), rural farmers (Temu, 1992), women (Johnston-Anumonwo, 1997), small-scale enter-prises (Steel and Webster, 1991), and public social spending (Castro-Leal *et al.*, 1998) have been well documented. With the exception of the Becker *et al.* (1994) seminal work, which examines the effect of SAPs on the urban dynam-ics of African cities, very few studies have focused specifically on the urban planning and management implications of SAPs. This chapter examines the problems that confront urban planners and managers in Ghana as they attempt to deal with the consequences of structural adjustment programs. It begins with an examination of the growing urban crises in the Accra Metropolitan Area, Ghana's largest urbanized region. This is followed by an assessment of the challenges that face urban development planners and managers as they try to cope with these crises. The chapter then concludes with a call for an appropri-ate urban growth management strategy to ameliorate the inevitable outcomes of the urban crisis.

15:1.1 The Growing Urban Crises in the Accra Metropolitan Area

The Accra metropolitan area is the nodal center of the Greater Accra region (refer Figure 15.1) and exhibits the typical attributes of a primate city. It ac-counts for 58 percent of industrial establishments in the country, 50 percent of the value-added in manufacturing, and 46 percent of industrial employees and medical doctors. Greater Accra is the most urbanized of all ten regions in Ghana with 84 percent of its residents living in urban places. In the 1960s, the average annual rate of urban growth was 6.4 percent, and in the 1970s it was 5.0 per-cent or almost twice as high as the national annual population growth rate of 2.8 percent. Accra's population is currently estimated at 1.7 million (3.5 times larger than in 1960) with in-migration accounting for about two-thirds of its growth rate.

This rapid increase in urban population has led to host of interrelated problems associated with housing affordability, fiscal mismanagement, infrastructural service delivery, and land management. According to United Nations estimates, Accra has one of the lowest "city development indexes" (CDI) in the world, a composite index that measures the extent to which cities perform with respect to environmental management, health, education, infrastructure, and level of economic output (UNCHS, 1998). Accra's CDI rating is 22.3 out of a high of 90.99 (Table 15.1).

Development analysts have argued that structural adjustment programs can slow down the pace of urbanization by utilizing devaluation and trade liberalization programs to stimulate export activity in agricultural and mineral-producing regions. This has the anticipated effect of attracting migrants to these regions thus slowing the movement to cities. On the other hand, however, successful structural adjustment programs that stimulate economic growth could very well trigger rapid urbanization (Becker *et al.*, 1994). In the case of Ghana, however, the economic reform programs have been heavily biased towards urban areas, specifically Accra, where most of the pilot projects associated with rehabilitation and local government reform have occurred. As a result, Accra continues to bear the economic and administrative burdens that accompany rapid growth.

More recently, however, uncontrolled urban sprawl, spurred by inappropriate institutional policies, has begun to manifest itself on the physical landscape. Accra, in fact, is overurbanized to the extent that its socio-economic and administrative structure can no longer sustain the rapid pace of urbanization. Instead of being a prime mover of innovation, change and development, Accra has reached a stage where it is functionally unable to provide the desired socio-economic leadership required to expand formal job opportunities, to transfer appropriate technologies and skills to the hinterland, and to provide an environment that assures a higher standard of living to residents by way of health, education and decent affordable housing.

One of the persistent problems is the sporadic process of uncontrolled sprawl in the fringes of the Accra metropolitan region. Urban sprawl referred to in this chapter refers to scattered, untimely, and poorly planned urban development that occurs in the fringes of urban areas. Table 15.2, for instance, reveals that the land development multiplier in Accra (12.5), an indicator of suburban development, is twice as high in Accra than the African average (5.0). Also, housing production in Accra exceeds the African average by a factor of 3. Examples of sprawling suburban communities that have recently emerged in the Accra Met-

Figure 15.1   **Regions of Ghana**

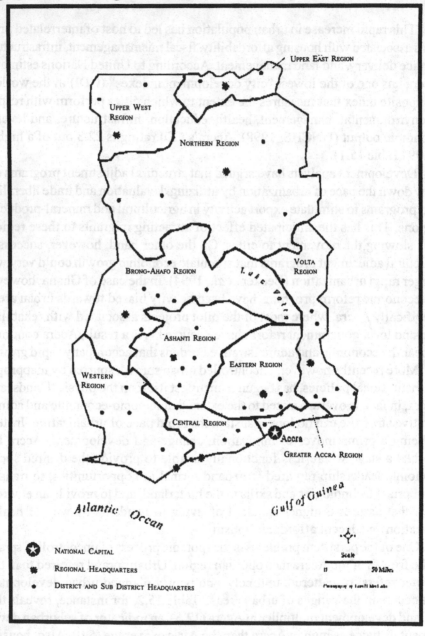

UPPER EAST REGION

UPPER WEST REGION

NORTHERN REGION

VOLTA REGION

BRONG-AHAFO REGION

Lake Volta

ASHANTI REGION

EASTERN REGION

WESTERN REGION

CENTRAL REGION

Accra

GREATER ACCRA REGION

Atlantic Ocean

Gulf of Guinea

NATIONAL CAPITAL

REGIONAL HEADQUARTERS

DISTRICT AND SUB DISTRICT HEADQUARTERS

Scale

15   50 Miles

ropolitan Area include Hatso, a mixed income community, Ashale-Botwe, an upper income community, and Adenta, a middle-income community interspersed with low-income households. These sprawled developments are not functionally related to the surrounding land use and they manifest themselves in the form of large expanses of low density, uni-dimensional development in the urban landscape. These new developments encroach upon lands important for environmental, farmland, and natural resource protection. Furthermore, they generate higher public costs for facilities and services.

Four major problems associated with urban sprawl in the Accra Metropolitan Area are: (1) the lack of an appropriate institutional response; (2) poor land management; (3) adverse economic impacts and inadequate urban service delivery; and the detrimental social and cultural implications.

15:1.2 Institutional Responses to Urban Sprawl

Planning institutions in Accra have hardly devised appropriate growth management mechanisms to guide the rate, character, and quality of new development.

Table 15.1 City Development Index for Selected African Cities

City	CDI
Accra, Ghana	23.21
Kumasi, Ghana	27.44
Tamale, Ghana	24.91
Dar es Salaam, Tanzania	27.20
Addis Ababa, Ethiopia	25.53
Bulawayo, Zimbabwe	63.99
Gaborone, Botswana	45.89
Nairobi, Kenya	45.56
Abidjan, Cote D'Ivoire	43.81

Source: *Compiled from UNCHS, 1998. Global Urban Indicators Database, United Nations Center for Human Settlements (Habitat).*

Sprawled residential development is characterized by unauthorized land development as most buildings are developed without prior approval from planning agencies. In Accra, the key agencies responsible for initiating, approving, and executing plans are the Department of Town and Country Planning, the Accra Planning Committee, and the Accra City District Council respectively. Any development undertaken by private individuals, groups, government or parastatal organizations in any part of the city without prior approval from the Accra Planning Committee and the City Council is unauthorized (Anipa and Aryeetey, 1988). In spite of this legal provision, rapid sprawl continues due to a number of factors.

First, there is a lack of coordination between housing affiliated agencies. State-run housing agencies, in particular, do not consult with planning committees for assistance on planning issues. A number of these agencies operate on the assumption that they are immune from regulations. This is due in part to the fact that there is inadequate information on the function and purpose of the Planning Committee, the Town and Country Planning Department, and the City Council in relation to land development.

Second, planning and housing-related agencies have failed to monitor residential developments and enforce legislation, due to a lack of resources, staff and inspectorate units. This has led to a violation of several building code and planning and subdivision regulations pertaining to setback requirements, land suitability, site coverage ratios, quality of building materials, and the provision of infrastructure. In the community of East Legon (see Figure 15.2), for example, several homes are located perilously close to high-tension cable wires. Furthermore, the majority of homes in the community are located on a floodplain and along the flight path of the international airport. In addition, building codes and regulations issued by the Ministry of Works and Housing tend to be costly and unrealistic, and discourage the use of indigenous materials.

15:2 Land Management

The problem of residential sprawl is further aggravated by inappropriate land management practices. The management of land is fundamental to any improvement in the quality of an urban environment and the ability of local governments to collect revenue to finance infrastructural improvements. Access to land facilitates the provision of housing, utilities, transportation and other relevant urban amenities. It also enables local governments and international agencies to carry out their policies more effectively. Between 1972 and 1982, the World Bank allocated 77 percent of its loans to site-service and slum upgrading

Table 15.2 Indicators of Urban Sustainability in the Accra Metropolitan Area, 1998

		Accra	Average
HOUSING	House Price/Income Ration	8.1	6.94
	Land Development Multiplier*	12.5	5.10
	Housing Production per 1000 population	25.0	7.50
Economic Conditions	% Informal Employment	69.6	56.1
INFRASTRUCTURE	% Households connected to		
	water network	45.7	37.6
	sewer network	12.0	12.7
	telephone network	35.9	11.6
TRANSPORTATION	% Work trips by:		
	private car	10.0	11.8
	bus/minibus	47.0	30.3
	foot	26.0	36.7
	bicycle	2.0	8.2
	Mean travel time to work	45min	37min
ENVIRONMENT	Solid Waste Disposal Methods		
	Landfill	3.0%	16.9%
	Open dump	97.0%	63.4%

* The Land Development Multiplier is the average ratio between the median land price of a developed plot at the urban fringe in a typical subdivision and the median price of raw, undeveloped land with planning approval in an area currently being developed.

Source: Compiled from UNCHS, 1998. Global Urban Indicators Database, United Nations Center for Human Settlements (Habitat)

schemes which were predicated upon securing land for the urban poor (Mabogunje, 1992).

A major problem with land in Accra is the inability of municipal governments to efficiently issue, register, and maintain records of land transactions. This has resulted in a proliferation of informal and illegal tenure arrangements. Customary land tenure systems, where the ownership of land is vested in a collective (family, village) while individuals enjoy rights of usage, are prevalent. It is the sacred and customary duty of traditional authorities (chiefs) to represent the interests of their respective communities. They make decisions on the amount and purchase price of land to be leased.

The problems associated with the customary land system are compounded by long delays and several inconveniences associated with the land permit system. The authorization process for land development is cumbersome, time-consuming, and very costly to developers. Six agencies are normally responsible for perfecting the completion of a title to land. It begins with the Land Commission processing the necessary documents, followed by the Town and Country Planning Department verifying that site plans comply with approved layouts and the Land Valuation division stamping the duty. The Internal Revenue Service then assesses the tax value of the property, the Deed Registry registers and issues the title, and the Survey Department concludes by processing the final title. This process can take any where from 6 months to 3 years. After delays with land title registration, it may take 1 to 3 years to obtain a building permit. It is not surprising, therefore, that developers prefer to "by-pass" the system by proceeding with site development before seeking authorization. This is a major reason for the uncontrolled and uncoordinated pattern and process of urban sprawl.

15:2.2 Economic Impacts and Problems with Urban Service Delivery

The economic and fiscal impacts of suburban sprawl are evidenced by the escalating public costs associated with services and facilities required by the new sprawling developments. A study by the World Bank (1984) demonstrates that there is a lack of a sufficient network of water pipes that extend to outlying communities. For example, the total daily water supply in Accra in 1987 was 3.2 million liters. If we assume a conservative average of 9 liters per day for 1.7 million people, this translates to a demand of about 15 million liters per day or about 20 percent of total demand. Newer developments in Adenta, Hatso and Ashale Botwe have septic tanks, which create problems associated with sanitation and waste disposal. Further costs are incurred as code violations and planning irregularities abound in these sprawled communities. The Accra Metro-

Figure 15.2 Internal Structure of Accra

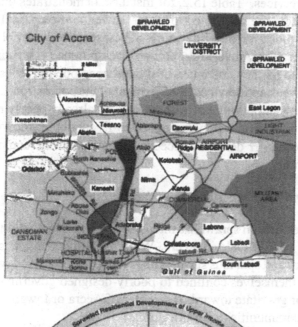

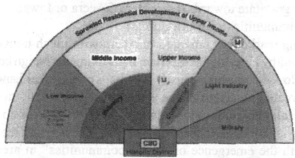

politan Authority, for example, does not engage in concurrency management to ensure that urban development is consistent with the provision of facilities. As a result, several dwelling units violate planning and building code regulations.

Unregulated and unmanaged sprawl has also meant that half of Accra's 1.7 million people live in areas with hardly any infrastructure. The World Bank's (1984) report on the rehabilitation status of Accra reveals that the lack of routine maintenance has caused drainage systems in the city to cease functioning, resulting in severe flooding conditions during the rainy season. The lack of maintenance and the frequency of floods have resulted in the deterioration and blockage of roads, thus creating extensive delays in the movement of goods and services and increasing the operational costs of vehicles.

Other problems relate to the waste disposal, electricity consumption, and communication services. Table 15.2, for instance, demonstrates that very few households in Accra are connected to water, sewer, and electrical networks. Furthermore, the collection of solid waste is at best sporadic with only 3 percent ending in landfills. The Accra Metropolitan Area produces an estimated 600 tons of solid waste per day and much of it is dumped openly. House to house service is once a week and services just 5 percent of the population. This results in a lot of dumping and burning of garbage into an already poorly maintained drainage system. Electrical supply is plagued by frequent breakdowns in a system where the consumer demands exceeds capacity, and the communication system is still erratic.

15:2.3 Social and Cultural Impacts of Urban Sprawl

The process of residential sprawl has intensified spatial segregation and social stratification on a large scale in Accra. While wealthy Ghanaians escape to enjoy the greenery and spacious surroundings of the suburbs, the lower income households find themselves confined to poorly-designed government estates, and the very poor gravitate towards the center of Accra or towards one of the peripheral slum communities such as Madina.

A dual housing market has emerged in Accra with plush housing estates that cater to the wealthy elite, including Ghanaians living in foreign countries on the one hand, and low income high density dwellings on the other hand. This has resulted in a bifurcated housing finance system that offers low-interest loans and longer amortization periods to the wealthy and offers little or no assistance whatsoever to low to moderate-income households. The housing market is further polarized with the emergence of "gated communities" in areas such as East Airport, East Legon, and McCarthy Hill. The comforts and benefits enjoyed by these residents make them immune from disruptions. This increasing spatial polarization diminishes any chance of them acting as an effective voice to improve the system.

Rising housing prices in Accra offer few options to low-moderate income households. Accra's house price to income ratio of 8.0 is higher than the African average of 6.9 (Table 15.2). In the inner city, few options exist in state housing estates where problems abound with inappropriate design characteristics, building code violations, poor ventilation, and high dwelling densities. Also, multi-storeyed family compounds offer little relief since they are fully occupied with dwelling densities as high as 15 persons per dwelling.

As the housing market has reached a saturation point in central Accra, residential development has encroached upon valuable agricultural land in the urban fringe and has had a detrimental impact on the lives of low-income families who practice urban agriculture as a means of survival and sustenance. Research on urban agriculture is limited. A few case studies in Kenya (Lee-Smith et al., 1987; and Freeman, 1991), Zambia (Rakodi, 1988) and Lome-Togo (Schilter, 1991), offer some perspectives and reveal that the practice offers income and employment alternatives particularly for women in the informal sector. Schilter's study of Lome, Togo appeals to the sensitivity of the government to maintain appropriate land reserves for agriculture in view of the continuous physical development of the city.

Residential sprawl in Accra also has the effect of disrupting cultural and traditional institutions. Accra is affiliated ethnically with the Ga Mashie people, the original 16th century settlers. Since the early beginnings of urbanization in the early 17th century, the city has become more ethnically diverse and several land transactions have occurred. The net result has been a dislocation of the traditional institutions of the Ga people, thus fuelling ethnic tensions. Most of the newly-developed land in the suburbs was purchased by financial and home-building institutions from the Ga chief who has the sacred duty of safeguarding the communal property rights of his constituents.

As the crises of residential sprawl and municipal mismanagement continue to unfold in the Accra metropolitan area, there is a need for municipal government authorities to respond immediately with more efficient, innovative and effective ways to manage growth. A number of challenges and development management options are reviewed below.

15:3 Challenges of Urban Development Management and Planning in Accra

The complexities of the urban crises in the Accra Metropolitan Area call for a comprehensive and integrated urban development and planning strategy. The World Bank instituted a series of urban development projects in the early 1990s to address these problems. The first Bank-assisted urban project, the Accra District Rehabilitation Project, focused on Accra only. The project emphasized the rehabilitation of the urban infrastructure and services, including roads, water supply, sewerage and sanitation, drainage, solid waste, markets, public transport, traffic management, and area upgrading. Another objective was to strengthen institutional capacity at the central and municipal levels, and improve the mechanisms associated with cost recovery and revenue mobilization. These initiatives were then extended to four other cities—Kumasi, Tamale, Tema, and Sekondi-Takoradi—under the auspices of the Priority Works Project, the Urban II Project, the Urban Transport Project, and the Water Sector Rehabilitation Project. The

World Bank and IMF structural adjustment program does emphasize reforms in local government, institutional capacity building, and privatization–all policies that have a bearing on urban revitalization. It is too early to assess the impact of these projects, however, the challenges presented below have some bearing these initiatives.

15:3.1 Developing a Comprehensive Urban Information System

Reliable information is essential for urban development management and planning. A comprehensive Urban Information System (UIS) that focuses not only on demographic and economic variables but also on a wide range of qualitative and quantitative measures serves as a necessary basis for informing the different phases of urban development management and planning. Figure 15.3 provides a schematic representation of a UIS and indicates the types of information required. For example, an urban profile is an important component of an information system. It represents qualitative and quantitative information on the basic demographic structure, settlement systems, and sector profiles such as land use, transportation, and the labor market. Housing demand, for example, is influenced by labor market changes or new transportation investment. The urban profile should compile existing information on these sectors and be continuously updated. Government agencies often collect data from a variety of sources such as special surveys and the census; however, these are often available in a fragmented manner. It is difficult to access public data in a consistent way, as they exist in a fragmented manner in the offices of different ministries. It is, therefore, important for the urban profile to provide guidance to the numerous departments that collect urban data, so as to improve their content and coverage. The urban profile should be updated annually to incorporate changes brought about by sector plan implementations within the region.

Land information systems and geographic technologies can be employed to develop a better recording system of land inventory, land transactions, and land assessment and valuation. Falloux (1989) has pointed out that only 1 percent of land parcels in Africa South of the Sahara have been comprehensively surveyed. It is imperative to link a comprehensive land survey to all other attributes of a land management system, namely population, housing, production cost structures, impact assessments and institutional structures as shown in Figure 15.3. A combination of geographic technologies that focus on aerial photos, satellite imagery, and geographic information systems (GIS) can be employed to update land parcel inventories, conduct land suitability analyses, and update information on existing urban infrastructure and facilities.

The effectiveness of any urban development management and planning strategy depends on the efficiency of the implementing institutions. An additional problem in Ghana is the shortage of specialized institutions, particularly in the private sector, such as banks, mortgage and insurance companies, and community development corporations. As a result, the responsibility of plan implementation rests with a public sector that is constrained by bureaucratic rules and regulations. Consequently, an important component of an urban information system in Ghana should be an inventory of institutions that are active in the urban sector. Information can be organized so that size, experience, activity levels, functional specialization, and performance criteria can classify institutions. Given the dominance of the public sector in planning activity, it would be useful to include information on private institutions and community groups with their goals, expertise, and attitudes.

Monitoring is the assessment of planning policies as they are being implemented. It consists of a continuous process of data assembly, screening, and reporting unintended problems and new opportunities that will cause the modification of the plan to meet its desired objectives. Impact assessments are important in measuring performance and achievement levels of urban programs. It determines the extent to which program objectives have been accomplished and the extent to which target groups such as the urban poor have benefited from such programs. Such information is always useful for in designing future plans and strategies.

Education and training are necessary prerequisites to the successful design and implementation of comprehensive urban information system. Also, in order to minimize cost, a network of information-sharing agencies can be established. Public and private agencies can then have the flexibility of retaining sensitive information and sharing those parts that may be of benefit to the public good.

15:3.2 Land Management

The availability and management of land are fundamental to any improvement in the quality of an urban environment. Municipal governments in Africa are, therefore, faced with the challenge of ensuring that records of land transactions are issued, registered, maintained, and expedited more efficiently. Land information systems facilitate the recording, measurement and processing of land parcel boundaries and property rights, and enhance the opportunity for increasing property tax revenues. In Ghana, the key challenge facing urban managers and planners are the recurring problems of tenure insecurity, land litigation, inadequate land surveys, and the dilemmas of resolving conflicts between customary and statutory rights to land. The Land Title Registration Law was enacted in 1986 to address these problems. Key provisions in the law include the

**Figure 15.3 An Urban Information System for Urban
Development Management and Planning**

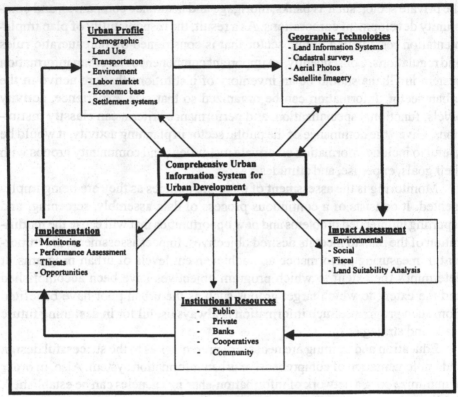

validation of physical boundaries on the ground and on the map; the adjudication and settlement of competing claims; the registration of adjudged interests; and the guarantee by the state that such recorded interest be indefeasible (Agbosu, 1990).

This new law is applicable to the Accra Metropolitan Area, which is being used as a pilot study. Three key administrative units are primarily responsible for land management: the Lands Commission Secretariat administers public lands and registers deeds; the Survey Department conducts mapping and cadastral surveys; and the Land Title Registry maintains records of land transactions. In spite of the enactment of new legislation to secure land tenure and minimize litigation, problems in land management still persist. A major challenge rests with the ability of the three agencies of land management to collaborate and coordinate effectively. Perhaps a "one-stop shop" that incorporates and streamlines all three functions is warranted. The Land Title Registry is still plagued by manual procedures and needs more efficient computerized methods to store and process titles. The Survey Department also requires computerized methods to expedite cartographic procedures and facilitate field surveys. In addition, the Lands Commission is still faced with the challenge of harmonizing land transfers between public, communal and private invests. These deficiencies need to be resolved to create an enabling environment for private sector land development and for comprehensive urban development and management.

15:3.3 Decentralization and Building Local Capacity

African municipal governments have historically been extremely centralized. The International Monetary Fund (IMF) has indicated that local governments in industrialized countries account for 57 percent of all government jobs, compared to only 6 percent in African countries (Stren, 1992). Consequently, loosely structured and poorly functioning municipal governments in Africa continue to be out of touch with the needs of local communities. The challenge, therefore, for urban communities in Ghana rests with the government's ability to decentralize municipal functions in order to strengthen institutional development, build local capacity, and empower disenfranchised communities to deliver and maintain urban services.

Ghana is among a number of African countries dedicated to the decentralization effort. The Ghana government enacted the Local Government Law in 1988 to promote popular participation and ownership of the machinery of government by shifting the process of governance from command to consultative processes, and by devolving power, competence and resources to the district level. The decentralization policy was extended to include the District Assemblies Common Fund Act of 1993, the National Development Planning Systems Act of 1994, and urban, zonal, and town council committees in 1994.

Decentralization was further motivated by the Program of Action to Mitigate the Social Costs of Adjustment (PAMSCAD). In an effort to sustain the Structural Adjustment Programs, the government instituted PAMSCAD in 1987 to protect vulnerable groups such as the urban poor from the negative consequences of the economic reform program. PAMSCAD embodied a degree of decentralization and emphasized community initiatives, training and placement services for workers, basic needs provisions, devolution of responsibility for project initiation and implementation, and capacity building at the local level.

Ghana's decentralization policy is an ambitious effort that shifts away from an emphasis on deconcentration to one on devolution. The former refers to the redistribution of decision-making authority and financial and management responsibilities among different levels of central government with some local input. Local agencies act as agents of the central government and expend resources allocated to them by central government authorities. Devolution, on the other hand, is the transfer of authority for decision-making, finance, and management to quasi-autonomous units of local governments. It transfers responsibilities for services to municipalities that elect their own councils, raise their own revenues, and have independent authority to make investment decisions (Wunsch, 1999). The devolution of authority empowers communities to create opportunities to sustain themselves. If successfully implemented, it alleviates the burden on government and allows communities to take stock and provide for their needs in a more efficient manner.

In light of the ambitious and multidimensional nature of Ghana's decentralization policy, the World Bank has suggested a more deliberate multi-year strategy of implementation that focuses on a few priority areas at a time. One such area is in local public finance reform where the system for central-local revenue transfers have been targeted. In addition, reforms in local government management, including budgeting, personnel, and operations and management have been prioritized for immediate attention.

A major feature of the public finance reform program is to establish resource sharing through the District Assemblies Common Fund, which advocates the transfer of not less than 5 percent of total government revenue to the districts. Districts receive allocations on the basis of their demographic and income profiles, and their revenue improvement characteristics. Municipalities such as Accra will consist of urban, zonal, and town or area councils and unit committees (Ministry of Local Government and Rural Development, 1999). Unit committees are consultative bodies at the grassroots level responsible for carrying out duties that are delegated to them by District Assemblies. This provides community residents with the potential of managing and operating the services that they require. This potential can only be realized, however, if residents receive the necessary education and skill-based training to increase their capacity to meet their own needs.

A major challenge in implementing this ambitious decentralization policy is the lack of an administrative and social infrastructure to strengthen the institutional capacity to manage and deliver urban services. One must question the government's ability to delegate authority to community-based and grassroots organizations in the formal and informal sectors of the urban economy, given the lack of trained human resources and inappropriate institutional mechanisms. As a result, can the Ghanaian government afford to delegate responsibilities by creating public enterprises, housing authorities, community development corporations, and special service districts to address the needs of disenfranchised and marginalized populations in the Accra Metropolitan Area? The fact remains that essential public services such as police, electricity, water, and land titling are not yet decentralized. Of equal importance in the successful implementation of a decentralized policy is the government's ability to articulate its agenda and follow through on a phased program of action. This can be accomplished if there is a shared and open discussion on the process, an adherence to local priorities, and an effort to uphold the principles of accountability.

15:3.4 Cost Recovery and Taxation

Municipal governments and community-based constituencies have relied heavily on central governments for revenues and in the process have denied themselves an opportunity to exercise any financial authority. An essential component of a fiscal decentralization policy is to transfer the responsibility of decision making and revenue generation to urban councils and other subdistrict levels. This implies taking on the responsibility of cost recovery through user charges and co-financing or coproduction, where users participate in providing urban services through monetary and labor contribution. The IMF and World Bank, through structural adjustment, have encouraged cost recovery through user charges as a major source of revenue for financing and maintaining basic urban services such as garbage collection, sanitation, and community facilities. In the Accra Metropolitan Area, user charges account for just 3.2 percent of local government revenues. This is because the central government has not motivated local residents and authorities to assume the responsibility for projects, thus denying them the opportunity to play a meaningful role in shaping the destinies of their own communities. Also, for political reasons, governments prefer to charge low rates for services thereby failing to maximize the full potential of cost recovery mechanisms.

Urban managers and planners must face the challenge of improving cost recovery mechanisms by improving services in local communities, mobilizing local resources, and encouraging residents to gain a sense of participation and accomplishment. Successful cost recovery occurs through self-help and coproduction as residents develop creative ways to pool their resources and talents.

Furthermore, since property taxes are such an important source of revenue, it is imperative that municipal authorities improve property tax assessments and collection procedures. This is predicated by the development of comprehensive land information systems, the periodic updating of valuation rolls, and improvements in land management procedures as discussed above.

15:3.5 Role of the Private Sector

Another relevant component of the decentralization process is economic decentralization, which involves shifting responsibility for functions from the public to the private sector. Privatization is a key emphasis of the structural adjustment reform program and it has direct implications for urban development management and planning. Privatization can range in scope from contracting or leasing out tasks and responsibilities to private sector firms, while municipal governments retain overall supervisory and regulatory control, to a complete transfer of responsibilities for providing urban services to private sector firms who set their own price (UNCHS, 1999). In other words, it ranges from the provision of goods and services based entirely on the free operation of the market to public-private partnerships where governments, community groups, voluntary organizations and the private sector collaborate or cooperate to provide services.

Privatization is beneficial if it relieves the financial and administrative burden on the government, satisfies unmet needs, reduces the cost of public services to consumers, improves efficiency by promoting competition, enhances innovation and new technology, and improves responsiveness to cost control measures (McMaster, 1991). The Ghanaian government, in collaboration with the World Bank and IMF, embraced the notion of privatization to help address shortfalls in urban infrastructure and services. The potential exists for community-based organizations and private companies to provide municipal services such as solid waste management, parking, management of public restrooms, improving neighborhood roads, and health and education services.

Municipal governments in Accra and Kumasi have received support from the World Bank to contract with local entrepreneurs to collect household refuse. Refuse is collected in both cities through door-to-door services and the provision of metal containers located at transfer stations for communal use. Door-to-door services are provided to only 5 percent of households in Accra, while in Kumasi only a third of refuse generated ends up in landfills. Also, there is inequality in the distribution of services as low-income households are largely ignored. Consequently, open dumping and burning are practiced widely. Post (1999) mentions that on average private contractors provide door-to-door service at a 12 percent lower cost than the Waste Management Department in Kumasi. In addition, default rates are three times lower for privatized operations than municipal operations.

A key determinant of the success of privatized operations is cost recovery. Some critics have questioned the ability and the willingness of low-income households to pay for urban services. Studies in Nigeria and Ghana, however, have shown that the poor are willing to pay for quality services. The Nigerian study revealed that private water vendors in Onistha delivered twice as much water as the public water system and that payments for vended water were 20 times more than the payments made for water from the utility (World Bank, 1990). Archer et al (1997) found that over 70 percent of households in Atonsu, Kumasi were willing to contribute financially to proper refuse collection. These studies confirm that the success of a cost effective privatized urban service is contingent upon the quality of service provided, the appropriateness of the technology employed and the efficient enforcement of fee collection procedures. In the case of solid waste management a combination of cost effective and simple refuse collection technologies such as power tillers, and wheelbarrows or push carts could result in higher community participation and waste disposal rates.

The successful privatization of urban services further depends on the Ghanaian government's ability to ensure a competitive environment by avoiding favoritism in awarding of contracts and preventing private providers from monopolizing urban services.

Furthermore, since privatization is still in its infancy in Ghana, municipal governments need to provide the necessary mechanisms to facilitate privatization by providing support services such as information referral networks, financial management, accounting and legal assistance, and performance monitoring guidelines. In addition, governments need to minimize the constraints that frequently frustrate the privatization process. These include: avoiding too much bureaucracy, avoiding delays in payments to contractors, preventing the indebtedness of essential municipal services in order to attract entrepreneurs, and shoring up institutional finances to facilitate financing for acquisitions.

15:4 Instituting an Effective Urban Growth Management Planning Strategy

It is clearly evident from the above discussion that the task of resolving the urban problems in the Accra Metropolitan Area is a complex one. It calls not only for a comprehensive and integrated approach that captures the essence of the challenges discussed above, but also an effective growth management strategy that embodies the principles of rational and cost effective planning. Effective monitoring and coordination of residential sprawl and its attendant problems, requires the institution of growth management techniques that call for a more rational and orderly form of development. The literature on growth management is more commonly associated with American states such as Florida, California, New Jersey, Maine, Vermont and Massachusetts. Hardly any consideration has been given to the applicability of such techniques within an African context. A number of appropriate techniques are worth considering in the context of the Accra Metropolitan.

352 IMF and World Bank Sponsored Structural Adjustment Programs in Africa

First of all, planning agencies in Accra need to ensure that concurrency management is enforced. This will ensure that suburban and exurban development takes place concurrently with the provision of adequate facilities. Due to poor monitoring and enforcement procedures, several developments in suburban communities like Adenta and Ashale Botwe have proceeded without a sound infrastructure in place. This is certainly detrimental to the safety, welfare, and well being of residents. There needs to be more interaction between private developers and planning agencies to ensure that proper procedures are followed and to increase awareness about the functions and responsibilities of public entities.

Planning agencies may also want to consider instituting *flexible urban growth boundaries* to restrict and redirect residential development to the extent that farmlands and natural resources are protected, and also to ensure that appropriate service and infrastructure standards are met prior to the approval of new development. *Clustered residential developments* and *planned unit developments* also need to be considered to promote more compact development that encourages smaller, more cost-efficient development which, in turn, facilitates the installation of facilities and preserves open space. Clustered developments are also more cost effective since they allow for medium density development and more opportunities for affordable housing. Furthermore, more compact developments cost less to maintain.

Open-space zoning is required to preserve agricultural areas to sustain the economies of farm families. Since urban agriculture is becoming more prevalent in Africa cities, there is a need to ensure that the livelihoods of such families, who play an important role in the urban economy, are protected. Open space zones could be supplemented with *Holding Zones* to delay development until more viable options are carefully considered and verified.

Besides these growth management techniques, other challenges face the metropolitan governing bodies of the Accra Metropolitan Area. For instance there is a need to enhance the administrative and technical capacity of the Accra Metropolitan Authority to monitor land development and coordinate the delivery and operation of services. A qualified and skilled labor force is required to carry out the day-to-day operations and regulations of urban management. Nowhere is this more evident than in the area of urban land management. Most African countries lack the institutional and administrative capacity to record land transactions. Without proper land registration procedures, African cities cannot administer property taxes, which are an essential revenue source for local governments. There is a need, therefore, to conduct appropriate workshops and technical training programs to upgrade the human resource base required to improve land management practices. Becker *et al.* (1994), in reference to a World Bank staff appraisal report in Ghana, call for a need to restructure salaries and incentives and develop special post-graduate programs to train both public and private sector managers. Unfortunately, such programs are lacking in Ghana and other African countries. The Economic Development Institute of the World Bank is already involved in organizing a number of workshops and seminars geared towards civil service improvement, capacity building for policy analysis, coordination and implementation of national development policies, and developing training strategies to strengthen institutional management (Adamolekun, 1989).

Another challenge involves the need to improve the financial capability of the municipal authorities in Accra to provide adequate urban services. Improving the administrative and technical capacity of urban governments would enhance the ability of urban governments to become financially viable and autonomous. In view of the difficulties involved with the administration of property taxes, most local governments continue to be dependent on external sources for finances, which are not always reliable. In addition to improving the capacity to collect property taxes, local governments need to improve their tax and fee collecting procedures. Adding more tax collectors to the municipal payroll can do this.

Finally, there is a need to develop appropriate and innovative solutions that are not only cost-effective, but also meet the needs of local constituents. The planning and municipal organs of the Accra Metropolitan Area need not always have to rely on western ideals and norms, and be pre-occupied with preserving out-moded colonial structures and procedures. According to Mabogunje (1992), urban Africa needs a strategy of governance that capitalizes on the institutions and processes which people are familiar with and one that entices people to participate actively in the management of cities. Kironde (1992) offers some guidance along these lines in the area of sanitation as one example. He suggests that urban governments begin to invest in ventilated improved pit latrines (VIPs) as an alternative to water-borne sanitation systems which are capital intensive and unaffordable for a majority of urban residents. It may be unrealistic to expect comprehensive sewer systems to be extended to all urban residents in the Accra Metropolitan Area. Although certain biases exist in regards to the health hazards of pit latrines, the ventilated improved systems, if properly constructed and maintained, may very well be the more realistic cost-effective alternative.

Several attempts have been made by international agencies and local African governments to deviate from applying rigid standards in the housing sector.

Gone are the days when slum eradication, rigid minimum building codes, and forced resettlement were seen as solutions to the housing problem. Donor agencies like the World Bank and United States Agency for International Development have forged partnerships with African governments to encourage local self help efforts. Site-service, core housing, and squatter upgrading programs are now regarded as viable alternatives. Site-service programs have been adopted in a number of African countries, including Zambia, Zimbabwe, and Kenya. Unfortunately, site and service programs do not always reach their intended beneficiaries. The poor sometimes end up re-selling their site to middle and upper income households.

African urban governments, such as Accra's, need to be more creative and innovative in their approach to urban problems. They also need a change in attitude to incorporate the utilization of local indigenous resources and to encourage more local participation and initiative in the urban revitalization effort. All this can be done within a framework that encourages an interdependent relationship between international, national, and local agencies.

References

Adamolekun, L. (1989). *Issues in Development Management in Sub-Saharan Africa*, Washington, DC: World Bank Economic Development Institute.

Agbosu, L. K. (1990). Land Registration in Ghana: Past Present and Future. *Journal of African Law*, 34(2): 104.

Anipa, J. and Aryeetey, E. (1988). Planning for a Distressed African City: The Accra Experience. *Planning African Growth and Development*, Accra, Ghana: Institute of Social, Statistical, and Economic Research.

Archer, E., Larbie, B. and Amin, A. (1997). *Privatization of Refuse Management in Atonsu, Kumasi*. Ghana, Research Paper, No.7. Kumasi, Ghana. University of Science and Technology.

Becker, C., Hamer, A. and Morrison, A. (1994). *Beyond Urban Bias in Africa*, Portsmouth, NH: Heinemann.

Castro-Leal, J., Dayton, J., Demery, L. and Mehra, K. (1999). "Public Social Spending in Africa: Do the Poor Benefit?". *The World Bank Research Observer*.Vol. 14, No. 1, February.

Falloux, F. (1989). *Land Information and Remote Sensing for Renewable Resource Management in Sub-Saharan Africa: A Demand-Driven Approach*. Washington, DC: World Bank.

Johnston-Anumonwo, I. (1997). "Geography and Gender in Sub-Saharan Africa" in S. Aryeetey-Attoh, *Geography of Sub-Saharan Africa*. NJ: Prentice-Hall.

Kironde, J. (1992). "Received Concepts and Theories in African Urbanization and Management Strategies: The Struggle Continues". *Urban Studies*, 29(8):1277-1292.

Lee-Smith, D., Manundu, M., Lamda, D. and Gathuru, P. (1987). *Urban Food Production and the Cooking Fuel Situation in Urban Kenya*. Nairobi. Mazingira Institute.

Mabogunje, A. (1992). *Perspectives on Urban Land and Urban Management Policies in Sub-Saharan Africa*. Washington, DC: World Bank.

McMaster (1991). *Urban Financial Management: A Training Manual*. Washington, DC: World Bank.

Ministry of Local Government and Rural Development (1999). *Comprehensive Development Framework: Decentralization*. Republic of Ghana.

Owusu, J.H. (1998). "Current Convenience, Desperate Deforestation: Ghana's Adjustment Program and the Forestry Sector". *The Professional Geographer*. 50, 4.

Post, J. (1999). "The Problems and Potentials of Privatizing Solid Waste Management in Kumasi, Ghana". *Habitat International*. Vol. 23, 2: 201-215.

Rakodi, C. (1988). "Urban Agriculture: Research Questions and Zambia Evidence". *Journal of Modern African Studies* 26(3):495-515.

Schilter, C. (1991). *Urban Agriculture in Lome: Agronomic and Socio-Economic Approaches*. Paris and Geneva: Karthala and IUED.

Steel, W. F., Webster, M. (1991). *Small Enterprises under Adjustment in Ghana*. Washington, DC: World Bank.

Stren, R. (1992). "African Urban Research Since the Late 1980s: Responses to Poverty and Urban Growth". *Urban Studies*. 29(3/4): 533-555.

Temu, P. (1992). "Food and Agriculture in Africa" in J. Nyang'oro and T. M. Shaw. *Beyond Structural Adjustment in Africa: the Political Economy of Sustainable and Democratic Development*. New York: Praeger.

UNCHS (1998). *Global Urban Indicators Database*. United Nations Center for Human Settlements (Habitat). Nairobi, Kenya.

United Nations Center for Human Settlements (Habitat) (1999). *Privatization of Municipal Services in East Africa: A Governance Approach to Human Settlements Management*. UNCHS: Nairobi, Kenya.

World Bank (1984). *Ghana: Accra District Rehabilitation Project*. Washington, DC: World Bank.

World Bank (1990). *Nigeria: Willingness to Pay for Urban Water Services*.

Washington, DC: Transportation, Water and Urban Development Department. World Bank.

Wunsch, J. (1999). "Decentralization, Local Governance and the Democratic Transition in Southern Africa: A Comparative Analysis". *African Studies Quarterly*: Volume 2, Issue 1.

16 Structural Adjustment and the Health Care System

JOSEPH R. OPPONG

16:1 Introduction

Structural Adjustment Programs (SAPs) have been the standard prescription of the International Monetary Fund (IMF) and the World Bank for the "internally-generated" economic stagnation and retrogression that characterized most African economies during the 1970s (Ould-Mey, 1996; Owusu, 1998). This stagnation was attributed to market distortions, which in turn were due to overwhelming governmental intervention in the domestic economy. Unwarranted state interference in the market, over-bloated public service, state ownership of manufacturing enterprises and investment in social welfare were essential symptoms of the debilitating and crippling disease that clogs the wheels of market efficiency (World Bank, 1981; Green, 1987). Consequently, removing these market distortions and curtailing or severely limiting government intervention was the only viable solution. Structural Adjustment Programs thus prescribe three groups of reforms: deflationary measures, which consists of removal of subsidies and reduction of public expenditures, institutional changes in the form of trade liberalization and privatization, and expenditure-switching measures in the form of currency devaluation and export promotion (Logan and Mengisteab, 1995).

The hallmark treatment of Structural Adjustment Programs includes: massive currency devaluation; price, interest rate and trade liberalization; credit ceiling and controls over money supply; freeze on wages and salaries; privatization of public enterprises; public expenditure reduction; withdrawal of subsidies (real and imagined); introduction of cost recovery measures on social services (particularly education and health); reduction of the size of the civil service through staff retrenchment, and increased revenue mobilization through introduction and enforcement of a range of direct and indirect taxes (Olukoshi, 1996). Abundant justification is presented for this harsh prescription regimen. For example, private enterprise, because of its private/non-state status, has been attributed quali-

357

ties such as political independence, economic rationality, efficiency, dynamism, resourcefulness, adaptability, innovativeness and so on–the supposed qualities of successful enterprises under competitive situations (Gibbon, 1996).

This chapter examines the effect of Ghana's structural adjustment program on the health care system. It argues that structural adjustment is producing irreversible changes in Ghana's health care system and new forms of "private" health care have emerged in response to the changing economic climate and context of health care provision. Using the example of Buruli Ulcer in the Western Region of Ghana, the chapter argues that decreased use of the formal health system makes it increasingly difficult to determine the disease burden for planning and implementing effective health interventions. Many people are postponing seeking health care primarily due to increased cost of treatment, often with disastrous results.

16:2 Structural Adjustment and Health

The health implications of structural adjustment policies in Africa have been a subject of great interest (Turshen, 1999; Stock, 1995; Stock and Anyinam, 1992; Oppong, 1997). Stock (1995) argues that cutbacks in health sector spending instigated by the IMF under Structural Adjustment Programs have led to large staff layoffs and significant salary reductions. This has produced mere consulting clinics without drugs and sometimes without medical staff because many doctors migrate to seek better economic conditions. Ironically, more Malian physicians are practicing in Paris than in Mali itself (Stock, 1995). Similarly, user fees and cost recovery programs, instead of improving quality of care and inducing increased utilization have caused people to delay treatment or do without it resulting in higher disease and death rates. Stock argues that the solution to the funding crisis in health lies, not in collecting small user fees from patients, but in focusing on the root causes of the crisis, particularly the effect of debt repayment on the ability of African governments to deliver even rudimentary health services. While debt forgiveness is not the entire answer, it is an obvious starting point.

Turshen (1999) provides a comprehensive, well-documented, scathing expose of the health implications of structural adjustment. Using startling statistics, e.g., the median year at death in Europe is 75 years but in Africa it is five years, Turshen laments the emergence of the World Bank as the single largest donor in the health sector in the world today. Given the already frightfully high rates of morbidity and mortality in Africa and the record of World Bank policy failures, the large-scale replacement of tested and proven WHO initiatives such as primary health care with the untested World Bank model of privatization is unjustifiable. Turshen argues that recent outbreaks of new infectious diseases such as Ebola, Marburg and AIDS demonstrate the need for the very same public health system the World Bank and IMF have systematically dismantled. Privatization, by increasing costs, will deter most of the ill from seeking care causing them to die quietly at home, and thus delay awareness and response to new epidemics. In the concluding chapter of the book, Turshen makes a strong case for linking health with its political and economic context. Because health is political in every sense of the word, involving government intervention to prevent illness, for example, health debates should involve a broad discussion of the political choices and not be restricted to the narrow question of financing personal medical care. Turshen makes the case that investments in a private health sector are inefficient and detrimental in Africa.

A general perception exists that adjustment programs have demanded draconian cuts in the expenditures that were already woefully inadequate, especially in the social sectors where public spending is often no more than one or two dollars per capita a year (Sahn, *et al.*, 1997). Moreover, as a result of currency devaluation, food prices are extremely costly, too costly to permit the poor to buy enough food to satisfy basic requirements.

Oppong (1997) captures this viewpoint quite well. He argues that reduced government ability to invest in health and reduced inability of individuals to invest in food, nutrition and health care, both attributed to Structural Adjustment Programs, not only paralyze public health systems, but also hurt nutrition and compromise health infrastructure. Citing the 1995 World Health Report, Oppong cautions that increased poverty in Structural Adjustment Program countries should be a major concern. The Report states:

> Poverty is the main reason why babies are not vaccinated, why clean water and sanitation are not provided, why curative drugs and other treatments are unavailable and why mothers die in childbirth. It is the underlying cause of reduced life expectancy, handicap, disability, and starvation. Poverty is a major contributor to mental illness, stress, suicide, family disintegration and substance abuse. Every year in the developing world 12.2 million children under five years die, most of them from causes that could be prevented for just a few US cents per child. They die largely because of world indifference, but most of all they die because they are poor (*World Health Report*, 1995).

Because health care depends on many imported inputs such as pharmaceuticals and vaccines, fiscal difficulties and unfavorable trade balances mean a limited ability to import these essentials as governments devote more funds to paying off national debts. Moreover, cutbacks in health expenditure do not affect urban and rural residents equally. Urban-based hospitals and programs, usually with significant political support, emerge unscathed while those programs benefiting rural residents receive the biggest cuts. Beside, such cuts as reduction in transportation or fuel budget in a period of escalating fuel and spare parts prices mean limited rural travel by health workers and, consequently, less access to better trained health workers for rural residents. In short, rural residents are forced to subsidize urban health care while living with little or no health care. Finally, high inflation and depreciated incomes compels many workers to supplement their meager incomes with second or third jobs, and "borrowing" official supplies for personal use becomes widespread practice (Oppong and Toledo, 1995). The combined effect of all these is to increase the disease burden while reducing the ability of individuals to fight off disease and diminish access to care.

Applied to HIV-AIDS, Structural Adjustment Program-induced reduction in government health expenditures may increase rates of HIV infection in three main ways. First, because sexually transmitted diseases (STDs) facilitate the spread of HIV, reduced government funding for diagnosis and treatment of STDs increases the risk of transmitting HIV (UNAIDS, 1999). Reduced funding for blood screening, poor hygienic practices in clinics (e.g., inadequate or no sterilization of equipment) due to funding cutbacks may also spread HIV in health care facilities. In fact, Minkin (1991) calls clinical settings in developing countries an often neglected and potentially the most dangerous location for HIV infection. Routine procedures such as blood tests, injection of antibiotics and vaccines, vaginal examinations, abortions and other invasive procedures performed under unhygienic conditions simply spread diseases. For example, Structural Adjustment Program induced cutbacks in Tanzania has produced an environment of health care provision where "protective gear is seen as a luxury" (Lugalla, 1995). Lamenting the situation, Lugalla (1995) reports the death of health workers in government hospitals due to exposure and infection from treating patients without protective gear such as gloves. At the Muhumbili Medical Center, the country's biggest referral hospital, three nurses died in 1995 from cholera they contracted after treating patients with the disease. In addition, cutbacks in health education expenditures reduce the efficacy of public education campaigns and awareness of the risks of infection and thereby, may increase the spread HIV.

Second, structural adjustment programs cause or intensify economic recessions increasing inequality and poverty rates. Poor people engage in more risky behavior and lack access to health care resources that would lower their risk of contracting HIV. They are more likely to be malnourished, which would increase their risk of contracting HIV if they are exposed to it. Moreover, they are more likely to migrate to urban areas, where the prevalence of HIV infection is higher. Furthermore, Structural Adjustment Programs marginalize women, making them more vulnerable to HIV. Erosion of women's real incomes and increasing poverty intensify gender inequality and weaken women's ability to negotiate sexual relations and practices.

Third, structural adjustment programs, through their focus on increasing incentives for greater trade openness, promote an export-oriented pattern of economic growth that may increase the chance of contracting HIV. Greater incentives for exporters lead to the creation of more dense trading networks between rural and urban sectors with the growth in export crop production, and an increase in urbanization where export-oriented industries are located. Improved rural-urban spatial interaction enhances the spread of diseases such as HIV.

16:3 The Changing Structure of Health Care

Structural adjustment programs produce significant structural changes in the health care delivery system in Sub-Saharan African countries. Besides requiring cuts in government expenditure on public health services, structural adjustment specifically promotes privatization of health care services (Turshen, 1999). Typical results include massive layoffs of government health workers, decay of public health infrastructure, and the imposition of user fees that compel people to postpone or forgo treatment with dire consequences. Combined with the escalating cost of private health care facilities, the deteriorating quality but increasing cost of government health services produces a burgeoning demand for affordable and accessible health care services. Thus, the overall context of health care delivery will undergo long-term structural change leading to the evolution of new forms of care and new groups of health care providers.

Implementation of Structural Adjustment Programs in Ghana led to cutbacks in health expenditure resulting in large layoffs, significant salary reductions (due to inflation), and closure of many facilities. For example, the devaluation of the Ghanaian currency, the cedi, raised the cost of drugs and essential medical supplies since most drugs and medical supplies used in the country have a very high foreign exchange component (Ghana Ministry of Health, 1999). In

1996, in support of its application to the IMF for additional loans, the Ghana government wrote:

> The quality of health care is suspect, resulting in low utilization of services, particularly by the poor. ... [D]rug shortages are still common, medical equipment often does not work, personnel are not effectively deployed, and staff morale suffers. ... Health services and management is [sic] particularly weak at the community, sub-district and district levels, where people should be making first contact with the health system (Ghana Ministry of Health, 1995).

Due to the wage and salary freeze and high inflation associated with Structural Adjustment Programs, many health workers were compelled to take second jobs to make ends meet, resulting in absenteeism and poorer health care. Many of the already inadequate number of doctors left the country to seek better economic fortunes in Europe and North America. In fact, the physician stock of Ghana decreased from 1,700 in 1981 to 800 in 1984 (Vogel, 1988). The imposition of user fees, locally "Cash and Carry", drastically reduced access to biomedical health care (Anyinam, 1989), compelling users to forgo or delay treatment and seek alternatives to biomedical services. The impact was more drastic in rural areas, however. While the fees led to an immediate decline of 25-50 percent in hospital and clinic visits in the Accra area, it was as high as 45-80 percent in some rural areas during the first eight months (Waddington and Enyimayew, 1989).

Low morale among health workers due to their increasingly deplorable conditions of service has been a perpetual problem. Nationwide strike actions by physicians and nurses in public health institutions have become a recurring reality of life. For example, both doctors and nurses walked away from their jobs in 1999 in support of their demand for higher wages and improved conditions of service. This led to increased caseloads at quasi-government hospitals such as the Military and Police Hospitals, where nurses were working twelve hours or more a day. The problem of morale among health workers is so widespread that the Ghana government's Comprehensive Development Framework document released in 1999 observed thus:

> Low staff morale, especially within the middle level providers is also seen as a key cause of the low service quality in the [health] sector. This maybe in direct response to the service conditions which to a lot of health workers are far below that of other sectors. Another contrib-

uting factor is the limited number of skilled staff that tends to create a high workload for most health workers. Supervision, monitoring and regulation of staff and service delivery, is also inadequate (Ministry of Health 1999, p. 5).

The report also noted the poor maintenance of public health care facilities citing leaking roofs, cracked walls, faulty plumbing and sewer systems and obsolete health equipment with no spare parts or preventive maintenance programs. Other problems included shortage of drugs and medical supplies. A recent news report noted "illegal" fees being charged by nurses at the Korle Bu Teaching hospital due to the shortage of such critical inputs as gloves (Ghana Review, March 1, 2000). Mothers had to pay extra fees after delivery to get their babies bathed.

Moreover, privatized health care–a deliberate goal of structural adjustment –received a new boost, providing health care for the affluent, particularly urban, populations. The combined contraction of government provided biomedical services and the increase in costly private biomedical services drove many people away from the formal biomedical system. For example, while 66 percent of ill persons in the rural savanna regions of Ghana consulted practitioners other than doctors in 1987-88, in 1991-92 it was 74 percent (Ghana Statistical Service, 1989; 1992).

Furthermore, poverty appears to have increased considerably throughout the country. In 1997, more than 6.5 million (about 30 percent) Ghanaians lived in absolute poverty, and the poverty of 1.8 million people (about 10 percent of the total population) was described as hard-core (World Bank, 1997). The proportion of the population classified as poor increased from 43 percent in 1981 to 54 percent in 1986 and 55 percent in 1997. While the proportion of urban poor increased from 30-35 percent in the 1970s to 50 percent in the mid-1980s, the rural rates increased from 37 percent in the late 1970s to 54-60 percent in the mid-1980s (Green, 1987). Increasing poverty in the midst of cost recovery in health and social services simply means restricted access to biomedical health services.

With no reductions in endemic diseases such as malaria and increases in poverty-related diseases stemming from unbalanced or insufficient diet (Turshen, 1999), Ghana's recent economic difficulties have produced an unceasing and ever-increasing demand for inexpensive and accessible health care (Oppong and Hodgson, 1998). Because of the marginalization of traditional medical practitioners (see Oppong and Williamson, 1996 for a more thorough discusssion) and the inadequacy of preferred biomedical resources, self-treatment became the leading source of health care for most residents particularly in the rural areas. The 1988 Primary Health Care Review (Adjei *et al.*, 1988) reported that 43.4 percent of rural mothers used self medication compared with 27.2 percent for urban. For most practical purposes, in Ghana, self-medication means obtaining some medicine (usually on the advice of a vendor or a "knowledgeable" relative) and applying it to the self-diagnosed ailment. In rural areas, these medicines are usually obtained from itinerant drug vendors. Consequently, a proliferation of new groups of providers (variously termed drug peddlers, fringe practitioners, bus-stop dispensers) has emerged to fill the void created by reduced access to biomedical care. One of these new groups of providers, Itinerant Drug Vendors (IDVs), comprise a broad group of mobile pharmacists and itinerant health providers who dispense both modern and indigenous pharmaceutical products, primarily on a cash basis in open markets, homes, villages and towns (Oppong and Williamson, 1996; Twumasi, 1988). Many of them are considered illegal and sometimes viewed as quacks by government health officials. Yet, they are frequently the only source of health care for those who do not use traditional medicine but are unable to afford biomedical health care (private or public). IDVs are a primary source of substances used for self-medication because they not only recommend medications, but usually dispense them right on the spot as needed: no prescription forms, no waiting, no traveling. Many sell traditional herbal preparations which they manufacture themselves. Some only sell biomedical pharmaceutical products obtained from pharmacies and drug store wholesalers, while others sell a mixture of biomedical pharmaceuticals and herbal medications. They provide free medical counseling and when a patient is unable to afford the full dosage of the required medication, IDVs readily sell partial dosages. They are affordable and readily accessible. In short, as Oppong and Williamson (1996) argue, IDVs have evolved from the changing context of health care in Ghana. The political economic context, physical and cultural environment, and prevailing diseases have created the conditions for the emergence and popularity of IDVs.

16:4 Structural Adjustment Programs and Increasing Disease Burden: The Case of Buruli Ulcer

Buruli Ulcer disease, caused by *Mycobacterium ulcerans*, has been identified worldwide, predominantly in tropical climates, and in several West African countries, it is considered endemic (Van der Werf, *et al.*, 1999; Asiedu *et al.*, 1998). According to the WHO, BU is rapidly becoming the third most prevalent mycobacterial disease, with an impact soon to surpass leprosy (WHO, 1998). Cote d'Ivoire experienced a three-fold increase in reported cases between 1987-1991, and a decade-long retrospective study documented dramatic increases affecting up to 22 percent of the population in some communities (Marston *et al.*, 1995). While these data are alarming, they most likely underestimate BU occurrence because there are no reliable tools for the surveillance and diagnosis of the disease other than clinical signs and symptoms (Dobos *et al.*, 1999). Moreover, the disease typically occurs in remote areas where residents have little contact with the health care system.

Buruli Ulcer produces chronic skin lesions of a necrotising cutaneous and subcutaneous nature that cause severe debilitation to the patient. Healing usually takes 4 to 6 months and involves extensive scar formation that frequently results in deformity, particularly in children, and stunted limb growth. In one study, 26 percent of patients were left with functional disability of a limb (Marston *et al.*, 1995). No cure for the disease currently exists, but death from BU is rare. Affected children frequently become school dropouts while adults suffer productivity losses with incalculably harsh socioeconomic consequences.

16:4.1 Buruli Ulcer Disease in Ghana

In Ghana, the first probable case was identified by Bayley in rural Accra in 1971 (Asiedu and Etuaful, 1998). Since then, BU cases have been identified in all ten administrative regions of Ghana. Van der Werf *et al.* (1989) described 96 cases of Buruli ulcer in the Asante Akim North district of Ashanti region. The disease burden in Ghana is not known but when systematic reporting was instituted in 1993, 1,300 cases were reported. Following this, the Ghana Ministry of Health organized a nationwide case search to establish the actual disease burden in mid-1999. Initial reports suggest a much higher than expected and rapidly increasing occurrence of the disease. The Western Region which had not reported any cases previously, for example, reported 265 cases after the national case search. The highest number of cases was from Ahanta West district (Dixcove) 70, followed closely by Wassa Amenfi (Asankragwa) with 55. The Ministry of Health team concluded that Asankragwa area was an endemic region and should be targeted for a detailed study to discover the true epidemiology of the disease. The team observed that patients living at Asankran-Bremang and Dunkwa did not have access to a health facility. Their inability to walk to the Asankrangwa Hospital (about 10-14 miles away) or pay the one thousand cedis fee (then about $0.30 US) for daily wound dressing kept them away. In short, the cash and carry system instituted directly under structural adjustment not only deprives the sick of much needed care, but makes it difficult to get a true estimate the burden of the disease. Without adequate knowledge of disease burden, effective intervention is severely constrained or rendered impossible. Cutbacks to public health mean limited resources for dealing with Buruli Ulcer disease.

The Cash and Carry system has also affected the quality of care provide. In June 1999, during a summer field study with medical geography students from the University of North Texas, we interviewed several physicians about the conditions of health care provision in the country. While these interviews were by no means random, they provide insights into the decreased quality of care problems attributable to the Structural Adjustment Program.

16:5 Qualitative Interview Results

A Physician from Tuberculosis Treatment Facility Accra at the Korle Bu Teaching Hospital observed:

> It is not that we physicians here do not know the proper regimens or procedures for treating our patients. We do, but are extremely limited by the poverty of the patients we treat. Let me give you an example. After examining a very sick patient, I requested that he should go and get a chest x-ray. The patient told me: "Doctor, I have no money to pay for that x-ray. Just give me some medicine, any medicine to help me. Please don't let me die". How do I provide effective treatment? I feel like I'm shooting in the dark, my hands are tied behind me because of this "Cash and Carry" system.

In a district hospital, another physician said:

> In the absence of appropriate laboratory tests due to patient's inability to afford them, I end up prescribing several (usually 2-3) different drugs, shotgun therapy, in the hope that one of the medications may be effective. Several times my patients tell me: "Doctor, I can't afford to buy all these medications. Please select the most important one for me so that I can get that one". In fact, sometimes, they ask me to reduce the dosage for them because they cannot afford the full dosage of even one drug.

Another district medical officer said:

How do I treat the effects of poverty and under-nutrition? I can't feed them three times a day!

As one direct result of the Cash and Carry system, patients have to provide their own supplies even for surgical procedures in public hospitals. Taking gloves, dressing materials, and even saline fluid for surgery is quite common. One doctor recounted a case where in the middle of abdominal surgery it was detected that the patients' supplies had been exhausted. During the procedure fecal matter from the intestines accidentally spilt into the abdominal cavity. The standard procedure under those conditions would be to wash out (lavage) the cavity with sterile normal saline fluid. When the surgeon requested the fluids he was told that the patient had bought an insufficient amount and therefore did not have enough fluids to cover the lavage. The abdominal cavity was subsequently closed after a partial lavage was done—leaving fecal matter behind.

One can only wonder in dismay and maybe anger at how frequently such troubling events occur in the health care system. How many unnecessary deaths occur each year because impoverished patients are unable to afford medical care? How many die because physicians hold back on treatments patients are unable to afford?

These effects of Structural Adjustment Programs are by no means unique to Ghana. Other African countries undergoing adjustment programs are facing similar problems. In Senegal, government spending on health from public funds is around US$ 1 per capita, and at least 40 percent of spending on health comes out of family budgets. In the 1980s and early 1990s, an economic downturn and a devaluation of the currency under Structural Adjustment Programs led to a collapse of much of the government-run infrastructure in health. The number of medical consultations fell by eight percent between 1978 and 1986, even though the population grew by 28 percent over the same period.

16:6 Conclusion

Ghana's structural adjustment program is producing irreversible changes in the health care delivery system. Government funded health care is systematically being replaced with privatized health care. Cutbacks on government expenditure on health care and other changes in the economy such as devaluation, increased poverty, and the Cash and Carry system have produced reduced use of health care facilities in the midst of increased need for health care services. As a result, the health status of many people is being compromised in the interest of a hypothetical future economic gain. There appears to be a general lack of a comprehensive properly integrated program for health care delivery and mainly ad hoc programs to deal with unexpected side effects of Structural Adjustment Programs. In fact the only one consistent thing about the health care policy implemented so far has been privatization at all cost. Unfortunately, because privatization favors and targets those who can pay, the large majority of Ghanaians, particularly rural residents and the poor, have no place to obtain care in privatized system. They will have to live with the remains of a demolished public health system since private health care concentrates in urban centers and is profit-maximizing. It is time to recognize that privatization does not address the problem of inaccessibility due to poverty, and privatized health care does not seek the well-being of the poor. It caters to those who can pay. Who speaks for the newly emerging hard-core poor and urban poor? Can the majority of Ghanaians count on their elected government to speak for them?

References

Adjei, S., G. Owusu, J. Adamafio and S. Akyeh (1988). *Primary Health Care Review, Ghana - 1988.* Accra, Ghana: Ministry of Health.

Anyinam, C. (1989). "The Social Cost of the IMF's Adjustment Programs for Poverty: The Case of Health Care in Ghana". *International Journal of Health Services*, 19: 531-47.

Anyinam, C. (1994). Spatial Implications of Structural Adjustment programs in Ghana. TESG 85(5):446-450.

Aseidu, K., Etuaful, S. (1998) "Socioeconomic Implications of Buruli Ulcer in Ghana: A Three-Year Review". *American Journal of Tropic Medical Hygiene*, 59:1015-1022.

Asiedu, K., Raviglione, M., Scherpbier, R., eds. (1998). Geneva, Switzerland. Global Tuberculosis Programme, World Health Organization.

Brydon, L. and K. Legge (1996). *Adjusting Society: The World Bank, IMF and Ghana.* London: Tauris Academic Studies.

Dobos, K., Quinn, F.D., Ashford, D.A., Horshburg, C.R., King, C.H. (1999). "Emergence of a Unique Group of Necrotizing Mycobacterial Diseases". *Emerging Infectious Diseases*, 5(3). <http://www.cdc.gov/ncidod/eid/vol5no3/dobos.htm>

Ghana Ministry of Health (1995). "Ghana Health Sector Support Program". Proposal prepared for the World Bank by the Ghana Ministry of Health. Accra: Ministry of Health.

Ghana Ministry of Health (1999). "Comprehensive Development Framework for Health. Accra: Ministry of Health". <http://www.ghanacdf.org.gh/>

Ghana Statistical Service (1989). *Ghana Living Standard Survey*. Accra: GSS.

Ghana Statistical Service (1992). *Ghana Living Standard Survey 1991*. Accra: GSS.

Gibbon, P. (1996). "Structural Adjustment and Structural Change in Sub-Saharan Africa: Some Provisional Conclusions" in Gibbon, P. and A. O. Olukoshi (1996). *Structural Adjustment and Socio-Economic Change in Sub-Saharan Africa*. Uppsala: Nordiska Afrikainstitutet 1-47.

Green, R. H. (1987). "Stabilization and Adjustment Programs". *Country Study 1: Ghana*. Anankatu, Finland: WDER.

International Monetary Fund (2000). "Ghana: Selected Issues Series: IMF Staff Country Report No. 00/02". <http://www.imf.org/external/pubs/ft/scr/2000/cr0002.pdf>

Logan, B. I. and Mengisteab, K. (1993). "IMF-World Bank Adjustment and Structural Transformation in Sub-Saharan Africa". *Economic Geography*, 69:1-24.

Lugalla, J. L. P. (1995). "Economic Reforms and Health Conditions of the Urban Poor in Tanzania". Paper presented at the American Anthropological Association Annual Conference, November 15-19, 1995, Washington, DC.

Marston B.J., Diallo, M.O., Horsburgh, C.R., Jr, Diomande, I., Saki, M.Z., Kanga, J.M., *et al.* (1995). "Emergence of Buruli Ulcer Disease in the Daloa Region of Côte d'Ivoire". *American Journal of Tropical Medical Hygiene*, 52:219-24.

Minkin, S. F. (1991). "Iatronic AIDS: Unsafe Medical Practices and the HIV Epidemic". *Social Science and Medicine*, 33 (7): 786-790.

Olukoshi, A. O. (1996). "Extending the Frontiers of Structural Adjustment Research in Africa: Some Notes on the Objectives of Phase II of NAI Research Program" in Gibbon, P. and A. O. Olukoshi. *Structural Adjustment and Socio-Economic Change in Sub-Saharan Africa*. Uppsala: Nordiska Afrikainstitutet, 49-86.

Oppong, J. R. (1997). "Medical Geography of Sub-Saharan Africa" in Samuel Aryeetey-Attoh (ed.) *Geography of Sub-Saharan Africa*. Upper Saddle River, NJ: Prentice Hall, 147-181.

Oppong, J. R. and M. J. Hodgson (1998). "An Interaction-based Location-allocation Model for Health Facilities to Limit the Spread of HIV-AIDS in West Africa". *Applied Geographic Studies* 2(1):29-41.

Oppong, J. R. and J. Toledo. (1995). "The Human Factor and the Quality of Health Care". *Review of Human Factor Studies*, 1(1):80-90.

Oppong, J. R. and D. A. Williamson (1996). "Health Care Between the Cracks: Itinerant Drug Vendors and HIV-AIDS in West Africa". *African Rural and Urban Studies* 3(2):13-34.

Ould-Mey, M. (1996). *Global Restructuring and Peripheral States: The Carrot and the Stick in Mauritania.* London: Littlefield Adams Books.

Owusu, J. H. (1998). "Current Convenience, Desperate Deforestation: Ghana's Adjustment Program and the Forestry Sector". *Professional Geographer*, 50(4): 418-436.

Owusu, J. H. (1998). "The Myth of Development Under Ghana's Structural Adjustment Program: Evidence from the Formal Wood Processing Sector" in S.B., S.K. Adjibolosoo and B. Ofori-Amoah (eds.), *Addressing Misconceptions About Africa's Economic Development: Seeing Beyond the Veil.* Lewiston, New York: The Edwin Mellen Press, 193-216.

Sahn, D. E., P. A. Dorosh and S. D. Younger (1997). *Structural Adjustment Reconsidered: Economic Policy and Poverty in Africa.* New York: Cambridge University Press.

Stock, R. (1995). *Africa South of the Sahara: A Geographic Interpretation.* New York: Guilford Press.

Turshen, M. (1999). *Privatizing Health Services in Africa.* New Brunswick, NJ: Rutgers University Press.

Twumasi, P. A. (1988). *Social Foundations of the Interplay Between Traditional and Modern Medical Systems.* Accra: Ghana Universities Press.

UNAIDS and WHO (1999). *AIDS Epidemic Update: December 1999.*

van der Werf, T., van der Graaf, W., Tappero, J., Asiedu, K. (1999). Seminar: *Mycobacterium Ulcerans Infection.* Lancet 354:1013-18.

Vogel, R. J. (1988). *Cost recovery in the health sector: Selected country studies in West Africa.* Washington, DC.: World Bank. Technical Paper No. 82.

Vogel, R. J. (1994). *Financing Health Care in Sub-Saharan Africa.* Westport, CT: Greenwood Press.

Waddington, C. J. and K. A. Enyimayew (1989). "A Price to Pay: The Impact of User Charges in Ashanti-Akim District, Ghana". *International Journal of Health Planning and Management*, 4:17-47.

World Bank (1995). *Country Briefs.* Washington DC: World Bank.

World Bank (1995) *World Development Report.* Washington DC: World Bank.

World Bank and the Government of Ghana (1999). *Comprehensive Development Framework for Ghana to Achieve Vision 2020.*

World Health Organization (1995). *World Health Report.* Geneva: WHO.

17 Adjustment Reforms in a Poor Business Environment: Explaining Why Poor Institutions Persist under Ghana's Reforms

Nicholas Amponsah

> We [Ghanaians] need two legs to walk: a strong and effective state and a strong private sector; we have neither and are not likely to have either anytime soon. We are like a cripple...with no legs, pushing himself around on a crude board with wheels, surviving only by begging and trying to look sympathetic to the potential alms giver.[1]

17:1 Introduction

One of the serious drawbacks of Ghana's remarkable structural adjustment reforms since 1983 is the lukewarm participation of the private sector, especially domestic Ghanaian private business in this program of shared-growth (Amponsah, 2000; Leechor, 1994; Herbst, 1993; Tangri, 1992). A major reason why domestic private business in Ghana has not responded favorably is institutional uncertainty (Amponsah, 1999; 2000; Herbst, 1993). Yet the reform regime has demonstrated its commitment to implementing these market reforms, which it fervently believes is necessary for Ghana's sustained economic turnaround. Leadership commitment in itself cannot be transformed into deeds, for there are fundamental prerequisites that make the market system work in all capitalist societies, most notably a credible institutional environment. Why does a poor business environment persist even under seventeen years of leadership commitment? Why has a regime so committed to liberalizing the economic environment to pave the way for private sector participation not found it expedient to craft the requisite rules of the game that enable the latter to actively participate?

This chapter utilizes survey data[2] on domestic private entrepreneurs and the political and bureaucratic elite to explain why institutional problems have persisted under Ghana's reforms said to be a shining example of successful market liberalization in sub-Saharan Africa (The World Bank, 1994). The study demonstrates that a wide gulf exists between domestic private entrepreneurs and the political and bureaucratic elite on the problems of institutional decay and the need for reforms. While the former sees existing institutions as dysfunctional and oppressive, the latter do not believe so or have taken them for granted. Accordingly, the reform leadership in Ghana, like many African countries, is yet to recognize the imperative of institutional reform as a necessary step for securing positive private sector response.

As the most successful case of adjustment reforms in Africa that yet faces the challenge of a poor institutional environment, which impedes private sector response, Ghana is an excellent place to explore the reasons why poor institutions persist even under market reforms. The following section discusses the paradox of leadership commitment to market reforms under dysfunctional political and economic institutions. It review's the literature on the fractured relationship between Ghana's reform regime and its domestic private entrepreneurs. The third section makes a comparative analysis of regime-society synergy by contrasting Ghana's case with East Asia. Section four examines the perspective of domestic private business in Ghana on the problems of institutional uncertainty with the view to comparing their perceptions with that of the political and bureaucratic elite in Ghana in section five. Section six draws conclusions.

The case of Ghana is quite paradoxical because of the persistence of its reforms within an economic environment in which poor institutions persist. This has been so because the reform regime thinks that market liberalization simply means fiscal restraint, divestiture of public firms and deregulation. These can take place without inducing private sector response if the business environment is uncertain. Amponsah (2000) notes that the institutional climate within which domestic entrepreneurs in Ghana operate is, indeed, a factor that accounts for their negative response to the incentives that market reform is purported to have brought. Private entrepreneurs in Ghana continuously express several concerns about the Ghanaian regulatory environment that constitutes obstacles to secure investment. They note that in their business transactions with officials not only are the rules and procedural regulations ambiguous, they are also not likely to be observed or enforced. They lament at the inadequacy of their rights to business and property. Why have these concerns persisted under sixteen years of leadership commitment to reforms? Does the reform leadership share these concerns? Only when the reform leadership shares these concerns of domestic private entrepreneurs in Ghana will there be concerted efforts to craft the requisite institutional foundations that facilitate the positive response of the latter in this process of shared-growth.

17:2 Leadership Commitment: Necessary, but Not Sufficient

The importance of the commitment of the elite of a reforming regime to the reform paradigm cannot be overemphasized. The elite, comprising important political decision makers, as well as top bureaucrats must not only be committed to the reform, they need to have a deeper understanding of the new logic of development processes implied in the market system. It means that the elite must have an unfettered recognition of the critical role private economic actors should play for the success of market reforms. There is yet another important prerequisite for market reforms to succeed. The reform elite must have a tacit understanding and recognition that the institution of private property, contractual rights and fairly enforceable legal rules and regulations are critical for the workings of the market system, and they have to show their preparedness to ensure that these prevail. Above all, they must recognize that they need the cooperation of the private business elite for their success under market reforms, while the latter also need them in order to be able to sprout. In other words, they need to know that to achieve economic success, the state needs private business, private business needs the state, and both need the market.[3] Therefore the goals and aspirations of the state elite and private business entrepreneurs must be congruent. Where the goals of these two critical actors are at polar opposites the objective of achieving success through reforms also becomes an uphill battle. Unfortunately, this has been the case in the relationship, perspectives and objectives of the political and bureaucratic elite, and domestic private entrepreneurs during Ghana's sixteen years of consistent structural adjustment reforms.

Three decades ago Esseks (1971) identified the fractured relationship between the Ghanaian State elite and the indigenous business. Today, the polarized relationship between the political elite and domestic private entrepreneurs persists (Tangri, 1992; Leechor, 1994; Hutchful,1995). According to Hutchful

(1995: 303), Ghana's "reform team had few structural, social, or political links with the private sector". He notes that several attempts to institute institutional reforms that will bring some flexibility into the system failed because of the lack on interest on the part of the reform elite. This attitude of the elite in Ghana and other African countries contrasts with that of the elite in the Highly Performing Asian Economies (HPAEs), who were able to engender the cooperation of their domestic business. It is worthwhile examining the role of the elite in the HPAEs in the process of implementing developmental programs of shared-growth.

17:3 East Asian and African Elite in Developmental Roles: A Comparative Assessment

Recent studies on the Asian miracle demonstrates how the political elite there took serious steps to woo the business sector through cooperation and the creation of adequate and effective institutional framework (Campos and Root, 1996). The correspondent situation in Ghana manifests a wide gulf between the private sector and the political elite (Hutchful, 1995; Tangri, 1992). Unless the reform elite in Ghana, and other African countries recognizes the critical role of their domestic private sector and erect appropriate and credible institutions that support private economic activity, the goal of achieving significant development through market liberalization will be a daunting task.

The role of the political elite in nurturing domestic private entrepreneurship is particularly critical for late developing nations. What Evans (1992) finds particularly remarkable about the fast growing East Asian economies is an elite who shares a concerted commitment to, and, support for, private business. He notes that individuals who were convinced of the value of local industrial production found themselves in positions of leverage in the state apparatus. The elite did not take positions in government for their own sake. They used their ideas to develop policies and institutions that enabled private business to become groomed and well established (Campos and Root, 1996). To be able to fulfill this task the elite must not only have a vision for reforms, it also needs a conviction that a credible institutional environment is a fundamental prerequisite for private business' success. Accordingly, critical members of the executive and the bureaucracy in the HPAEs played important advocacy roles for private business with which they had shared aspirations. In particular, they crafted institutions necessary for "reducing the risk and uncertainty entailed in entering a new sector or a new kind of endeavor" (Evans, 1995:80).

The above characterization of the role of the state elite in the HPAEs seems to be in contrast to what occurs in the Poorly Performing African Economies (PPAEs), as the case of Ghana clearly indicates. In 1989, after some six years of sustained liberalization reforms in Ghana, private industrialists still felt alienated by the state elite. They expressed their main concerns, and modest ones at that, to the reform regime. They urged:

> the need for officials at the highest echelons of government machinery to acknowledge the efforts of the private sector through their pronouncements and actions and to give due recognition of the role being played by the private sector in Ghana's economic recovery effort...[Private entrepreneurs] observed with great concern that the harassment of investors by organs of the revolution/security agencies have not changed (Tangri, 1992:99).

These sentiments, expressed nearly a decade ago, seem to echo the findings of the survey for this study. It is thus worthwhile to revisit this issue to see the extent to which these private sector concerns have been taken serious and effectively addressed by the reform elite. These concerns, reflecting government's contempt for private entrepreneurship and consequent harassment of private business people, has much affinity with the problems of institutional uncertainty in Ghana.[4] In fact, according to Tangri, representatives of private business in Ghana have made it clear that illegal harassment of business by government continued to "create uncertainty regarding the security of investment and property rights" (Tangri, 1992: 99).

Amponsah (1999: Chapter 5) investigated the institutional climate in which Ghanaian private entrepreneurs operated with the goal of gauging why they exhibited a lukewarm response to reforms. The study demonstrated that because private entrepreneurs are not certain about the sanctity of their property and business, they are not inclined to plunge their resources into investment. If private entrepreneurs share negative views about the institutional environment within which they operate how do the political and bureaucratic elite in Ghana perceive the same phenomenon? The analyses that follow contrasts the perspectives of these two critical actors, i.e., private business on the one hand, and the state elite on the other, on what critical factors there are that need to be resolved for Ghana to ensure a measure of significant investments and improved economic performance. First, I test to see how important private entrepreneurs consider the institutional environment in their investment decisions. Following that I also test the expectations of the political and bureaucratic elite on these same issues with the view to verifying whether these two critical actors shared analogous perspectives on what needs to be done for development goals to be attained.

17:4 How Private Entrepreneurs in Ghana see the Institutional Environment

Some analysts of African development believe that the failure of Africa to industrialize and develop is due to the inherent weakness of Africans in entrepreneurial acumen (Hart, 1995). This view also shared by the Ghanaian reform elite, attributes economic failure to the general lack of technological and entrepreneurial acumen and ability to innovate (Lall, 1995). According to Lall, if industries in the less developed countries of Africa are not sprouting, as they should, the answer lies with the low levels of technological development and industrial adaptability. Where industrial entrepreneurs lack "the information and skills, technical, organizational and institutional, that allow productive enterprises to utilize equipment and information efficiently", industrial progress would certainly be stalled (Lall, 1995: 261). In fact, the idea that poor technological capacity and entrepreneurial acumen, accounts for the low development of industrial enterprises in the less developed countries of Africa has permeated most studies on African development (Berman and Leys, 1994:8).

Such belief notwithstanding, recent studies make it clear that this deterministic view of entrepreneurial development is erroneous. In a World Bank study of enterprise development in Ghana, Aryeetey *et al.* (1994) found that initial lack of resources such as abundant capital or sophisticated technical facilities did not always matter as much for entrepreneurial growth as we are made to believe. This study is so relevant for our purpose that it seems worthwhile recounting extensively. They note:

> The successful Small and Medium-sized Enterprises (SMEs) typically have built up their capacity gradually by reinvesting profits, sometimes by bank overdrafts (occasionally loans), supplier's credits or foreign assistance. Gradual growth appears to enable the entrepreneur to adapt to the market and master management tasks. For example, a successful engineering firm that now employs about twenty people started as a backyard operation producing school equipment, shifted to simple work benches and equipment for wood processing, then tried making a corn mill, used both technical innovation and strong after-sales service to build a good reputation, obtained a loan to import used equipment, developed a wide range of products, and is

now looking to expand to a permanent structure and specialize in a limited number of items (Aryeetey *et al.*, 1994:111).

Morris (1998:59) succinctly argues that "entrepreneurship is neither innate to certain people and societies nor a random or chance event. Rather, it is determined by the environmental conditions (encompassing the environmental infrastructure of political, economic and legal institutions), operating at a number of levels", in various societies. To what extent do domestic private entrepreneurs and the political and bureaucratic elite considers issues of institutional certainty as the major precondition for sustained economic diversification and development as Morris powerfully argues? To be sure, these institutional imperatives for development would be attained only if the major actors in the process of shared-growth reach a consensus on their critical importance and accord them priority in their programs.

We developed a five-variable questionnaire related to institutional certainty and its impact on investment prospects to explore the views of domestic private entrepreneurs and the political elite on the institutional environment within which Ghana's adjustment reforms are being undertaken. In Table 17.1, the views of domestic private entrepreneurs on four of the variables are summarized. In all, 448 private entrepreneurs and 45 political and bureaucratic elite were interviewed.[5] Some of the private entrepreneurs were randomly selected from a registry of private enterprises countrywide listed by the Ghana Private Enterprises Foundation in Accra. Others were selected purposively after an initial enumeration.

The pervasiveness of problems associated to insecure land tenure in Ghana cannot be overemphasized. Yet all governments since colonial times take them for granted or simply gloss over them. Rimmer (1992:39) sums up the problems of land tenure in Ghana since the colonial times this way:

> Proprietary boundaries–communal, family and individual–were often vague if not quite indeterminate, and purchasers might therefore buy land from sellers whose ownership was questionable; sometimes, indeed, the same land was sold by more than one vendor, and sometimes land was sold precisely because the title to it was uncertain. Litigation over land rights was therefore commonplace, and in the absence of legislation applying the statute of limitations, no case was finally ever settled. The resulting confusion and uncertainty were often deplored.

Question one attempts to elicit the views of private entrepreneurs on the

Table 17.1 The View of Ghanaian Private Entrepreneurs on the Institutional Climate

	Do you agree that....?							
	Q1. existing laws on land ownership are effective enough for secure business operations in Ghana?		Q2. existing property and contract laws are adequate and effective for secure business operations?		Q3. domestic private investors in Ghana are not responding to SAP due to mistrust for the legal environment?		Q4. policy instruments contained in SAP provide adequate incentives for private investment and increased productivity?	
Responses	Frequency	%	Frequency	%	Frequency	%	Frequency	%
Yes	42	10.0	130	30.6	292	65.2	93	21.8
No	347	82.8	223	52.9	140	31.3	274	64.2
Don't know	30	7.2	72	16.9	16	3.6	60	14.1
Total	419	100	425	100	448	100	427	100

extent to which they think a secure land ownership rights exists in Ghana. This is very important as the World Bank duly recognizes the need for land security to serve as an incentive for individuals to improve and develop their lands especially in agriculture (Stein,1994: 1836). Entrepreneurs were asked whether the existing laws on land title and ownership were adequate or effective enough for secure business operations. The answer was a resounding, "no". The overwhelming majority (82.8 percent) said they felt the existing laws on land ownership did not provide adequate guarantees for secure business investment. Only ten percent of private entrepreneurs surveyed thought the existing land security rights were adequate. A major obstacle that has prevented generations of Ghanaians from benefiting from the value in land tenure, including the ability to deploy land as collateral still persists as domestic business proprietors deplore.

The next issue of great importance in assessing the institutional environment is the certainty of an individual's rights to contract and do business securely. Again, as the responses indicate (question 2) the majority of private entrepreneurs in Ghana (52.9 percent) felt that the existing laws on property and contractual rights were not effective enough for secure business operations. Furthermore, we asked private entrepreneurs what in their view is holding back potential local private investors such as themselves from investing even under sustained efforts of the Ghanaian regime to implement an IMF/World Bank adjustment program aimed at giving much prominence to private economic actors (question 3). The question was, do you think private investors such as you are not responding to Ghana's 15 years of Structural Adjustment Programs with increased investments because of deep-seated mistrust for the existing legal and economic institutions? The aim was to gauge the extent to which the perceived insecurity of the institutional environment was an important factor in their investment decisions. The results of this poll showed that the overwhelming majority, 65.2 percent, was positive that the general lukewarm attitude towards investment was, among other things, due to the mistrust entrepreneurs have for the legal institutional situation. In other words, private entrepreneurs in Ghana have little trust in the manner in which the state system specifies what constitutes the rights of individuals to conduct business, or in the way they are protected or enforced.

The fourth question asked domestic entrepreneurs the extent to which they thought the policy instruments contained in the Structural Adjustment Program provide enough incentives, including those related to the legal institutional dimensions, to induce potential private investors to commit their resources to productive investment ventures in Ghana. This question is particularly important in light of the view held in donor circles that, adjustment programs, by reforming previous ruinous statist policies and removing various market distortions, should induce prospective private entrepreneurs to accelerate investment. Gyimah-Boadi (1995:311) is surprised that "despite the steps taken to liberalize the economy and *improve the general climate for business* (emphasis mine) private investment did not rise significantly". Since such a view is popularly held even in academic circles, we considered it necessary to find out from private business themselves how the economic reform program has actually contributed to the improvement of the general institutional climate. We asked respondents, do you think the policy instruments that have been implemented under the Structural Adjustment Program for the past 15 years provide enough incentives to attract private investment?

The results of the survey on this issue contrasted with the neoclassical view held in donor circles that a liberalized economy would lead to an improved economic environment. The majority of Ghanaian private business entrepreneurs surveyed (64.2 percent) felt that the policy instruments of the adjustment reforms, i.e., trade liberalization, currency rationalization or devaluation, divestiture of state economic enterprises and privatization, in themselves do not provide the requisite incentives necessary to induce increased investment from private sector actors. If the private sector actors shared these views, what does the political and bureaucratic elite supposed to shape policies for the former, think about these same critical matters. The next sections examine the perspectives of the political and bureaucratic elite on these same issues.

17:5 How the Political and Bureaucratic Elite in Ghana see the Institutional Environment

It is clear that domestic private business considers the institutional environment as vital but that this is unfavorable or insecure in Ghana. How do the other critical actors on the other side of the spectrum, the political and bureaucratic elite, perceive these same issues? The perspective of the political and bureaucratic elite–the reform regime–on these institutional issues is very critical because it will determine the direction of their actions and the effectiveness of their outcomes. In fact, the process of industrial transformation for late developing nations, especially under market reforms, is essentially how to make the market work to promote national entrepreneurship. Therefore, the need for the reform elite to have a deeper understanding of the basic concerns of local capitalist entrepreneurs becomes more critical.

If private business entrepreneurs and the political and bureaucratic elite share contrasting views on the conditions necessary for reforms to succeed, the task of ensuring improved economic conditions through these reforms will most likely be a daunting one. It is important that there be some amount of congruence in the views of these two critical actors on the problems that exist and what needs to be done. Hutchful (1995:303) notes the importance of bridging the gap between private business and state elite in the process of adjustment reforms. He rightly observed that the reform leadership in Ghana "had few structural, social, or political links with the private sector", and this accounts for the limited success of reforms. It is in light of this that we surveyed the political and bureaucratic elite on how they viewed the institutional situation in Ghana and the importance they accord it.

The political and bureaucratic elite surveyed included certain leading politicians, some of whom are representatives of local or national legislative assemblies, i.e., district assembly representatives and parliamentarians. We also interviewed some directors of certain key ministries involved in policy analysis, formulation and implementation, such as Trade and Industry and Mobilization. Others interviewed included some of those in the top echelons of certain parastatal bodies responsible for private business promotion, such as the Private Enterprises Foundation and the Association of Ghanaian Industries. A few members of the top echelons of the political parties were also interviewed. The responses from the political and bureaucratic elite are summarized in Table 17.2.

How did the HPAEs ensure the sprouting of domestic entrepreneurship and ultimate industrial transformation? Lee (1997) provides a clue through an empirical analysis of the *"maturation and growth of infant firms"* in South Korea. The political leadership in South Korea, he argues, was not only committed to the cause of private entrepreneurs; it backed this commitment with concrete actions in support of domestic private firms. He cites as an example the establishment of an institutionalized procedure for private industrialists to receive government support. Regime leaders in South Korea laid down the "rules of the game" for receiving government support clearly. Their institutionalized scheme of support for private business worked this way. Domestic firms would be protected from unhealthy foreign competition and be subsidized, but under specific rules of agreement. Internally, these domestic firms would have to accept competition, and their performance standards in terms of increased export output was the sole yardstick for continued protection and subsidy. Lee writes:

> Exports was a "focal point" of government-business relations, *with players on both* sides knowing that export performance would be used as a principle for adjudicating *unforeseen consequences* (Lee, 1997: 1278).

Table 17.2 The View of the Political and Bureaucratic Elite in Ghana on the Institutional Climate

	Do you agree that....?							
	Q1. existing laws on land ownership are effective for secure business operations?		Q2. existing property and contract laws are adequate and effective for secure business operations?		Q3. domestic private investors in Ghana are not responding to SAP due to mistrust for the legal environment?		Q4. policy instruments contained in SAP provide adequate incentives for private investment and increased productivity?	
Responses	Frequency	%	Frequency	%	Frequency	%	Frequency	%
Yes	22	48.9	24	53.3	13	28.9	37	82.2
No	17	37.8	20	44.4	31	68.9	5	11.1
Don't know	6	13.3	1	2.2	1	2.2	3	6.7
Total	45	100	45	99.9	45	100	45	100

In their desire and eagerness to stimulate the growth of domestic entrepreneurship regime leaders in Korea as elsewhere in the HPAEs employed institutionalized principles and not whimsical rent-seeking techniques (*important lesson number one*). The elite of the HPAEs also understood the concerns of private industrialists while the latter also knew what the state required of them (*important lesson number two*). In the process of shared-growth, therefore, the importance of congruence in ideas between the elite and private entrepreneurs over the prerequisites necessary for economic productivity cannot be overemphasized.

Does the political elite in Ghana share similar understandings of the problems arising from the institutional climate, or do they have different perspectives to those of private business entrepreneurs? The responses to these questions by the political elite have serious consequences for any future task of working towards the improvement of the institutional climate. Indeed, it has been accurately observed that in most developing countries, in the course of pursuing any development objectives, conflicts often occur that fundamentally impede the entire process thus frustrating the set goals. Davis (1970: 53) notes that this situation occurs "precisely as a consequence of an incongruence in the aspirations, attitudes, motivations and corresponding behavior of each group comprising the social structure".

As the responses from Ghanaian private business indicated land ownership in Ghana as in several parts of Africa is shrouded in confusion and complication. Why has this problem persisted under various regimes since colonial times? Rimmer (1992: 39) notes "the Commissioner of Lands in the 1950s took the view that litigation on land in native courts was "not excessive" relative to litigation on other matters", and hence it did not consider it as significant enough to warrant policy reforms. This view is still held both in official circles and the academia. According to Ninsin (1989:165) land ownership in Ghana is "essentially usufructuary and NOT proprietary; for it terminates with the extinction of the original owner's family ... the usufructuary right in the lands of the communal group is restricted to its members: it is not transferable to non-members of the group...In other words, the right of the member of the communal group in the group's lands *does not confer on him freehold title*". This kind of land ownership obviously contains a great deal of insecurity in terms of title ownership that might pose a threat to potential investors. Among the land question problems Ninsin, records situations involving:

> peasants being chased away from the lands they have been cultivating for decades; payment of exorbitant ground rents to landlords;

ejections, brutal assaults, illegal arrests and other forms of intimidation by landlords; bloody confrontations between contending parties (Ninsin 1989: 179).

In 1986 the reform regime in Ghana responded to the issue of land title security with a law, the *Land Title Registration Law 1986 PNDCL 152*. In a memorandum to the law the government recognized that:

> Systematic land tenure research in Ghana has revealed radical weaknesses in the present system of registration of instruments affecting land under the Land Registry Act 1962 (Act 122). The chief among them is litigation, the common sources of which are the absence of documentary proof that a man in occupation of land has certain rights in respect of it; the absence of maps, and plans of scientific accuracy to enable identification of parcels and ascertainment of boundaries; and lack of prescribed forms to be followed in case of dealings affecting land or interests in land (Ninsin 1989: 178).

In question one we asked the political and bureaucratic elite whether they agreed that existing laws on land ownership in Ghana are adequate for secure business operations? Earnestly, a slight majority (48.9 percent) felt that the existing laws on land ownership were not adequate enough for secure business operations. Only 37.8 percent thought the existing laws on land were effective, while the rest (13.3) had no idea. And, in fact, one official was quick to point out the recent ethnic conflict between the Kokumba and Nanumba of Northern Ghana over farming land dispute in which many lives were lost. Nonetheless, many of them commented that they do not think that the ineffective laws on land in themselves have any significant impact on business operation, especially, manufacturing entities. As one local assembly representative in Sunyani intimated, in her view land problems may exist in relation to rural subsistent farming issues, but certainly not a national problem since not much farming is being commercialized in rural Ghana yet. Besides they believed the government's legislation, Land Title Registration, is capable of resolving any inadequacies.

On the other hand, most of the political elite surveyed believed that existing laws, rules and procedural regulations on business property and contractual obligations are adequate and effective enough for secure business operations in Ghana (question 2). On this issue 53.3 percent of the political and bureaucratic elite in Ghana thought that those property rights and contractual obligations existing in Ghana are adequate and effective. This is in sharp contrast with the view expressed earlier by business entrepreneurs.

Several studies conclude that the regime leaders of Ghana, unlike their counterparts in the HPAEs, have no confidence in indigenous Ghanaian entrepreneurs as a force that can bring about industrial transformation.[6] The political elite often demonstrates contempt and suspicion if not outright hostility towards private business. Because the elite is not enthusiastically disposed towards private business it is not inclined to put in place institutional mechanisms that will augment private business ventures. Question three asked the political and bureaucratic elite whether they thought Ghanaian private investors are not responding to the Structural Adjustment Program with increased investments because of deep-seated mistrust for the existing legal and economic institutions? The results showed that the political elite shares a different perspective on this issue. Unlike private entrepreneurs, the majority of the political elite (68.9 percent) indicated that they do not think the lukewarm response of Ghanaian private investors to reforms is due to mistrust for the institutional environment, which they believe, is secure.

In the view of the elite the legal and institutional climate should not constitute any obstacle to private investors in Ghana. An official at the Private Enterprises Foundation for instance suggested that though there may be several obstacles, they were all dwarfed by the macroeconomic failures such as the unsustainable inflation and the continued depreciation of the Ghanaian currency. Another official at the Trade and Industry Ministry alluded to this same issue as the most probable reason for low private investment in Ghana. The elite believes that the low response of Ghanaian private business entrepreneurs to market reforms is rather due to their innate weakness. This view was expressed by an important political figure, the Minister for Trade and Industry in Ghana. As she put it the inherent "structural weakness" of Ghanaian entrepreneurs is the reason why they "have not been able to take advantage of the liberalized trade regime and compete with imports".[7] Thus, the idea that indigenous private business is inherently incapacitated and an unlikely candidate to contribute effectively towards meaningful national development seem to be an entrenched official thinking. This belief is more strongly held because the elite are convinced that the Structural Adjustment Program has created an incentive framework, which should induce capable entrepreneurs to increase their productive activity as responses to question four indicate.

As noted earlier, Ghanaian private entrepreneurs have misgivings on the incentives that Structural Adjustment Programs are supposed to bring. What does the elite also think about the potency of the policy instruments of the Structural Adjustment Program in inducing increased economic activity. This issue is important because the foundation of the adjustment program rests on the assumption that the application of the market friendly macro-economic policy instruments would provide incentives that would unleash markets and stimulate private investors to commit resources to investment. In question four we asked the political and bureaucratic elite whether they thought the policy instruments contained in the Structural Adjustment Program provide adequate incentives and guarantees that should entice prospective domestic private investors to invest in Ghana? As the responses indicated most elite (82.2 percent) affirmed the adequacy of the incentive structure entailed in the adjustment reform package. Only 11.1 percent of them thought the Structural Adjustment Program did not provide enough incentives for private investors. Clearly, there is no consensus between the critical actors in the reform process. And this lack of consensus plays a part in the apparent rift between Ghanaian private business and the reform regime. "There remains", ... writes Gyimah-Boadi (1996:85), "the problem of promoting domestic private investment in the face of poor relations between the Rawlings regime and key elements of the local capitalist class". With such divergent perspectives between the two critical actors in the process of shared growth, the prospects of meaningful development appear a daunting task. Not surprisingly, the two critical actors also share different perspectives on the prospects of investment.

17:6 Prospects of Investment in Ghana: The Views of Private Business and Political Elite

The divergent perspectives of private entrepreneurs and the political elite on the institutional climate are also reflected in their expectations on investment. The final survey sought to elicit the expectations of these two critical actors on the prospects of investment in Ghana under the Structural Adjustment Program. This is important because the nature of their expectations on domestic private investment would determine how fervently they strive to find solutions to whatever problems that impedes it. Question five asked, under the present political and economic regulatory environment, do you think it expedient for an entrepreneur to plow back a significant part of accrued profits as re-investment to expand the operations of the industry (in the event the one's business was greatly successful)? In Figure 17.1 the views of the political elite and private entrepreneurs are compared.

In fact, in 1994 the *Courier* asked Ghana's Minister of Trade and Industry, Mrs. Emma Mitchell a similar question. The *Courier* asked, "What difficulties do you face in trying to stimulate industrial development in your country (Ghana)?" In reply she noted that the goal of her ministry was to stimulate the emergence of a more internationally competitive industrial sector, especially one that utilizes domestic resources with capacity to export (The Courier, 1994: 26-29).

The Minister rightly identified a number of macroeconomic constraints. Among the primary obstacles that militated against the attainment of the goal of industrial development according to her was limited credit for private business, which makes venture capital hard to come by. Another problem, she noted was the continued depreciation of the Ghanaian currency the cedi, which has eroded the capital base of many industrialists. She also reiterated the exposure of the weak Ghanaian businesses to unrestrained competition as the result of trade liberalization. What can we say is the official position on the issue of institutional climate? On the issue of institutional concerns she said:

> Other problems relate to the general institutional and legal constraints
> that affect production, particularly for exports. But the government
> has recognized these problems and has taken firm measures, notably
> in the most recent budget, to address them (The Courier, 1994).

As is obvious, reference to the institutional constraints in official circles

Figure 17.1 Expectations on Private Investment in Ghana under Structural Adjustment Programs: The Perspectives of Private Entrepreneurs and the Political and Bureaucratic Elite

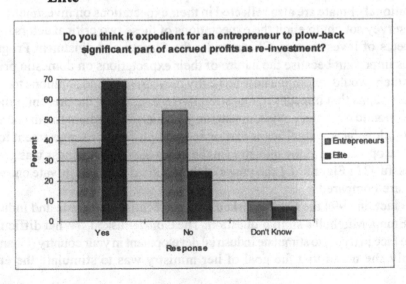

implies issues involving bottlenecks in the macroeconomic framework for which it is believed the government of Ghana has done a lot to resolve ostensibly through budgetary statements. The view of Ghanaian political and bureaucratic elite, thus, seems to be that as long as government has divested itself from direct economic activity and provided incentives to putative private entrepreneurs, the way has been paved for vigorous private sector activity. Not surprisingly, a majority of them (68. 9 percent) thought that under incentives that a liberalized economic environment has ushered in Ghana it was expedient for prospective private entrepreneurs to seriously commit their resources to investment, Figure 17.1. In the view of a member of Ghana's parliament, the fact that the country was under constitutional rule should convince prospective investors to expand their economic activities. Only a small minority of the elite (24.4 percent) felt that the economic environment in Ghana was not secure enough in Ghana for brisk economic ventures.

The expectation of private business on investment prospects sharply contrasts with that of the elite as illustrated in Figure 17.1. Only a minority of them (36.2 percent) felt that the economic environment in Ghana under reforms was conducive enough for them to commit their resources to investment. The majority of 54.4 percent said they were not positively inclined to reinvest their profits under the existing institutional environment they considered insecure. This is quite untypical of Ghanaian entrepreneurs who are noted to re-deploy profits as reinvestment (Aryeetey *et al.*, 1994).

17:7 Conclusion

What must be done before market liberalization can succeed in countries such as Ghana where the framework for private markets is lacking because of several decades of statism? Starr (1990: 31) notes that under such circumstances, "the basic foundations of a private sector need to be constructed". He argues that "this process is not to be understood as a mere relaxation of state controls. On the contrary, it requires an active effort by the state to design new laws and institutions, to ensure security of property rights and rights of voluntary association". This task cannot be performed without the consensus of the state elite and private actors on the importance of this prerequisite. Unfortunately, in Ghana, the two critical actors share contrasting views on the fundamental issues that must be resolved in order for market liberalization to work. In fact, a number of the members of the elite surveyed constantly made reference to issues such as inflation, which they thought more seriously impacted on investment prospects than institutional bottlenecks such as land and property rights issues. They also pointed to the continuous depreciation of the Ghanaian currency, the "cedi", as the result of it being floated as part of the currency rationalization package of the Structural Adjustment Program reforms. These matters, they felt, were the more important issues that warranted government attention. For these and other reasons, therefore, the more critical issue of the legal institutional situation, which private entrepreneurs consider very seriously, is less important for the elite.

Clearly, therefore, a major explanation for the continued institutional uncertainty can be attributed to the disdain of the Ghanaian political elite towards domestic capitalists on the one hand, and the corresponding mistrust of the former for the latter. Under such wide and divergent perspectives the chances of securing the consensus necessary for building appropriate institutions are slim. These factors explain the reason why the reform leadership have not yet recognized institutional credibility as a necessary precondition for the success of reforms, or have simply taken them for granted. The reform regime misconstrues lifting state controls for a safe economic environment. For Ghana to make meaningful gains from its committed reforms, regime leaders need to rethink the building of consensus around institutional reforms.

Notes

1. This was a typical Ghanaian official's characterization of the dilemma of low investment and productivity under market liberalization reforms. See Thomas M. Callaghy, "Lost Between State and Market: The Politics of Economic Adjustment in Ghana, Zambia and Nigeria", in Joan M. Nelson, ed., *Economic Crisis and Policy Choice: The Politics of Adjustment in the Third World* (Princeton, N. J.: Princeton University Press, 1990), 283.

2. The survey was originally conducted in the Fall of 1997 and the Spring of 1998 as part of the field research for a Ph.D. dissertation. See Nicholas Amponsah, *Institutional Credibility and Entrepreneurial Development: Explaining Ghana's Mixed structural Adjustment Results, 1983-1998* Ph.D. Dissertation, Claremont Graduate University, 1999.

3. This kind of symbiotic relationship between the state and private business is noted to be one of the major explanations for Korea's dynamic industrial transformation. See Peter Evans (1995), Embedded *Autonomy: States and Industrial Transformation.* Princeton, N. J.: Princeton University Press, 57.

4. Ghana's adjustment reform has followed the classic Smithian paradigm in terms of its liberalization, openness to trade and efforts to ensure the workings of the invisible hand.

5. For the detailed survey questionnaire, see Amponsah (1999), *Institutional Credibility* Appendix A.

6. See for example Tangri (1992) Government-business Relations; Esseks (1971) Government and Indigenous Private Enterprise in Ghana.

7. Ghana's Minister for Trade and Industry alluded to this view in an extensive interview on the difficulties the Ghanaian regime faces in trying to stimulate industrial development, especially from indigenous Ghanaian industrialists. See *The Courier* (1994).

References

Alderman, H. (1994). "Ghana: Adjustment's Star Pupil?" in *Adjustment to Policy Failure in African Countries*, edited by D. E. Sahn. Ithaca: Cornell University Press.

Amponsah, N. (1999). "Institutional Credibility and Entrepreneurial Development: Explaining Ghana's Mixed Structural Adjustment Results, 1983-1998". Ph.D. diss., Claremont Graduate University.

Amponsah, N. (2000). "Ghana's Mixed Structural Adjustment Results: Explaining the Poor Private Sector Response". *Africa Today* 47 (1) 9-34.

Aryeetey, E., A. Baah-Nuako, T. Duggleby, H. H., and Steel, W.F. (1994). *Supply and Demand for Finance of Small Enterprises in Ghana*. Washington, DC: The World Bank.

Berman, B.J. and Leys, C. (1994). *African Capitalists in African Development*. Boulder: Lynne Rienner Publishers.

Campos, J.E. and Root, H.L. (1996). *The Key to the Asian Miracle Making Shared Growth Credible*. Washington, DC: The Brookings Institution, 1996.

The Courier (Journal of the African-Caribbean-Pacific and European Union) (1994). "Country Report, Ghana: Striving to Keep up the Momentum". 44 (March/April): 20-45.

Davis, I. (1996). Ghana's Encouraging Elections: The Challenges Ahead. *Journal of Democracy*, 8. 2: 78-91. *Social Mobility and Political Change*. London: Pall Mall, 1970.

Esseks, John D. (1971). "Government and Indigenous Private Enterprise in Ghana". *Journal of Modern African Studies* 9. 1: 11-29.

Evans, P. (1992). "The State as Problem and Solution: Predation, Embedded Autonomy, and Structural Change" in *The Politics of Economic Adjustment: International*.

Evans, P. (1995). *Embedded Autonomy: States and Industrial Transformation*. Princeton, N. J.: Princeton University Press.

Gyimah-Boadi, E. (1995). "Explaining the Economic and Political Successes of Rawlings: The Strengths and Limitations of Public Choice Theories" in *The New Institutional Economics and Third World Development*, edited by J. Harris, J. Hunter and C. M. Lewis. London: Routledge.

Gyimah-Boadi, E. (1996). "Ghana's Encouraging Elections: The Challenges Ahead". *Journal of Democracy*, 8. 2: 78-91.

Hart, E. I. (1996). "Liberal Reforms in the Balance: The Private Sector and the State in Ghana, 1983-95". Ph.D. diss., UMI: Princeton University.

Herbst, J. (1993). *The Politics of Reform in Ghana, 1982-1991.* Berkeley: University of California Press.

Hutchful, E. (1995). "Why Regimes Adjust: The World Bank Ponders its 'Star Pupil'". *Canadian Journal of African Studies* 29 (2): 303-317.

Lall, S. (1995). "Industrial Adaptation and Technological Capabilities in Developing Countries" in *The Flexible Economy: Causes and Consequences of the Adaptability of National Economies*, edited by Tonny Killick. London: Routledge.

Lee, J. (1997). "The Maturation and Growth of Infant Industries: The Case of Korea". *WorldDevelopment* 25 (8): 1271-1281.

Leechor, C. (1994). "Ghana: Front-runner in Adjustment" in *Adjustment in Africa: Lessons from Country Studies* edited by Ishrat Husain and Rashid Faruqee. Washington, DC: World Bank.

Morris, M. H. (1998). *Entrepreneurship Intensity.* Connecticut: Quorum Books.

Ninsin, K .A. (1989). "The Land Question since the 1950s" in *The State, Development and Politics in Ghana* edited by E. Hansen and K. Ninsin. London: CODESRIA Books.

Rimmer, D. (1992). *Staying Poor: Ghana's Political Economy, 1950-1990.* Oxford: Pergamon Press.

Starr, P. (1990). "The New Life of the Liberal State: Privatization and the Restructuring of State-Society Relations" in *The Political Economy of Public Sector Reform and Privatization*, edited by E. N. Suleiman and J. Waterbury. Boulder: Westview Press.

Stein, H. (1994). "Theories of Institutions and Economic Reform in Africa". *World Development.* 22 (12): 1833-1849.

Tangri, R. (1992). "The Politics of Government-Business Relations in Ghana". *Journal of Modern African Studies* 30 (1): 97-111.

The World Bank (1994). *Adjustment in Africa: Reforms, Results and the Road Ahead.* Washington, DC: The World Bank.

18 Progress in Adjustment in Ghana: Is Growth Sustainable?[1]

ROBERT ARMSTRONG

18:1 Introduction – Progress in Adjustment

Ghana launched its far-reaching Economic Reform Program (ERP) in 1983. After more than a decade of economic decline and political instability, three shocks had brought the economy close to collapse: a prolonged drought, a marked deterioration in the terms of trade, and the unexpected return of more than a million Ghanaian workers expelled from Nigeria. The crisis changed attitudes and perceptions both in Ghana and in the World Bank. Bank staff formed a strong partnership with a core group of Ghanaian officials responsible for preparing and implementing the ERP. A key factor in sustaining the reforms since 1983 has been the continuity and competence of this core group.

The main achievements of the ERP are well known. Trade was liberalized and inflation drastically reduced. Extensive price and distribution controls were dismantled. Cocoa and other exports recovered. GDP growth has averaged about 5 percent a year over the past decade. Poverty has been reduced and social indicators have improved. The Ghanaian strategy was to "go for growth" above other objectives, and the World Bank supported this strategy. This led to the designation of the rate of G.D.P growth as the dominant indicator of the success of the strategy, for both the government and the bank. World Bank loans of more than $2 billion, including more than $1 billion for adjustment operations, have helped Ghana's economic progress over the past decade. But a study by Operations and Evaluations Department (OED) of the World Bank warns that the progress will not be sustained, unless Ghana speeds up the implementation of a large unfinished agenda of policy reform.

The current strategy has not fostered the response needed from the private sector. Nor has it raised agricultural productivity. The Bank needs to frame its assistance strategy to help Ghana achieve participatory development and self-reliant, private sector-led growth.

18:2 Long-term Concerns

While Ghana has been hailed as a success story, prospects for sustaining satisfactory rates of growth and poverty reduction are uncertain:

• *Agricultural growth is* much slower than necessary and feasible, and may be slower than population growth (see Table 18.1). Over the decade 1983-94, the Bank emphasized structural reform of prices, marketing, and enterprise ownership. But Ghana made slow progress in eliminating the parastatal monopoly in cocoa marketing and in developing more efficient markets. Little was done to improve the productivity of the main food crops or to generate dynamic innovation in the small farm sector.

• *Fiscal problems* have resurfaced. Deficits are larger than is consistent with low inflation and adequate credit to the private sector. Fiscal problems, combined with excessive credit to public enterprises, still depress private investments and savings, and underlie the resurgence of inflation in 1993-95.

• *Savings and investment* Total and private savings and private investment rates are extremely low. Recently the private investment rate was only 4 percent of GDP, and the national savings rate was only 1 percent of GDP (see Table 18.1).

• *Aid* Ghana's recovery in the 1980s was led by the public sector, supported by external assistance. In 1988-92, annual net receipts of official development assistance averaged more than 10 percent of Ghana's GDP. Much of the aid was quick-disbursing and used to finance current expenditures. But in today's stringent aid climate, Ghana cannot count indefinitely on large concessional inflows.

• *Environmental problems* are serious. Soil fertility loss, soil erosion, and deforestation have been estimated to cost Ghana 4 percent of its GDP annually (see "Ghana 2000 and Beyond". World Bank, 1992).

• *Poverty* Since the adjustment program started, important "quality of life" indicators, including child malnutrition, infant mortality, literacy, and access to clean water, have improved. Between 1988 and 1992 the proportion of Ghana's population in poverty fell somewhat, from 36 to 31 percent. Almost all the improvement in incomes came from economic growth; income distribution remained fairly stable. The income gains benefited most regions, and especially rural areas, but poverty increased in Accra. Living standards among the poorest groups remain seriously low. Most of Ghana's poor are food crop and export crop farmers with incomes about one third of the national average. Targeting of social spending has not improved.

18:3 Bank Assistance Strategy

In the 1980s, the Bank's assistance strategy was relevant to Ghana's needs in giving priority to stabilizing, liberalizing, and rehabilitating the economy. The strategy closely accorded with the government's strategy of "going for growth" through a public sector-led recovery program. Both lending and nonlending services were efficacious and cost-effective. They put appropriate emphasis on getting prices, balances, and incentives right, on rebuilding moribund infrastructure, and on mobilizing aid. The soundness of most of the Bank's policy advice was founded on good economic and sector work. In the early 1990s, however, the Bank's assistance program did not adapt sufficiently-either in its priorities or instruments-to Ghana's changed conditions. Complacency led to a loss of momentum in addressing the unfinished agenda of adjustment. Neither the Bank's nor the government's strategy has adequately incorporated the goals of poverty reduction, raising agricultural productivity, institutional reform, and environmental sustainability.

Over the decade 1983-93 as a whole, the Bank's most effective instruments were its macro- and policy-oriented economic work, macro-level policy dialogue, the first two structural adjustment operations, lending for infrastructure rehabilitation, and aid mobilization efforts. Lending for agriculture and in the social, financial, and industrial sectors had less satisfactory results, as did efforts to foster coordination among development agencies. The Bank and other agencies sometimes pushed more assistance on Ghana than could effectively be managed by the public sector. And not enough attention was paid to the disadvantages of large aid flow, including new forms of rent seeking through patronage relationships. The least effective instrument was technical assistance aimed at fostering institutional reform.

For the Bank's assistance to be as relevant, efficacious, and cost effective as possible, the Bank and government should rethink their priorities. Externally supported spending on infrastructure and social sector development can keep economic growth going for a time. But to sustain growth and poverty alleviation will require much more vigorous responses from the private sector and from agriculture than have been achieved so far. This calls for a breakthrough in the investment climate and business environment for private business. It also calls for a new agriculture and human resources development strategy.

18:4 The 1990-1993 Country Strategy Paper (CSP)

The 1990 Country Strategy Paper (CSP) was the first for Ghana. By that time, Ghana had attained the status of a model client. In the Bank's annual report on ratings, Ghana had received very high ratings for several years. The government's ability to pursue its reform agenda during the years 1986 and 1990 was enhanced by the fact that this period was relatively uneventful in terms of major political changes. The 1990 CSP envisaged that a broader sharing of power on the political front would help foster support for further economic reforms to maintain the pace of development. The 1990 CSP was a more sober document than other reports on Ghana at a time when the country was being extolled as a success story, and when some bank staff were turning to accelerate Ghana's growth to the level of the fast growing countries of East Asia. The 1990 CSP defined 3 priority objectives, first, to intensify efforts to promote private sector development, recognizing that this would require the government to take a more "hands off" approach to managing the economy. The second one was to improve public sector management and implementation capacity. Thirdly, the paper proposed a few measures to encourage the government to address human resource development, poverty, and environmental issues that were seen to determine the economic and social sustainability of the reform program.

The 1993 CSP was prepared in the first half of 1993 shortly after elections. The 1993 CSP was significantly influenced by the completion in 1992 of "Ghana-2000 and Beyond: Setting the stage for accelerated growth and poverty reduction". This was a major Bank report produced with large Ghanaian participation, which emphasized need for action in three areas in order to accelerate growth above the 5 percent rate achieved during the previous decade. These areas included, private sector development and export promotion, public sector management, and human resource development. In these respects the 1993 CSP appropriately put increased emphasis on the long-term issues.

18:4.1 Elements of Strategy

Conditions for private enterprise To persuade the private sector that the increasingly market-friendly policies are unlikely to be reversed, Ghana needs to provide unequivocal support for private sector development, carry out in-depth administrative reforms, and undertake accelerated programs of privatization. Ghana's overstaffed and poorly functioning public administration and parastatal enterprises impede private sector development in several ways: through giving excessive credit to parastatals, which crowds out the private sector from access to finance; preferential treatment of public firms; de facto monopolies; rent-seeking, obstruction, and harassment by public officials; inefficient delivery of infrastructure and other public goods and services; and through sending the signal that statist policies and intimidating practices are still tolerated. Fostering private sector development is an area in which the International Finance Corporation should play an increasingly active role.

A "nexus strategy" to address the links among rapid population, growth, declining agricultural productivity, and environmental deterioration Ghana's unexploited land and forests have been dwindling fast, and rising population density has led to shorter fallow periods and reduced soil fertility. The combination of rapid population growth, declining agricultural productivity, and environmental deterioration needs priority attention in the Bank's lending program and in economic and sector work, policy dialogue, and aid coordination.

Agriculture The reform of inefficient agricultural parastatals is too slow. More in-depth analytical work is needed to identify ways and means to increase private investment in agriculture-particularly in nontraditional export crops-and to raise agricultural productivity in ways consistent with environmental sustainability.

Education Ghana launched a comprehensive educational reform program in 1986, to provide a more vocationally and practically oriented curriculum and wider access to basic education, especially for rural families. With strong government commitment and heavy support from the Bank and donor agencies, the program provided all the inputs thought to be necessary. But test results are very poor, suggesting that students are learning very little. This has provoked a new stocktaking by government and donors. Other countries, too, have found that successful educational reform can prove more complicated than expected. There seems to be agreement that Ghana's reforms allowed too little time for debate, curriculum design, piloting, and teacher training.

Capacity building, civil service reform Ghana's underused human capacities and the poor ethos of its bureaucracy are arguably more important constraints on the functioning of government than is the scarcity of skills. Thus far, the government and the Bank have emphasized increases in the supply of skills through externally financed training programs, and neglected the institutional context and governance conditions that determine how effectively skills are used. Too often, technical assistance to Ghana has been dominated by short-term considerations, rather than geared to long-term capacity building and institutional learning.

The need now is for a coherent and comprehensive strategy for reforming public administration. This strategy should address the root causes of underutilization of Ghana's trained people, overstaffing and low productivity in the civil service and public enterprises, and the poor performance of most technical assistance projects (not only the Bank's). The Bank should encourage the government to convene a local consultative group to produce an action plan for making technical assistance more effective.

Divestiture and restructuring of public enterprises Privatization of Ghana's large and inefficient public enterprise sector has been slow. Only 54 of 300 nonfinancial public enterprises had been fully divested by 1993. Some important privatization measures have been taken since then, but Ghana needs to draw up a clear action plan for enterprise divestiture and rationalization.

For public enterprises, many pricing and procurement decisions are still taken by government and efficiency improvements have been limited. Capacity utilization is poor. Many enterprises need to be liquidated. Bank staff attributes the slow pace of privatization to insufficient political support-partly because of Ghana's ideological heritage and partly because of vested interests. The Bank's approach to privatization may have been too formulaic and not sufficiently heedful of important political and social dimensions.

Public expenditure management Expenditure allocation has improved markedly since the early1980s. The Bank made major contributions through its economic work on public investment and public expenditure reviews. But weaknesses in budget execution, accounting, and evaluation systems still need priority attention. A huge public sector wage bill-for nearly 600,000 employees in a country of 16 million people-underlies the difficulties in controlling the fiscal deficit. There is an urgent need to improve public expenditure monitoring and control mechanisms.

Poverty To foster social and political sustainability, Ghana's strategy needs to ensure that the benefits of growth, and of external assistance, are broadly shared. Implementation problems prevented the Program of Actions to Mitigate the Social Cost of Adjustment (PAMSCAD) and the Social Dimensions of Adjustment Program from achieving significant results.

18:5 Reflection on Outcomes

Ghana achieved notable success over the past decade in macroeconomic policy reform as measured by various macro-level indicators (GDP growth, export recovery, removal of price distortions, and so on), but far less in terms of institutional and structural change and as measured by various micro level indicators (school test results, soil fertility, productivity of civil servants, and so on). The

Table 18.1 Ghana: Sustainability Indicators

	1984-87	1988-92	1993
Savings, investment, aid dependency			
Private investment as % of GDP	4	7	4
National savings as % of GDP	5	7	1
"Genuine savings rate,"[a] as % of GNP	-12	-4	-1[b]
ODA disbursements as % of GDPP	6	11	9[b]
ODA disbursements as % of GDP			
Nexus indicators			
Total fertility rate	7	6	6
Agricultural growth rate	3	2	3
Environmental indicators			
Annual loss from deforestation, erosion, soil degradation, as % of GDP		4	
Deforestation rate (annual % of total forests)		1	
Export growth			
Traditional	11	5	20
Nontradition	32	8	0
Government revenue as % of GDP	12	13	17
Percent of population below poverty line	36[c]	31[b]	

a. *Obtained by subtracting depreciation of fixed capital and depletion of natural resources from the nation savings rate.*
b. *1992*
c. *1998*

successes have been in rehabilitating infrastructure, getting the prices right, freeing up foreign exchange and other markets, reforming taxes and reducing subsidies, initiating cost recovery measures, and putting in place some important elements of an enabling environment for private sector development.

The record on poverty alleviation is mixed. Owing to a lack of baseline data for the early 1980s, it is difficult to establish changes in poverty up to 1987/88, when such baseline data became available. But the available evidence suggests that poverty decreased somewhat in the mid-1980s and that the proportion of the population in poverty has dropped marginally from about 43 percent in 1987/88 to about 42 percent in 1991-92.

Ghana also has a mixed record on stabilization, although budget deficits were reduced considerably between 1983 and 1992. Inflation remained high and too much of the burden of stabilization was put upon monetary policy, with crowding out effects on the private sector. The least satisfactory performance has been in agriculture and manufacturing output growth, in institutional reform (including the civil service and public enterprise divestiture and reform), in investment project implementation, and most importantly, in the "sustainability indicators". The disappointing response of private savings and investment is reflected in these sustainability indicators. Clearly the heavily aid-dependent, public sector-led growth of the 1980s and early 1990s will not be sustainable indefinitely. As will be discussed below, the interrelated issues of dependence, economic governance, and institutional development themselves form a nexus of "sustainability issues" that comprise core issues for the Bank's future assistance strategy.

Thus, the performance of the Ghanaian economy is a qualified success story. The attention and praise given by spokesmen for the Bank and other donors to the successes at the macro level may have served to mask some shortcomings at the sectoral and micro-levels. While GDP growth was at or near the targeted 5 percent annual rate, the composition of this growth was markedly different from what was projected/targeted, with services (largely trade) growing much faster and agriculture and manufacturing growing much slower than targeted. Moreover, there seems to have been a "halo effect", and some apparent successes at the design stage have not been translated into successes at the implementation stage.

In sum, different stories can be told about Ghana depending upon the choice of indicators, and this choice may focus attention on which actions and outcomes will be most closely monitored and measured in future.

Ghana's performance was affected by events and shocks beyond the control of the government, but which called for a response by the government and donors. Table 18.2 shows the nature of the shocks experienced by Ghana over the 15 years 1978-92. (Note that a positive sign for the shocks refers to adverse shocks rather than to beneficial windfalls.)

The terms of trade deterioration was quite substantial in the years just preceding the ERP and in recent years. During the mid-1980s Ghana's terms of trade improved slightly until 1987 (104, with 1984 = 100), but the index has deteriorated steadily since then (to 74 in 1992 and 66 in 1993). Falling cocoa prices were a main determinant of this decline, as they declined from 90-110 cents per pound in the period 1984-87 to between 50-56 cents over the years 1989-93.

Ghana was able to offset these external shocks mostly through its access to increased concessional external financing. Since a substantial share of this external financing was in the form of quick disbursing non-project aid, the government retained considerably more financial degrees of freedom than had it been forced to undertake import intensification or economic compression. But if the terms of trade deterioration is permanent, one cannot ignore also the undesirable effect of raising the country's aid dependence and the consequences of the "Dutch disease".

Natural disasters and exogenous circumstances are other external shocks. In Ghana, with more than 40 percent of GDP originating in agriculture, weather conditions significantly affect GDP; performance can vary severely owing to periodic droughts such as in 1981-83 and again (to a lesser extent) in 1990. Nigeria's expelling of over one million Ghanaians in 1983 resulted in about a 10 percent increase in Ghana's population that year. Conversely, the overvaluation of the CFA franc (currency of the Communauté Financiere d'Afrique) relative to the Ghanaian cedi and hard currencies up to 1993 was a positive exogenous development whereas the CFA devaluation in 1994 constituted a shock.

Table 18:2 External Shocks and Performance Response Measures
(as percent of GDP)

	Period Average		
	1978-82	**1983-87**	**1988-92**
External Shocks			
Terms of trade	3.4	-0.3	2.8
Export volume	0.7	-0.1	0.5
Total	4.1	-0.4	2.3
Performance response measures			
Additional net external financing	1.6	-0.8	2.3
Export promotion	-0.4	0.0	0.9
Import intensity	2.8	1.0	-0.5
Economic compression	0.0	-0.6	-0.3
Total	4.1	0.4	2.3

Source: *World Bank data*

There was relative peace between Ghana and its neighbors, political stability within the country and no major natural disasters to cope with between 1984 and 1993. Periodic tensions with Togo and Cote d'Ivoire did not cause major disruptions to the economy. There was however a man-made disaster in February 1994 when tribal strife in the northern region resulted in deaths of more than a thousand persons, in large movements of refugees fleeing this strife, and in the interruption of all developmental activities in the region. The causes and consequences of this strife need to be better understood if the Bank's strategy is to be as "relevant" as may be necessary to help avert recurrences in future. Such events clearly highlight the need for donor strategies to be grounded in a good understanding of a country's political economy.

Because the aid inflows largely offset the exogenous shocks over the period under review, the latter cannot be blamed for those aspects of Ghana's performance that were less than satisfactory. Yet, the impact on the economy on the behavior of the various agents in the economy, are very different depending upon whether Ghana's foreign exchange resources come from export earnings or from aid. This requires that the Bank pay close attention to how changes in macro variables affect micro-level incentives and behaviors, as in the case of cocoa earnings versus aid transfers. It also means that positive windfalls (such as the discovery of oil, a surge in cocoa prices, or very plentiful supplies of external assistance) typically have some distorting, downside effects that need to be monitored and minimized insofar as possible. These effects may, in some cases, outweigh the benefits when a "behavioral approach" and a long-term view are taken. In Ghana's case, the benefits of plentiful aid have more than compensated for the windfall losses from the terms of trade. But one type of distorting effect that large aid flows can create is that known as the "Dutch disease" effect.

18:6 Changes Needed

The Bank's assistance strategy needs to be framed in a long-term perspective of at least a decade.

The problems just outlined need urgent attention, but they will not be solved by quick or simple policy measures or by piecemeal investments.

Broaden and deepen "ownership" In the 1980s, the Bank dealt with a small group of leaders and technocrats accountable to an unelected head of state. In the 1990s, Ghana has an increasingly active parliament (notwithstanding its one-party composition) and new forms of decentralized organization and accountability. The political reforms may slow decision-making and policy implementation in the short term. But insofar as they broaden the "ownership", and thus improve the sustainability of policies, they will serve to deepen and to speed development over the longer term. The Bank needs to build the analysis of political economy factors more explicitly into its country assistance strategy.

Interact with broader constituencies As part of its role in helping to elucidate policy options, the Bank should interact with broader constituencies in and outside Ghana. It can help provide information and analysis on Ghana's economic and social situation and prospects by disseminating its documents more widely, by encouraging the government to discuss Bank and jointly authored reports with broader audiences, and by undertaking more outreach activities in regard to nonlending services. In the process, Bank staff should listen more to views and concerns of other analysts of Ghanaian issues.

Focus on governance The Bank's policy dialogue with the government, and the design of its assistance strategy, should emphasize governance factors that tend to diminish the confidence of the private sector and hence private savings and investments.

Help improve aid coordination policies and practices Aid is well coordinated in some areas but not others, and the government should take a much greater role itself. In particular, sectoral and subsectoral coordination needs to be strengthened in a context of agreed sectoral strategies and action plans.

Bank management, responding to the study, agreed that Ghana's policies should seek to encourage private sector growth by accelerating privatization, increasing competition in the financial sector, liberalizing the cocoa and petroleum sectors, and sustaining macroeconomic stability; support institutional development by improving public expenditure management and restructuring the public service; and promote agricultural growth while protecting the environment. As to OED's recommendation that the unfinished adjustment agenda be implemented faster, management noted the tension between the goals of rapid implementation and Ghanaian ownership. The Bank plans to support measures that government itself develops for public service reform, provided they can be expected to achieve the goal of a leaner and more effective public service; this strategy is consistent with supporting a program that is both Ghana-owned and sustainable. On consultation with broader constituencies, management noted its plans for public dissemination of the results of the Bank's economic and sector work, and for supporting workshops for parliamentarians and the public at large. On public expenditure management, management noted that a financial management technical assistance project now in preparation will support implementation of expenditure monitoring and accounting systems in 16 ministries. Oil aid dependency and coordination, management noted that donor' impact and influence on government's choices remains large, even if declining. The government now coordinates aid in some sub sectors, and it led the development of the public expenditure program, which has become the basis for all donor support. Future Bank economic and sector work will examine the potential for reducing aid dependency. As to the skill mix of Bank staff, management noted that it is paying special attention to institutional development skills in its recruitment policy.

The Committee on Development Effectiveness of the Bank's board of executive directors, discussing the study, highlighted the following issues: the need for greater borrower participation in preparing the Bank's country assistance strategies; differences between management and OED on the proper pace of institutional development; aid dependency and ways of improving domestic resource mobilization; ways to strengthen aid coordination; and the need to be realistic about the time needed to implement reform programs and achieve sustained supply responses.

18:7 Conclusions – Key Strategic Issues and Recommendations

In order for the Bank's assistance to Ghana in the years ahead to be as relevant, efficacious, and cost-effective as possible, this report recommends that the Bank rethink some of its priorities; give more attention to institutional development, economic governance, and sustainability issues; redefine some of its comparative advantages; reformulate its performance indicators; and refine its process of country assistance strategy formulation. The Bank's strategy and instruments were effective, in the 1980s, in helping the Ghanaian government to stabilize, liberalize, and rehabilitate a moribund economy in crisis. Particularly in the mid-1980s, the Bank's program of assistance to Ghana warrants high ratings for the relevance of objectives, efficacy of implementation, cost-effectiveness, and staff performance. The Ghana government warrants a high rating for borrower performance during those years.

In recent years, however, the performance on both sides has been more mixed. Progress has continued in getting the prices right, in rehabilitating the infrastructure, in rationalizing the revenue structure, and in public investment programming. These are areas where both the government's and the Bank's strengths and comparative advantages can be built upon. Performance has however been disappointing in private sector development, privatization (at least until 1994 when more privatization activities were launched) and public enterprise restructuring, civil service reform, expenditure control, agricultural development, educational achievement, environmental control, and institutional development.

Ghana's agenda of unfinished adjustment is a long one and the prospects for sustaining a satisfactory rate of per capita growth and poverty alleviation are uncertain. Agricultural growth remains well below what is feasible (and may well be no better than the rate of population growth), private savings and investment rates remain extremely low, educational testing results in the primary schools are very low, and there are serious environmental problems. This being the case, proclaiming Ghana to be a "success story" may not be accurate nor in the country's best interest.

The objectives and instruments of the Bank's country assistance strategy for Ghana today are not so well adapted to current conditions as they were in the 1980s. The hardest parts of Ghana's adjustment agenda are yet to come, and the objectives and instruments have to be adjusted accordingly. More urgent attention needs to be given to several longer-term issues if the growth and gains achieved are to be sustained, or accelerated. The binding constraints are on the institutional and managerial side. Key issues that need to be more explicitly addressed in the Bank's country strategy are: What can the Bank do to improve the sustainability prospects? And what pace of change, on which key fronts, is needed to minimize the risks that Ghana's progress will not be sustained? Most of the truly important issues (as distinguished from the urgent issues that tend to command attention) can be called the "sustainability issues".

Two main categories of sustainability issues are the "leading sector" and aid dependency. During the 1980s, recovery had to be public-sector and foreign-aid led. Aid supported expenditures on infrastructure and social sector development can keep the growth going for a time. But sustained growth and poverty alleviation will depend upon a far more vigorous private sector response than has so far been achieved. Ghana's heavy dependence on foreign aid also has its disadvantages, and the future strategy should provide for a reduction in this dependency. The binding constraints and the prevailing political economy are now quite different from those in the 1980s, the latter being characterized by a transition toward democracy following elections in 1992 of a civilian government under a new constitution. The current challenge for the Bank is how best to help Ghana manage a transition to a more sustainable path while a transition to democratic political forms and processes is underway. The sustainability of the political and social transitions depends partly on how the economy fares, and vice versa. The Bank's strategy must consequently be a "political economy" strategy.

A new strategy should be based on a shared vision of what kind of society the Ghanaian government and people wish to bring about. It should be a central role of the Bank to help Ghana elucidate the options, and then to forge the needed new instruments and institutions. If in the 1980s a main thrust of the strategy was to "get the prices right", a main thrust for the 1990s should be in helping Ghana to "get the institutions and economic governance conditions right". The new strategy also needs to accelerate the pacing and revisit the sequencing of future policy and program actions in the light of the failure of the current strategy to foster the needed private sector response or to raise agricultural productivity. One of the report's conclusions is that unless Ghana's adjustment strategy becomes less "gradualist" than in the past decade, growth may slow down to the extent that political and social stability would be endangered.

Two critical issues are how to stimulate the needed private sector response and how to reverse the declining productivity in agriculture without harming the environment. The trends in several environmental indicators (for example, deforestation, declining soil fertility resulting from shortened fallows) need to be reversed. The government strategy (as of 1994) was still based on continued gradualism in adjustment, including civil service reform, privatization, public sector restructuring agrarian reform, and family planning. This gradualism was facilitated, during the last decade, by the substantial aid provided by the donors, especially in the form of quick disbursement policy, much of which financed current expenditure. And as noted earlier, the gradualist strategy did lead to growth although not necessarily of a sustainable nature. But the binding constraints have changed. What is needed now is a strategy that achieves a breakthrough in the investment climate and business environment, and improves the confidence of private sector investors and actors. Lack of confidence now appears to be the binding constraint to private sector development. Accelerated programs of privatization, administrative reform, and "right-sizing" of government appear to be *sine qua nons* of increased confidence by the private sector that Ghana's increasingly market-friendly policies are unlikely to be reversed. Accelerating adjustment will undoubtedly be difficult in the face of the developments on the democratization front that may tend to slow decision making, but it appears as a necessary condition for sustainable growth.

The Bank has a well-established comparative advantage in helping Ghana to correct its distorted macroeconomic and sectoral policies and to rehabilitate its infrastructure. The Bank now needs to further develop and improve its capabilities and change its comparative advantages to become more effective in key areas where its past assistance has been least effective, such as in fostering private sector development (an area in which the IFC needs to play an increasingly active role, including through the African Enterprise Fund), in promoting institutional development and capacity building, in helping the government to become smaller and more efficient, and in elucidating policy analyses and options to promote widened and more enlightened public debates on key policy issues among constituencies formerly excluded from such debate and decision making.

Noting that many of the findings, conclusions, and recommendations could usefully be applied to other countries and regions, the committee subsequently sent the Bank's President a list of generic issues for discussion with Bank management. These included:

- The Bank's propensity to underestimate the time required, especially in countries with weak institutional capacities, to implement reform programs and achieve sustained supply responses;
- the need for country assistance strategy to reflect the country's political economy;
- the need to consider costs as well as benefits of rapid trade liberalization, to design programs flexibly, and to monitor their effects so as to enable midcourse corrections;
- the need for advice and support in the financial sector to address root causes, not just symptoms, of problems;
- the need for an adequate method of defining and measuring the minimal acceptable degree of "ownership" needed for success in individual operations, and for consistent, thoroughgoing efforts to promote partnership and borrower participation in Bank operations;
- the need for high quality, judiciously timed, economic and sector work and for thorough analysis of proposed operations;
- skill mixes, staffing, and work location issues.

In conclusion the paper will do some justice by highlighting further ways ahead for Ghana that flow from this text. These are as follows:

For the design of Bank strategy:

- Focus the strategy on sustainability, institutional development, and economic governance.
- Monitor performance against specified sustainability and micro-level indicators.
- Develop a "nexus strategy" and sectoral action plans.
- Make the strategy "more strategic" and longer term.

For achieving objectives:

• Build on Bank's comparative advantages (E.S.W, policy dialogue, aid coordi-
  nation).
• Reach out, and listen to broader constituencies within and outside Ghana.
• Focus the dialogues on hard choices and options.
• Develop a more strategic approach to institutional development and technical
  assistance.
• Get at the root causes of inadequate expenditure management.
• Do not oversell Ghana as a success case.
• Beware the downsides of aid dependency.
• Develop new aid coordination policies and practices.
• For making Bank instruments and processes more effective:
• Get in-house ownership of, and accountability for, the Bank's own strategy.
• Put more 'evaluation' into country assistance strategies.
• Changes the skill mix of staff to better meet the skill requirements of the
  strategy.
• Reassess the mix of Headquarters vs. Resident Mission assignments.

Note

[1] This chapter is a modified version of the precis of Armstrong, R. P. (1996). Ghana Country
Assistance Review published by the World Bank. The evaluation covers the 1983-1993 period. It
is reproduced here with the permission of the World Bank, the copyright owners. The modification was made by Onai Chitiyo.

References

Armstrong, R. P. (1996). *Ghana Country Assistance Review: A Study in Development
    Effectiveness*. Washington DC: World Bank.
McCarthy, Desmond F. and Dhareswar, A. (1992). *Economic Shocks and the Global
    Environment*. WPS 870. Washington DC: World Bank.

19 Alternative Methods for Evaluating Structural Adjustment Programs

Francois K. Doamekpor

19:1 Introduction

Reform programs in several countries have reduced the dominant role once played by central governments. Through these programs, the extent of government economic activity has been drastically reduced, including the dominant role once played in certain key sectors of the economy. In some of the countries, these changes have culminated in smaller governments and more vibrant economies. In spite of the gains derived from these programs, determining how well the reformed public sector is performing remains a primary focus of practitioners, policymakers and interested parties.

This chapter is devoted to a review of some of the methods for measuring government conduct of work. It is written to provide a rationale and broad overview of some of the measures used to gauge performance. It is not intended to prescribe country-specific techniques or approaches for assessing organizational performance. As a result, no country data were used. This was done on purpose to facilitate a general but an in-depth presentation of an integrative model that incorporates various performance measures and indicators.[1]

Historical studies of governmental performance in both developed and developing economies are unanimous in their support for a reformed public sector. Comparative empirical studies that focused on performance in both public and private organizations indicated that performance in most public organizations lagged significantly behind performance of private organizations.

Several reasons have been advanced to account for non-performance in public organizations. These include the "displacement thesis" and the commonly used hypothesis of productivity lag. Other seminal ideas that explain the lack of public sector performance are derived from Parker and Hartley's (1991) hypothesis that public sector organizations performed better as they moved towards autonomy or adopted commercially minded approaches to management. Others, such as Goodman and Loveman (1991) view government as an unnecessary and costly drag on an otherwise efficient system.

To address some of these problems, governments in both developed and developing economies instituted several improvement programs which included privatization, restructuring of processes and systems, productivity improvement, program cost and paperwork reduction, to mention but a few. These efforts were aimed at improving efficiency and productivity, reducing public sector size and bureaucratic bottlenecks and promoting competition. In fact, proponents of reform programs argue that any government program or agency will operate more efficiently and at lower cost under private management (Worsnop, 1992).

The purpose of this chapter is not to declare gains derived from various reform programs. It is also not meant to argue the adoption of particular reform programs or claim that certain programs worked better than others. Instead, it focuses primarily on how public sector programs or activities can be measured. This is derived from the hypothesis that, while improvement programs are important, frameworks or models for evaluating performance are equally relevant. In effect, how a government program or activity is measured is as important as how well the program or activity does.

To achieve this purpose, this chapter is organized into three interrelated parts. The first part begins with a search for a definition for a public sector organization or agency. It attempts to answer two critical questions:
• What is a public sector organization or agency?
• How different is it from a private sector organization?

The reasons for this distinction are three-fold. First, it eliminates the potential for misinterpreting organizational arrangements and the concomitant impacts of these arrangements on outcomes. Second, organizational structure and purpose are shaped by ecology, which plays a critical role in growth and development. Organizational ecology molds production and delivery systems and determines ability to achieve desired goals and objectives. Third, organizational purpose is a significant determinant of behavior and a distinguishing criterion among types of social systems. A discussion, therefore, of organizational arrangements, goals and ecology will provide an appropriate framework to discriminate among types of social systems and determine the uniqueness of a public sector organization or agency.

The second and third parts will provide an overview of the various types of measurement criteria for evaluating performance in both private and public organizations. The third part especially, suggests a comprehensive approach to gauge the performance of a government program or activity. The chapter ends with a conclusion.

19:2 Part One: Public and Private Organizations

19:2.1 Organizations

The discussion over what constitutes an organization is a continuing debate. Pioneers in the field of management especially, those belonging to the classical school, such as Henri Fayol, F.W. Mooney, Max Weber and Lyndall Urwick, to mention but a few, provided the foundation for studying and understanding organizations. Their seminal work, referred to as principles of classical management,[2] provided the basis for the use and application of management approaches and techniques in organizations. The bureaucratic organization therefore, is an outgrowth of the thinking of the classical management school. Further, Frederick Taylor's scientific management work design theory provided additional thrust for perfecting operations in organizations.

It is worth noting, however, that organizations are not machines. They are not mechanical structures and should not evoke mechanical imagery. They are constantly changing and reacting to environmental needs. Morgan (1997) notes that organizations are cultures, political systems, brains and instruments of domination, to mention a few. In fact, space would not allow a complete presentation of the current status of the field and its history. For our purpose however, an organization will be defined as a coordinated system of inputs with a pre-defined goal or set of goals. It is non-mechanical, adaptive and shaped by environmental exigencies.

An important question at this point is whether organizations operating in the public sector are different from those found in the private sector. The next section is devoted to a discussion of the distinction between public and private organizations.

19:2.2 Public versus Private Organization

Instead of discussing what constitutes the public sector and what separates it from the other, emphasis will be placed on a review of the literature on the publicness of organizations. This is because; there is unanimous agreement on the distinction between both sectors.[3]

In an introductory note to his book entitled, "All Organizations are Public", Bozeman (1987) noted that, "some organizations are governmental, but all organizations are public". He went on to argue that, "...publicness is a key to understanding organization behavior and management—not just in government organizations but also in virtually all organizations". In Bozeman's view, the most formidable obstacles to the resolution of the publicness puzzle is the ambiguity of such terms as public organization, public service and even governmental. Publicness, therefore, is not the equivalent of government. I share this latter view with Bozeman and contend that all organizations are public, in so many ways. The key question is why are public organizations similar to their private counterparts? Relatedly, why are they different from one another?

19:2.3 The Publicness of Organizations–Why are all Organizations Public?

To a large extent, all organizations are public, because their objectives, purposes and *raison d'être* are shaped by the collective will of consumers and citizens, either consciously or unconsciously. This collective will is expressed through organized social action aimed at changing corporate or organizational behavior or policy. It could also be aimed at achieving a well-defined set of social goals. These actions ultimately affect organizational arrangements including the production, delivery and supply of both private and public goods and services.

The collective will knows no boundaries and exert influence on either public or private organizational values. What is more important, conscious or unconscious actions directed at organizations by members of society trigger a level of scrutiny requiring a greater degree of openness and accountability. For instance, private organizations are constrained, like their public counterparts, by resources and turn to the general public for new or additional capital. Private organizations procure long-term capital through the sale of shares to citizens or the public. Similarly, public organizations turn to citizens to raise operating revenue through taxation. To raise operating resources, both private and public organizations are compelled to depend on citizens for survival. This dependence on the public purse for survival requires extensive public scrutiny and openness. Hence, owners or contributors of capital and taxpayers could restrict organizational growth and development by denying much needed operating resources. In a democratic setting, members of the public could induce fundamental changes in organizational arrangements and purpose by exercising their respective voting rights during elections or regular and special meetings.

Against the background provided above, I argue that all organizations are subjected to a degree of scrutiny that precludes the constancy of the public equals government relationship. An organization operating in either the private or public sector is compelled to be influenced by public action.

It is in this vain that I agree with Bozeman's principle that the publicness of an organization refers to the degree to which that organization is affected by political authority. I might add that, in a democratic society, all organizations are public in as far public scrutiny, ownership and political authority are concerned.[4] At the same time, significant differences exist between a private organization and a public organization. The next section seeks to explore the distinction between the two organizational types.

19:2.4 The Goals of public and Private Organizations

In spite of the "publicness" of all organizations, significant differences exist in values and purpose. A private organization's primary purpose is to increase the initial wealth of citizens or investors. In the context of a market system,[5] private organizations seek to maximize gains on invested capital while attempting to retain the initial wealth invested. Values within such organizations promote attributes that lead to the attainment of the profit maximization goals.

Defining the purpose of a public organization, on the other hand, is similar to interpreting through a computer the topography of a distinct planet observed with the aid of a modern satellite. Not that the topography is invisible or blurred but evokes an imagery of its own. Not that the purposes of public organization is non-existent, but is often broad, complex, unwieldy and in some instances all-encompassing. As a result, judgment about effectiveness and achievement of desired outcome is, at best discretionary. This ambiguity places further emphasis on the importance of organizational goals in public sector organizations. In essence, of what relevance is a "purpose" or a "goal" to an organization and how important is that purpose or goal?

19:2.5 The Goal Orientation Model

To answer the above question, I borrow for this purpose, the goal-model for organizational analysis. The use of a goal- or purpose-model to determine ability to achieve desired outcomes or measure public organizational effectiveness is a necessary first step in the effort to measure performance. In its simplest form, the purpose-model provides a formula for the public and consumers to know how well organizations are doing. Parsons (1981) noted that the notion of goals for organizations is a distinguishing element, separating organizations from other social systems. In this context, relating purposes or goals to outcomes of organizations facilitates an assessment of performance. Furthermore, the goal notion may provide a rationale for adopting multiple measures for gauging public sector performance.

As Warwick (1975) noted:

> ...It is not enough to pack a briefcase with concepts and measures
> developed in other settings, unload them in a public agency and
> expect them to encompass all of the worthwhile reality to which they
> are exposed.

In a nutshell, the purpose of a public organization distinguishes it significantly from a private organization. This distinction is critical to the achievement
of the chapter's objective and begs a series of questions: (1) Do organizations
operating in the public sector behave differently and have objectives that are
different from their private counterparts? (2) Should public organizations be
evaluated differently? My answers to both questions are in the affirmative in
light of the discussion presented above which provides a foundation for propos-
ing different sets of measures for scrutinizing activities in public organizations.

19:3 Part Two: Measurement Criteria for Private and Public Organizations

19:3.1 Evaluation of the Private Organization

Private organizations operate for the purpose of making or maximizing profits.
They constantly seek to augment profit levels by controlling adjustable variables
until they reach a point where profits could no longer be increased. To the
private firm, profit is both quantifiable and measurable. For instance, profits (G)
at output level Q, (G_Q); can be defined as the difference between total revenues
and total costs. Total revenues $(TR_{(Q)})$ are given by:

$$TR_{(Q)} = P_{(Q)} * Q.$$

In effect, a firm sells an output level (Q), at a given market price of P per
unit, $P_{(Q)}$. All things being equal, total revenues will be equal to the product of
price and output.

Total costs $(TC_{(Q)})$ for output level (Q) will be the product of the unit cost
(α) and quantity produced and sold (Q). Hence $TC_{(Q)}$ will be defined as:

$$TC_{(Q)} = a_{(Q)} * Q.$$

The unit cost $(\alpha_{(Q)})$ is also the marginal cost, representing the extra cost for producing one extra output Q. With the revenue and cost functions, profits are defined algebraically as:

$$\Gamma_{(Q)} = (P_{(Q)} * Q) - (\alpha_{(Q)} * Q).$$

Alternatively, accounting profits are defined as total revenues less costs, such that:

$$\Gamma_{(Q)} = TR_{(Q)} - TC_{(Q)}.$$

A differentiation of the above alternate function with respect to output (Q) may provide the output level and necessary condition for profit maximization. Thus,

$$\frac{d\Gamma}{d_{(Q)}} = \Gamma'_{(Q)}$$

and,

$$\Gamma'_{(Q)} = \frac{dTR}{d_{(Q)}} - \frac{dTC}{d_{(Q)}} = 0$$

For a private organization to maximize profits, it must operate at an output level at which marginal revenue is equal to marginal cost. Further, for a first order profit maximization condition, the differentiated total revenue and cost functions will be set equal to each other, such that:

$$\frac{dTR}{d_{(Q)}} = \frac{dTC}{d_{(Q)}}.$$

Under competitive market conditions, the left side of the profit maximization[6] function is the marginal revenue, which equals the right side, known as the marginal cost.

Finally, for true profit maximization conditions to occur, the sufficiency or second order conditions must also prevail. These conditions require that marginal profit would be decreasing at the optimal level of output. At best, current or past performance of the private organization can be evaluated by focusing on profitability.

Although not an exhaustive list, additional measures of profitability besides the profit function include the following:

- The profit margin ratio.[7]
- Return on assets.
- Return on equity.

Table 19:1 provides a brief explanation of each of the profit measures indicated above.

The following financial measures are not without their weaknesses. First, the profit function discussed earlier does not incorporate risk. As such, it is unable to capture relative risks associated with varying investment opportunities. Second, current profit levels (which could be either unimpressive or, excessively high) may not properly represent long-term profitability. In the short term, it could produce misleading data about future prospects. Third, profit

Table 19:1 Measures of Profitability – Private Sector Organizations

1. Net profit margin[8]

$$\frac{\text{Net Income}}{\text{Total Operating Revenues}}$$

Measures profitability as a percentage of total operating revenue. It is an indirect measure of profitability since it is not based on actual investment or capital investment in assets.

2. Net Return on Assets

$$\frac{\text{Net Income}}{\text{Average Total Assets}}$$

Measures profitability relative to investment in total assets. It measures as well management performance.

3. Return on equity

$$\frac{\text{Net Income}}{\text{Average Shareholders' Equity}}$$

It is the ratio of net income divided by the average common stockholders' equity. It is a relative measure of performance, which could be used for comparative analysis.

levels or ratios omit the incorporation of forgone gains, and fail to ultimately compare both elements in profit estimation.

This elementary process of focusing exclusively on adjustable variables including unit price or marginal cost and quantity to achieve desired objectives simplifies the task of measuring performance in private organizations. The process also provides a universal formula for measuring and enhancing profit levels. In addition, the sufficiency conditions for profit maximization involving pricing and combinations of inputs can reasonably be estimated. Therefore, consistent monitoring and coordination of organizational inputs could enhance effectiveness or the achievement of desired outcomes.

Within this context, one would argue that, in spite of the publicness of all organizations, the primary focus of the private organization on profit distinguishes it significantly from a public organization,[9] and renders the task of organizational evaluation less onerous.

19:3.2 Evaluation of the Public Organization

In the previous section, it was easier to provide an overview of some universally accepted frameworks for measuring performance of private organizations. This was the case in view of the openness of the purpose of the private organization, which is clear, specific (to generate profits that maximize the wealth of owners) and measurable. One might be tempted to apply the same simplistic approach to determine effectiveness of public organizations.

I hasten to point out two types of organizations operating within the government sector. The first type incorporates organizations often referred to as government-owned enterprises. The second includes organizations, bureaucracies and programs that are managed by the government and are often the outcomes of general laws and market failure or, environmental exigencies.[10]

19:3.3 Government-Owned Organizations or Public Enterprises

Some governmental programs are operated like their private counterparts and are sometimes referred to as government-owned or public enterprises. Other synonyms for such organizations are government corporations, off-budget enterprises, parastatal organizations, and state-owned organizations, to mention but a few.

For the most part, state-owned organizations share similar characteristics with private sector organizations. The government provides the startup capital. They are self-sufficient (or supposed to be self-sufficient), independent or autonomous and are not subject to the annual budget review process for reappropriation. These organizations are creations of statutes for the purposes of achieving pre-defined goals or objectives. Their purposes are stipulated in the statutes that created them, thus simplifying to some extent the evaluation of performance. What is more important, they compete on the free market, and engage in profit-oriented activities. Because of their legal statuses they are permanent instruments of public policy. Activities of these organizations range from postal services to power and telecommunications. Recent reform[11] programs in most countries seem to have curtailed their growth and restricted them to a handful of economic activities.[12]

In addition to profit orientation, some government-owned organizations are designed to accomplish social and economic objectives. As a result, they have multiple objectives that require the application of multiple measures for gauging performance. Therefore, efforts that rely on profitability as measure of success or effectiveness might not generate the desired result. It is only a necessary first step.

19:3.4 Measuring Performance of the Government-Owned Organization

As a first step, government-owned enterprises must be examined in the context of their profit orientation. Even if the purpose to generate profit is secondary, that objective must be assessed using universally acceptable measures similar to those applied to profit-oriented organizations.

However, there is a problem associated with using the profit function for government-owned organizations. The profit function is affected by subsidies, tax exemptions and applications of below market rates of interest. These factors create distortionary effects that generate misleading results. For instance, in the case of government subsidy, unit cost pricing under the marginal cost concept might not yield accurate results unless adjustments are made to reflect the presence of subsidy. Similarly, the tax-exempt status of the enterprise creates a problem for determining the true profit level. A further discussion of the treatment of these problems is beyond the scope of this chapter.

The second step in the evaluation of these enterprises involves the assessment of their social and economic objectives. This step requires a more comprehensive format linking goals to outcomes, while ensuring efficiency, accountability and equity. A discussion leading to the presentation of a framework for accomplishing this objective is presented in the next few pages.

19:4 The Bulk of Government Activity

The bulk of government work can be summarized to include, general government, public safety, public works, health and welfare, education, culture and recreation, conservation and environment, the last but not least economic development. These traditional functional areas are managed in organizations that respond to societal needs and depend on external resources for survival. They are molded by the ever-changing needs of their environments, and managed not with the intent of making profit. Almost all organizations falling into this category depend on annual budget authorizations for survival. Because of their non-profit orientation they cannot be evaluated using the economic profit concept. Therefore, the application of the profit function for gauging organizational performance will be grossly inadequate.

19:4.1 Organizational Characteristics and Goal-orientation Approach

It is a well-known fact that there are no universally acceptable measures for examining performance in the traditional activity areas of government. In the past few years however, efforts to develop useful measures have yielded some positive outcomes. As a result, the literature on the subject has been enriched by research work and reports from practical experience.[13]

It is also well known that performance measures and indicators are needed to improve managerial performance, ensure efficiency, effectiveness, quality, and equity or fairness. They are also needed to enhance the quality of decisions that are made about resource allocation.

While measures and indicators of performance are urgently needed, they constitute only one aspect of the framework for conducting comprehensive assessment of a traditional public organization or, program. I share the view that a holistic approach to gauging performance might provide a better result and understanding of the functioning of the traditional public organization or program.

To achieve this objective, I borrow from the open systems theorists (Miller, 1978; Boulding, 1956), and argue that a traditional public organization (or for that matter, program) can be defined as a system by drawing a boundary around its activities. Earlier works (Katz and Kahn, 1978; Emery, 1969; Beer, 1980) emphasizing the application of the basic systems theory to studies in psychology, organizational analyses and management, to mention but a few, portrayed organizations as living organisms. By extending these biological principles to the study of traditional organizations, one might argue that these organizations are analogous to "living systems", and as "living systems", are greatly influenced by their environment.

One virtue of the application of the open systems theory is that organizational purpose is better understood in the context of ecology. A second virtue is that it facilitates the description and analysis of the mutual relationship between organizations and their environment. Further, this interdependence serves as a source for identifying processes and structures.

19:4.2 An Integrated and Holistic Model

Against the background provided above, I propose an integrated and holistic framework, which considers five interdependent elements in the evaluation of traditional public organizations. These elements are divided into five Categories and summarized below:

• Category 1–Organizational Characteristics
• Category 2–Organizational Goals and Objectives
• Category 3–Organizational Performance Measures and Indicators
• Category 4–Organizational Resource Acquisition and Utilization
• Category 5–Organizational Impact and Value Added

The above concepts are not entirely new but provide a formula that integrates all aspects of an organization. The proposed framework is not a model for a general evaluation of the public sector in its entirety.[14] Instead, it favors a case-by-case analysis of organizations and for that matter programs[15] operated by governments.

The framework was previously discussed under the Open Balance Sheet concept (Doamekpor, 1996) and has been revised and abbreviated for this purpose. Its five interrelated categories are summarized in Table 19:2. The various elements are also summarized in the next few paragraphs.

• Category 1–Organizational Characteristics

Category 1 describes the organization by providing the rationale for its creation or, origin. It provides information on the purpose of the organization, its clientele or target population including policy or legislative goals. Legislative goals are the same as theoretical goals. They are different from organizational goals and often couched in legal terminologies that require further translation for interpretation.

In sum, the essence of the first category is to justify the existence of the organization or program. Information in this category could also be organized to provide historical background information relevant to decision making or pro-

Table 19:2 Elements of the Open Balance Sheet Framework

Category 1–Organizational characteristics:
- Type (Description)
- Size (Budget and Manpower)
- Functional Responsibility
- Clients (Target Beneficiary)
- Origin
- Theoretical Goals or Statutory Goals

Category 2–Organizational goals and objectives:
- Organizational Goals
- Organizational Objectives
- (Timely, Parsimonious, Specific, Measurable, Quantifiable)
- Organizational Activities

Category 3–Organizational Activities, Performance Measures and Indicators:
- Need Measures
- Workload Measures
- Efficiency Measures
- Effectiveness Measures
- Responsiveness Measures
- Equity Measures

Category 4–Organizational Resource Acquisition and Utilization:
Resource Use:
- Sources
- Applications
- Financial Statement Analysis

Category 5–Organizational Impact and Value Added:
- Social impact
- Economic impact

---

gram revision.

• Category 2–Organizational Goals and Objectives

The second category contains information on organizational goals and objectives. Although the statement of goals appears to be simple, for most organizations defining precisely organizational goals is a tug of war. Yet, organizations are goal driven and clearly defined goals provide the thrust for achieving desired maximum acceleration.

Objectives are derivatives of goals. They are means by which ability to achieve desired goals can be assessed. In the context of this model, they are simple to understand, specific, quantifiable and measurable.

Therefore, a framework for assessing organizational performance must state distinctively both goals and objectives. In a practical sense, using goals and objectives, a simple evaluation methodology might consist of a comparison of actual performance and projected or targeted activity level.

Category 2 information also includes functions or activities necessary to achieve desired organizational or program goals. These activities consist of strategies designed to attain objectives. Goals, objectives and activities are therefore interrelated and linked to the theoretical or legislative goals.

• Category 3–Organizational or Program Performance Measures and Indicators

The third category consists of measures for examining various aspects of an organization, other than financial, social and economic. For the purpose of this chapter, six measures have been identified, namely; need, workload, efficiency, effectiveness, responsiveness and equity.[16]

The concepts presented in this framework are composite measures and define several other elements. A need measure defines the problem or need to be addressed. A workload measure is derived from a need measure and defines how much an organization can achieve, given existing capacity level. It describes the organization's output level or, what it has been able to do.

Efficiency[17] measures examine resource utilization and relate output with input. It is simply, the relationship between a program performance output and resource input. Effectiveness measures an organization's ability to achieve desired objectives. In its simplest form, an effectiveness measure can be derived from an organization's objective. It can also be defined as the relationship between desired outcome and actual outcome.

The last two measures, responsiveness and equity are interrelated. Responsiveness measures an organization's ability to meet the real needs of beneficiaries and citizens while equity addresses questions of fairness. An organization cannot be responsive without adequately covering questions of equity.

• Category 4–Sources and Uses of Resources

All organizations use resources to accomplish goals. In traditional public organizations, resources are acquired through legislative processes and sometimes the survival of organizations depend upon the instruments that led to their creation. Organizations that depend permanently on government funding must provide statements of accounts focusing on sources and uses of funds.

• Category 5–Organizational Impact and Value Added

The purpose of the fifth category is two-fold. First, it seeks to analyze the direct impact of the organization or program on beneficiaries and determines the indirect benefits derived by others in society. Second, it seeks to suggest the kind of information needed to assess social and economic impacts.

Not all organizations have social or economic goals. But, by their nature, they generate social and economic effects that benefit society, either directly or, indirectly. These net positive effects can be translated into greater well being for society. For instance, all traditional public organizations recruit and retain employees. By so doing, social problems that relate to unemployment are curtailed.

Further, these organizations generate what might be termed valued added consisting of wages and salaries including other payments channeled through various systems within the economy. Besides wages and salaries, organizations pay rent (not in all cases), contribute to retirement funds of employees, pay interest on both short- and long-term loans, to mention but a few. These payments generate further economic activity from which society benefits and must be accounted for in the evaluation process.

In summary, for a typical traditional public organization, the above general framework is most appropriate. For programs in general, the Category on performance indicators will vary greatly and depend on the type of activity under investigation. For instance, all categories proposed, except the performance measures category, could be used to evaluate the performance of an economic management program. Relevant performance measures or indicators would be derived from policy guidelines that relate to the program and institutional measures.

19:5 Conclusion

There are several reasons why measures for evaluating public organizations, programs or activities should be different from those applied to private sector organizations. Some of those reasons have been addressed in this chapter. Other reasons for doing so are provided elsewhere. Evidently, the uniqueness of the public sector, its institutional arrangement and complexity of rules and goals have significant impacts on planning, execution and achievement of desired objectives. Equally important is the need for accountability, equity and effectiveness. Needless re-stating that requirement for accountability and equity is much broader and innate for public organizations than private sector organizations. Furthermore, a clearer standard of transparency is needed to ensure attainment of desired objectives.

The discussion presented in this chapter follows a basic yet integrative formula for evaluating organizational or program performance for both private and public. Organizations with profit orientation must be assessed in the context of the profit function; otherwise, the integrative open balance sheet format would be appropriate.

Notes

[1] Performance measures and indicators are sometimes used interchangeably, although their technical definitions are different. Measures define program output and desired outcomes or results. They relate directly to desired outcomes or results. Indicators, on the other hand, are surrogates for expected performance results and are used when measures are difficult to determine or define.

[2] The purpose of this section is not to review the literature on organization theory. Rather, it is aimed at providing a description of an organization. For further reading, see Morgan (1997), Blumberg (1987), Hummel (1977), to mention but a few.

[3] The discussion on the uniqueness of the public sector is not new. According to McKinney and Howard (1979), the public sector is different as to purpose, financial base, context and constraints. The government sector, public sector or sometimes referred to as the public domain, provides a forum for establishing collective values and imposes restrictions on organizations that are the outcomes of laws requiring not only openness but also compliance and accountability. Stiglitz (2000) identifies two distinguishing features between both sectors. First, people who provide leadership in the public sector in a democracy are elected. Owners of capital choose their private counterparts. Second, governments have certain rights of compulsion that private

organizations do not possess. The private sector, on the other hand, is constrained only by market forces and regulations imposed by government resulting from market failure. For further reading on the distinction between both sectors, see David McKevitt and Alan Lawton (1994); Richard Stillman (1996) and Jan-Eric Lane (1995).

[4] For further reading on the publicness issue, see Rainey, Backoff and Levine (1976); Allison (1976); Lynn (1981).

[5] The market system imposes invisible constants that affect values and profit levels. The private organization must contend with the supremacy of the consumer and competition imposed by others. Price, quantity and quality are adjustable variables for the private organization. Consumer choice could also be influenced through aggressive marketing. The two exogenous variables are therefore, consumer supremacy and competition.

[6] The profit maximization concept presented in this section is discussed in the context of a perfectly competitive market. Profit is maximized in a competitive market when marginal revenue equals marginal cost. Profit maximization also requires that the firm will expand output as long as the first order marginal revenue exceeds marginal cost. Profit maximization conditions under monopolistic, oligopolistic and natural monopolistic conditions are excluded to retain the sharpness in focus.

[7] There are several other measures for examining other aspects of a private organization. For instance, there are measures for determining short-term solvency, activity level, financial leverage and overall worth of the organization. The discussion on these other measures is beyond the scope of this chapter. However, an introductory text in corporate finance might provide additional insight.

[8] Gross ratios are available for all three profitability measures. A gross ratio includes earnings before interest and taxes as numerator. For instance, gross profit margin equals earnings before interest and taxes divided by total operating revenues.

[9] Yutchman and Seashore (1967) hold a contrary view. They contend that effectiveness is a function of an organization's ability to procure needed inputs from its environment for survival. Therefore, a focus on a goal-oriented approach to organizational analysis may not yield enough data to discriminate among various types of social systems.

[10] Non-profit and voluntary organizations are not discussed in this chapter.

[11] The reform programs in most of the countries were aimed at improving the efficiency, profitability and productivity of government-owned organizations. They were also aimed at reducing their financial and economic dependence on governments, while ensuring accountability and ability to measure and monitor performance.

[12] For further reading see, Mary M. Shirley, "Managing State-Owned Enterprises" (Washington, DC: World Bank, 1983); Steven Greenhouse, "France Embraces Popular Capitalism", New York Times, June 8, 1987; J.A. and D.J. Thompson, "Privatization: A Policy in Search of a Rationale", The Economic Journal, Vol. 96 (March 1986); Financial Times, March 27, 1987; Price Waterhouse, "Privatization: The Facts" (London, 1987).

[13] Sylvia, Sylvia and Gunn (1997); Tigue and Strachota (1995); Doamekpor (1996); OECD Occasional Papers No.3 (1994); Governmental Accounting Standards Series, No. 093-A (1992); Levin (1983).

[14] Studies that have examined the overall effects of public organizations on output, especially government-owned organizations, did not take into account the varied and multiple objectives of such organizations. For example, using aggregate value added data and a variant of Cobb Douglas production function in the context of a residual analysis, Doamekpor (1998) examined the effects of these organizations on growth of output. The results indicated that contributions of these organizations, under the OLS method, were insignificant. Other studies (Plane, 1992; Swanson and Wolde-Semait, 1988) arrived at the same conclusion. The use of aggregate data on all homogenous public sector traditional organizations might not yield a positive result.

[15] A program is defined as a set of inter-related activities or projects, organized with the purpose of achieving desired objectives or outcomes.

[16] It will be impossible to exhaust the list of measures required to assess the performance of all programs or organizations. These measures require time to develop and are very often the outcomes of institutional policy and guidelines.

[17] Efficiency has several connotations. There are also several types of efficiency— social efficiency, economic efficiency, and total factor efficiency, to mention but a few. It is used here to refer to the simple relationship between an organization's output and input. It is sometimes used as a synonym for productivity.

References

Allison, J.T. (1979). "Public and Private Management: Are They Fundamentally Alike In All Important Respect?". Paper presented at the Public Management Research Conference, The Brookings Institution, Washington, DC November 19-20.

Boulding, K.E. (1956). *The Image.* Ann Arbor: University of Michigan Press.

Bozeman, B. (1987). *All Organizations Are Public: Bridging Public and Private Organizational Theories.* San Francisco: Jossey-Bass Inc.

Blumberg, R. L. (1987). *Organizations in Contemporary Society.* Englewood Cliffs: Prentice-Hall, Inc.

Doamekpor, F. K. (1996). "Examining Performance in State-Owned Organizations". *Journal of Social, Political and Economic Studies*, Vol. 21, No. 2, Summer.

Doamekpor, F.K. (1998). "Contributions of State-Owned Enterprises to the Growth of Total Output". *International Economic Journal*, Vol. 12, No. 4, Winter.

Goodman, B. G. and Loveman, G.W. (1991). "Does Privatization Serve the Public Interest?". *Harvard Business Review*, November-December.

Governmental Accounting Standards Series, No. 093-A (1992). "Preliminary Views of the Governmental Accounting Standards Board on Concepts Related to Service Efforts and Accomplishments Reporting". December 18.

Hummel, R. (1977). *The Bureaucratic Experience.* New York: St. Martin's Press.

Jan-Erik, L. (1995). *The Public Sector: Concepts, Models and Approaches.* Newbury, CA: Sage Publications Inc.

Levin, H.M. (1983). *Cost-Effectiveness: A Primer.* Newbury Park, CA: Sage Publications.

Lynn, L. E. (1981). *Managing the Public's Business.* New York: Basic Books.

McKevitt, D. and Lawton, A. (1994). *Public Sector Management: Theory, Critique and Practice.* Thousand Oaks: Sage Publications.

Miller, J.G. (1978). *Living Systems.* New York: McGraw Hill, New York.

Morgan, G. (1997). *Images of Organization.* Thousand Oaks, CA: Sage Publications.

Parker, D. and Hartley, K. (1991). "Do Changes in Organizational Status Affect Financial Performance?". *Strategic Management Journal,* Vol. 12, 631-641.

Parsons, T. (1981). *Social Systems in the Sociology of Organizations: Basic Studies* edited by O. Grusky and G. A. Miller. New York: Free Press.

Parsons, T. (1994). "Performance Management in Government: Performance Measurement and Results-Oriented Management". *OECD Occasional Papers* No. 3, Paris: OECD.

Rainey, H.G., Backoff, W. A. and Levine, C.H. (1976). "Comparing Public and Private Organizations". *Public Administration Review,* Volume 36 (2).

Stiglitz, J. (1988). "Economics of the Public Sector". New York: W.W. Norton Company, Inc.

Stillman, R.J. (2000). *Public Administration: Concepts and Cases.* Seventh Edition. Boston: Houghton Miffling Company.

Sylvia, R. D., Sylvia, K.M., Gunn, E.M. (1997). *Program Planning and Evaluation for the Public Manager* (second edition). Prospect Heights, Ill: Waveland Press.

Tigue, P. and Strachota, D. (1995). *The Use of Performance Measures in City and County Budgets.* Chicago, IL: Government Finance Officers Association.

Warwick, D.P. (1975). *A Theory of Public Bureaucracy.* Cambridge, MA: Harvard University Press.

Worsnop, R. L. (1992). "Privatization". *Congressional Quarterly Incorporated.* Vol. 2, No. 42, November 13.

Yutchman, E. and Seashore, S. (1932). "A System Resource Approach to Organizational Effectiveness". *American Sociological Review,* 32. December.

20 Africa Under World Bank/IMF Management: The Best of Times and the Worst of Times

KWADWO KONADU-AGYEMANG

> The IMF does not care whether you are suffering economic malaria, bilharzia or broken legs. They will always give you quinine *(Kenneth Kaunda. Quoted in Cheru, 1989:37)*.

> Our programs are like medicine. Some of the medicine has harmful side-effects, and there are real questions about what the dosage ought to be. The best that can be hoped for is that we are prescribing more or less the right medicine in more or less the right dosage *(Michael Mussa, Chief Economist, IMF)*.

20:1 Introduction

The preceding chapters have examined the ramifications of Structural Adjustment Programs in Ghana over the past 16 years. As is evident in all the chapters, these programs have showered mixed blessings on the country. Indeed, as a result of implementing the Structural Adjustment Programs, Ghana is experiencing the best of times (compared to the abysmal years of the late 1970s and early 1980s) at the macro level considering major economic indicators. As all the contributors to this volume acknowledge, Ghana has achieved unprecedented GDP growth rate of 4-6 percent a year during the past 16 years, high by sub-Saharan African standards. The country's industrial capacity that virtually collapsed in the late 1970s has been revamped, while infrastructure development has boomed. The economy's over-dependence on cocoa as the principal source of foreign exchange has been changed also, and non-traditional exports now account for a sizable proportion of the country's total earnings. Also tourism

427

that existed only in name before 1985, is now among the top four foreign exchange earners for Ghana. Fiscal management has improved and Ghana's credit rating is at an unprecedented high. Increased confidence in Ghana's economy by foreign financiers also has enabled the country to attract foreign loans and investment.

But it is also self-evident that the country has suffered, and is still suffering, from the crippling effects of market driven policies. Unemployment and underemployment have risen to unprecedented levels, due primarily to retrenchments in both public and private sectors. Access to health, education and other services that have traditionally been provided by the state has been curtailed. The massive currency devaluation and huge foreign borrowings also have created high levels of debt servicing that have in turn affected the ability of the country to import essential fuels, medicines, equipments and spare parts, and to invest in social services. Sixteen years of adjustment programs have not empowered the poor to break out of the vicious circle of poverty. Poverty levels among staple food producers in the northern savanna regions have remained unchanged, and urban poverty is increasing (Oxfam, 1999; GSS, 1995). Indeed, some reports indicate that the attainments of Structural Adjustment Programs notwithstanding, the average poor person will have to wait 30 years to see their incomes raised above the poverty line; the poorest of the poor will have to wait 40 years (Oxfam, 1999). Furthermore, in an effort to maximize the quantities of mineral and forestry resources exploited for export, the environment has been raped and water resources polluted (Owusu, 1998). All these have resulted from policies that seek to generate economic growth at all cost. Considering all the litany of ills that have afflicted Ghana since the adjustment programs were put in place, it may be safe to conclude that Ghana is also experiencing the worst of times (Konadu-Agyemang, 2000).

The verdict on the impact of Structural Adjustment Programs on Ghana presented in the preceding chapters may not be unique to the country, but may also aptly describe the bitter-sweet socio-economic situation in the 30 or more African countries that have embraced Structural Adjustment Programs in one form or another during the past 20 years. In all the countries in the region that have adopted the programs, Structural Adjustment Programs have resulted in slow and inequitable patterns of growth (Oxfam, 1999; Adedeji, 1995; Stewart, 1995). Areas which possess natural resources and/or have the capacity to produce the cash crops that feature in the export trade may have done quite well at the expense of less endowed regions, thus resulting in large scale socio-economic and spatial disparities (Anyinam, 1994; Konadu-Agyemang, 2000). The adjustment programs have also contributed to massive cutbacks in government investment in social welfare, health and education throughout the continent. A survey of annual spending trends for 16 countries in sub-Saharan Africa undergoing IMF programs, found that 75 percent cut public spending on education. This is in a region with 47 million children out of school (Oxfam, 1999: 2). Overall, Sub-Saharan African countries undergoing adjustment programs have reduced per capita spending on education by 1 percent per annum. In several countries, including Zambia and Zimbabwe, reductions in per capita spending have been very large, amounting to over 20 percent. In others, such as Tanzania, spending has stagnated at very low levels of $1 per pupil per year for educational materials (Oxfam, 1999; Stewart, 1995).

The programs also have created massive indebtedness in the region, and sizable portions of their foreign exchange earnings are allocated to debt servicing, especially to the IMF. While the Fund accounts for less than 5 percent of aid flows to the region, it receives back in debt repayments $300m more than it provides in new loans (Oxfam, 1999; 2000). In Tanzania, for example, government expenditure on debt servicing amounts to four times government spending on primary education. This is in a country where 1.2 million children are out of school, where there is only one book for every 20 pupils in rural areas, and where illiteracy rates are rising. In 1996 and 1997, debt repayments absorbed over one-third of Zambian government revenue - more than the combined health and education expenditure (Oxfam, 1999; 2000; Stewart, 1995). These transfers have been sustained in the face of pressing social problems: over 70 percent of the population lives below the poverty line, some 665,000 children are not in school, and child mortality rates are increasing. Mozambique has been spending $7 per person on debt-servicing compared to $3 on health-care, while 190,000 children aged under five die each year as a result of infectious diseases that could be prevented through low-cost public investment in health-care infrastructure (Oxfam, 1999; 2000; Stewart, 1995).

The negative impacts of IMF/World Bank economic restructuring programs are evident not in Africa alone. Indeed, in East Asia, the programs have prolonged and deepened a recession, which has led to dramatic increases in poverty and deteriorating education indicators (Khor, 1998; Ravichandran, 1998). In Indonesia for example, an additional 20 million people have been driven below the poverty line, with 1.3 million children dropping out of school. In Thailand, the school dropout rate increased three-fold in 1998, with 676,000 children leaving school (Oxfam, 1999; Bandow and Vasquez, 1994). The same story can be told about Latin American and Caribbean countries that have been "sapped" by the Bretton Woods Institutions (Stewart, 1995).

While the motives for seeking economic growth and stability are vital to poverty reduction (Oxfam, 2000) and are indeed laudable, it is a paradox that the very policies of the Bretton Woods institutions are also hindering progress towards the human development goals adopted by the international community for 2015. These include halving income poverty, reducing by two-thirds the number of under-five child deaths, and universal primary education. The IMF's efforts to achieve budget stability through stringent controls on public spending have hampered the efficient use of aid for poverty reduction, while at the same time leading to cost-recovery practices, which have excluded poor people from basic services such as health-care and education (Oxfam, 1999).

Structural adjustment programs are undermining recovery prospects, compounding inequalities, undermining the position of women, and failing to protect access to health and education services.

20:2 The Problem with IMF/World Bank Programs

The problem with the Bretton Woods "doctors" is that they seem to offer a common diagnosis, and therefore identical prescription, for all the ailing economies that consult them. So far as these institutions are concerned, the economic development problems of all developing countries are caused by a litany of ills that almost invariably include state interference in the operations of the price mechanism, over-bloated civil service, over-spending on social services (Ould Mey, 1996; Cheru, 1989). The attempt to put all developing countries together in one neat compartment often leads to a wrong diagnosis. In the words of an Oxfam report,

> Misdiagnosis has been a consistent problem. Countries turning to the IMF face complex development challenges, but Fund staff tend to see a single affliction: namely, overspending. They proceed to prescribe a "one treatment cures all" medicine in the form of budget austerity, high interest rates, and restrictions on public spending. For many countries, this amounts to a strategy of treatment through asphyxiation. Whole economies have been thrown into reverse gear, while social and economic infrastructures have collapsed (1999: 3).

It may be argued that while adjustment programs *per se* may be considered desirable, it is the manner in which the programs are designed and implemented that is responsible for all the "side effects of the IMF medicine". In an Oxfam report entitled *The IMF: wrong diagnosis, wrong medicine*, several pertinent design issues are raised. These include over-emphasis on short-term monetary goals, insufficient consideration of the implications of the programs for poverty, hampering of efficient use of aid flows, failure to accommodate human development goals, and failure to generate a sense of ownership (Oxfam, 1999).

A key flaw in the adjustment program is the growth-at-all-cost mentality. As discussed in chapter two, any development programs that aim to increase the GDP and GNP at all cost without regard to the human suffering that occurs in its wake, is bound to be problematic. While nobody is discounting the significance of buoying up the GDP/GNP, it should not be done at the expense of the environment, access to employment, health or education. Program designers will have to incorporate cost benefit analysis in all the programs, and efforts made in advance to remedy undesirable consequences. While programs like PAMSCAD have been introduced to mitigate some of the side effects of the restructuring programs, they are often put together as an afterthought, and turn out to be too little, too late.

The only way of giving a human face to the programs (see Cornia et. al, 1988) and making them "people friendly" is to include the citizens in the design, implementation and evaluation. So far public participation has not featured highly in most of the programs. Citizen participation involves the sharing of power over development and planning decisions with members of the community (Arnstein, 1969). The adjustment programs are often drawn by the staff of the sponsoring agencies in Washington in consultation with the public servants in the finance ministries of the adjusting countries. However, these public servants, and the ministers they serve, tend to be too elitist and often out of touch with the people at the grassroots. Moreover, since they exercise clout and power, they do not suffer the consequences of the harsh economic measures as much as their fellow countrymen and women. To help achieve a more efficient and equitable system, program economic planners should seek opportunities for a wider public involvement to achieve a more thoughtful and sensitive approach in making economic restructuring decisions. The process adopted to involve the public must be sincere and properly implemented since mere tokenism may lead to confusion and damage the public image of the planning and implementing authority.

Programs tend to work better if the beneficiaries play key roles in their formulation, implementation and evaluation. It is only recently that attempts have been made to seek public input through the Structural Adjustment Participatory Review Initiative (SAPRI). The SAPRI brings together 250 nongovernmental organizations (NGOs) and other civil society groups to work together to review the impact of adjustment lending and policy advice in selected countries, and aims to improve understanding about the impacts of adjustment policies as well as about how the participation of local, broad-based civil society can improve economic policymaking. However, this is a postmortem exercise, and moreover does not make room for enough grass root participation.

20:3 Conclusion

In conclusion, it is fair to say that that the adjustment programs have failed the poor and made life more difficult not only for Ghanaians, but for all the developing economies that have adopted them as vehicles for salvaging their ailing economies and improving the quality of life for their citizens. The programs are not "delivering the accelerated and more equitable growth on which poverty reduction depends. They are not protecting the access of poor people to basic services" (Watkins, 1999). This is not surprising, for while the ostensible purpose of the Structural Adjustment Programs is to help the developing countries grow their way out of debt and to improve the quality of life of their citizens, the first principal objective of the World Bank and IMF "is the creation and maintenance of a world capitalist system in which multinational corporations can trade, invest and move capital without restrictions from national governments" (Swift, 1991:14). In short, the adjustment program itself is part of the poverty problem. Without fundamental and far-reaching reforms, that will place the people's interest first, it will remain a barrier to the realization of the human development goals set by the international community for 2015 (Watkins, 1999).

Now let us hear the conclusion of the whole matter from the IMF's star pupil, President Jerry John Rawlings, himself. In his speech on the 8th anniversary of the December 31 1981 *coup d'etat*, he admitted the failure of Structural Adjustment Programs in Ghana thus:

> Fellow countrymen and women, *I should be the first to admit that the Economic Recovery Program has not provided all the answers to our economic problems. In spite of all the international acclaim it has received, the effects of its gains remain to be felt in most households. Many families continue to experience severe constraints on their household budgets.* There are many who have found it difficult during the past holidays to manage a modest celebration with a chicken for a meal, and a new dress for a child. Meanwhile, we're now thinking of how to meet our rents, the next term's school fees and other routine expenses. These make it hard for us to appreciate any significant gains we have made under ERP (Quoted in West Africa, January 15-21, 1990 edition. Italics supplied).

The pathetic 1990 picture of Ghana's conditions under the adjustment programs painted by Jerry Rawlings himself has not changed. If anything the conditions have become worst. As it was in 1990, so it was in 1997, and so it is in the year 2000 and ever shall be that Structural Adjustment Programs as they are currently composed are not in the interest of Africans. The whole menu of Structural Adjustment Programs contains striking internal contradictions, and will not, cannot, and indeed are not meant to benefit the poor any more than the original colonization of Africa was meant to. Although Structural Adjustment Programs may have succeeded to some extent at the macro-level, according to the World Bank and IMF criteria, the overall impacts at the micro-level have bee largely negative.

References

Adedeji, Adebayo (1995). "An African Perspective on Bretton Woods" in *The UN and the Bretton Woods Institutions: New Challenges for the Twenty-first Century*. Edited by Mahbub ul Haq, Richard Jolly, Paul Streeten and Khadija Haq. New York: St. Martin's Press.

Arnstein, S. (1969). "A Ladder of Citizen Participation". *Journal of American Institute of Planners* 35 (July) 216-224.

Bandow, D. and Vasquez, I. (1994). *Perpetuating Poverty: The World Bank, the IMF, and the Developing World.*

Cheru, F. (1989). *The Silent Revolution in Africa: Debt, Development and Democracy*. London: Zed Books.

Khor, M. (1998). *IMF Policies make patients sicker.* Panang: Third World Network.

Killick, T. (1995). *IMF Programs in Developing Countries: Design and Impact.* London: Routledge.

Konadu-Agyemang, K. (2000). "The Best of Times and the Worst of Times: Structural Adjustment Programs and Uneven Development in Africa – The case of Ghana". *The Professional Geographer*, 52 (3) 469-483.

Loxley, J. (1991). *Ghana: The Long Road to Recovery 1983-1990.* Ottawa: North South Institute.

Ould-Mey, M. (1996). *Global Restructuring and Peripheral States: The Carrot and the Stick in Mauritania.* Lanham, MD: Rowman and Littlefield.

Owusu, J. H. (1998). "Current convenience; Desperate deforestation: Ghana's Adjustment Program and the Forestry Sector". *The Professional Geographer* 50(4) 418-36.

Oxfam (1999). *IMF: Wrong diagnosis, wrong medicine.* Oxford: Oxfam.

Oxfam (2000). *Debt Relief and Poverty Reduction.* Oxford: Oxfam.

Ravichandran, K. (1998). "IMF Prescribed Wrong Medicine for Asian ills". *The Financial Express*, January 11.

Stewart, F. (1995). *Adjustment and Poverty: Options and choices.* London: Routledge.

Watkins, K. (1999). Speech at IMF Global Forum, Washington, 24 September, 1999.

World Bank (1995). *World Development Report.* New York: Oxford University Press.

Index

435

Printed in the United States
by Baker & Taylor Publisher Services